Engineering Graphics with SolidWorks 2012

David C. Planchard & Marie P. Planchard CSWP

ISBN: 978-1-58503-706-3

SDC

PUBLICATIONS

Schroff Development Corporation

www.SDCpublications.com

Schroff Development Corporation

P.O. Box 1334

Mission KS 66222

(913) 262-2664

www.SDCpublications.com

Publisher: Stephen Schroff

Trademarks and Disclaimer

SolidWorks® Corp. is a Dassault Systèmes S.A. (Nasdaq: DASTY) company that develops and markets software for design, analysis, and product data management applications. Microsoft® Windows, Microsoft Office® and its family of products are registered trademarks of the Microsoft Corporation. Other software applications and parts described in this book are trademarks or registered trademarks of their respective owners.

Dimensions of parts are modified for illustration purposes. Every effort is made to provide an accurate text. The authors and the manufacturers shall not be held liable for any parts or drawings developed or designed with this book or any responsibility for inaccuracies that appear in the book. Web and company information was valid at the time of the printing.

Copyright© 2012 by D & M Education LLC

Examination Copies

Teacher evaluation copies are available with classroom support materials and initial and final SolidWorks models. Books received as examination copies are for review purposes only. Examination copies are not intended for student use. Resale of examination copies is prohibited.

Engineering Graphics with SolidWorks 2012 is written to assist technical school, two year college, four year university instructor/student or industry professional that is a beginner or intermediate SolidWorks user. The book combines the fundamentals of engineering graphics and dimensioning practices with a step-by-step project based approach to learning SolidWorks with an enclosed 1.5 hour Video Instruction DVD. Learn by doing, not just by reading!

The book is divided into two parts: Engineering Graphics and SolidWorks 3D CAD software. In Chapter 1 through Chapter 3, you explore the history of engineering graphics, manual sketching techniques, orthographic projection, isometric projection, multi-view drawings, dimensioning practices and the history of CAD leading to the development of SolidWorks.

In Chapter 4 through Chapter 8, you apply engineering graphics fundamentals and learn the SolidWorks User Interface, Document and System properties, simple parts, simple and complex assemblies, design tables, configurations, multi-sheet, multi-view drawings, Bill of Materials, Revision tables, basic and advanced features. Follow the step-by-step instructions in over 70 activities to develop eight parts, four sub-assemblies, three drawings, and six document templates. Formulate the skills to create and modify solid features to model a 3D FLASHLIGHT assembly. Chapter 9 provides a bonus section on the Certified SolidWorks Associate CSWA program with sample exam questions and initial and final SolidWorks models.

Electronic Files

INTRODUCTION

Engineering Graphics with SolidWorks 2012 and Video Instruction DVD is written to assist technical school, two year college, four year university instructor/student or industry professional that is a beginner or intermediate SolidWorks user. The book combines the fundamentals of engineering graphics and dimensioning practices with a step-by-step project based approach to learning SolidWorks with the enclosed 1.5 hour Video Instruction DVD. Learn by doing, not just by reading!

The book is divided into two parts: Engineering Graphics and SolidWorks 3D CAD software. In Chapter 1 through Chapter 3, you explore the history of engineering graphics, manual sketching techniques, orthographic projection, isometric projection, multi-view drawings, dimensioning practices and the history of CAD leading to the development of SolidWorks.

In Chapter 4 through Chapter 8, you apply engineering graphics fundamentals and learn the SolidWorks User Interface, Document and System properties, simple parts, simple and complex assemblies, design tables, configurations, multi-sheet, multi-view drawings, Bill of Materials, Revision tables, basic and advanced features. Follow the step-by-step instructions in over 70 activities to develop eight parts, four sub-assemblies, three drawings, and six document templates. Formulate the skills to create and modify solid features to model a 3D FLASHLIGHT assembly. Chapter 9 provides a bonus section on the Certified SolidWorks Associate CSWA program with sample exam questions and initial and final SolidWorks models. Passing the CSWA exam proves to employers that you have the necessary fundamental engineering graphics and SolidWorks competencies.

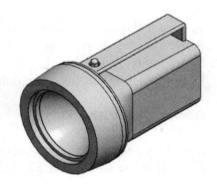

Review individual features, commands, and tools for each project with the book's 1.5 hour Video Instruction DVD and SolidWorks Help. The chapter exercises analyze and examine usage competencies based on the project objectives. The book is designed to compliment the SolidWorks Tutorials located in the SolidWorks Help menu. Each section explores the SolidWorks Online User's Guide to build your working knowledge of SolidWorks.

Desired outcomes and usage competencies are listed for each project. Know your objectives up front. Follow the step-by step procedures to achieve your design goals. Work between multiple documents, features, commands, and properties that represent how engineers and designers utilize SolidWorks in industry. The authors developed the industry scenarios by combining their own industry experience with the knowledge of engineers, department managers, vendors, and manufacturers. These professionals are directly involved with SolidWorks everyday. Their responsibilities go far beyond the creation of just a 3D model.

About the Cover

Displayed on the front cover is a SolidWorks FLASHLIGHT Assembly document. The FLASHLIGHT Assembly contains eight components, and four sub-assemblies.

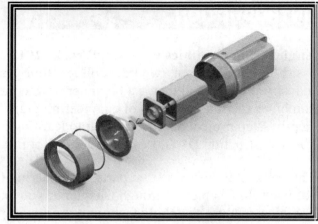

The following features are used to create the FLASHIGHT Assembly: *Extruded Base, Extruded Boss, Extruded Cut, Fillet, Chamfer, Revolved Base, Revolved Thin Boss, Revolved Thin Cut, Dome, Shell, Hole Wizard, Circular Pattern, Swept Base, Swept Boss, Loft Base, Loft Boss, Mirror, Draft, Rib*, and *Linear Pattern*.

The FLASHIGHT Assembly uses the following tools: *Insert Component, Hide/Show, Suppress/UnSuppress, Mate, Move Component, Rotate Component, Exploded View*, and *Interference Detection*. The FLASHIGHT Assembly drawing contains a Custom Drawing Template, Annotations, Bill of Materials, Balloons and more.

About the Authors

David Planchard is the founder of D&M Education LLC. Before starting D&M Education, he spent over 27 years in industry and academia holding various engineering, marketing, and teaching positions and degrees. He holds five U.S. patents and one international patent. He has published and authored numerous papers on Machine Design, Product Design, Mechanics of Materials, and Solid Modeling. He is an active member of the SolidWorks Users Group and the American Society of Engineering Education (ASEE). David holds a BSME, MSM with the following Professional Certifications: CCAI, CCNA, CCNP, CSWA, and CSWP. David is a SolidWorks Solution Partner, an Adjunct Faculty member and the SAE advisor at Worcester Polytechnic Institute in the Mechanical Engineering department.

Marie Planchard is the Director of World Education Markets at DS SolidWorks Corp. Before she joined SolidWorks, Marie spent over 10 years as an engineering professor at Mass Bay College in Wellesley Hills, MA. She has 14 plus years of industry software experience and held a variety of management and engineering positions. Marie holds a BSME, MSME and a Certified SolidWorks Professional (CSWP) Certification. She is an active member of the American Society of Mechanical Engineers (ASME) and the American Society for Engineering Education (ASEE).

David and Marie Planchard are co-authors of the following books:

- **A Commands Guide for SolidWorks® 2012**, 2011, 2010, 2009 and 2008

- **A Commands Guide Reference Tutorial for SolidWorks® 2007**

- **Assembly Modeling with SolidWorks® 2009,** 2008, 2006, 2005-2004, 2003 and 2001Plus

- **Drawing and Detailing with SolidWorks® 2010**, 2009, 2008, 2007, 2006, 2005, 2004, 2003, 2002 and 2001/2001Plus

- **Engineering Design with SolidWorks® with Video Instruction 2012**, 2011, 2010, 2009, 2008, 2007, 2006, 2005, 2004, 2003, 2001Plus, 2001 and 1999

- **Engineering Graphics with SolidWorks with Video Instruction 2012**, 2011, and 2010

- **SolidWorks® The Basics with Multimedia CD 2009**, 2008, 2007, 2006, 2005, 2004 and 2003

- **SolidWorks® Tutorial with Video Instruction, 2012**, 2011, 2010, 2009, 2008, 2007, 2006, 2005, 2004, 2003 and 2001/2001Plus

- **The Fundamentals of SolidWorks®: Featuring the VEXplorer robot, 2008** and 2007

- **Official Certified SolidWorks® Associate Examination Guide, Version 3; 2011, 2010, 2009,** Version 2; 2010, 2009, 2008, Version 1; 2007

- **Official Certified SolidWorks® Professional (CSWP) Certification Guide with Video Instruction DVD, 2011, 2010**

- **Applications in Sheet Metal Using Pro/SHEETMETAL & Pro/ENGINEER**

Acknowledgments

Writing this book was a substantial effort that would not have been possible without the help and support of my loving family and of my professional colleagues. I would like to thank Professor John Sullivan and Robert Norton and the community of scholars at Worcester Polytechnic Institute who have enhanced my life, my knowledge, and helped to shape the approach and content to this book.

The author is greatly indebted to my colleagues from Dassault Systèmes SolidWorks Corporation for their help and continuous support: Jeremy Luchini, Avelino Rochino, and Mike Puckett.

Thanks also to Professor Richard L. Roberts of Wentworth Institute of Technology, Professor Dennis Hance of Wright State University, and Professor Jason Durfess of Eastern Washington University who provided insight and invaluable suggestions.

Finally to my wife, who is infinitely patient for her support and encouragement and to our loving daughter Stephanie who supported me during this intense and lengthy project.

Contact the Authors

This is the third edition of this book. We realize that keeping software application books current is imperative to our customers. We value the hundreds of professors, students, designers, and engineers that have provided us input to enhance our book. We value your suggestions and comments. Please visit our website at **www.dmeducation.net** or contact us directly with any comments, questions or suggestions on this book or any of our other SolidWorks books at dplanchard@msn.com or planchard@wpi.edu.

Note to Instructors

Please contact the publisher **www.schroff.com** for additional classroom support materials: PowerPoint presentations, Adobe files along with avi files, term projects, quizzes with initial and final SolidWorks models and tips that support the usage of this text in a classroom environment.

Trademarks, Disclaimer, and Copyrighted Material

DS SolidWorks Corp. is a Dassault Systèmes S.A. (Nasdaq: DASTY) company that develops and markets SolidWorks® software for design, analysis and product data management applications. Microsoft Windows®, Microsoft Office® and its family of products are registered trademarks of the Microsoft Corporation. Other software applications and parts described in this book are trademarks or registered trademarks of their respective owners.

The publisher and the authors make no representations or warranties with respect to the accuracy or completeness of the contents of this work and specifically disclaim all warranties, including without limitation warranties of fitness for a particular purpose. No warranty may be created or extended by sales or promotional materials. Dimensions of parts are modified for illustration purposes. Every effort is made to provide an accurate text. The authors and the manufacturers shall not be held liable for any parts, components, assemblies or drawings developed or designed with this book or any responsibility for inaccuracies that appear in the book. Web and company information was valid at the time of this printing.

The Y14 ASME Engineering Drawing and Related Documentation Publications utilized in this text are as follows: ASME Y14.1 1995, ASME Y14.2M-1992 (R1998), ASME Y14.3M-1994 (R1999), ASME Y14.41-2003, ASME Y14.5-1982, ASME Y14.5M-1994, and ASME B4.2. Note: By permission of The American Society of Mechanical Engineers, Codes and Standards, New York, NY, USA. All rights reserved.

Additional information references the American Welding Society, AWS 2.4:1997 Standard Symbols for Welding, Braising, and Non-Destructive Examinations, Miami, Florida, USA.

References

- SolidWorks Users Guide, SolidWorks Corporation, 2012

- ASME Y14 Engineering Drawing and Related Documentation Practices

- Beers & Johnson, <u>Vector Mechanics for Engineers</u>, 6th ed. McGraw Hill, Boston, MA

- Betoline, Wiebe, Miller, <u>Fundamentals of Graphics Communication</u>, Irwin, 1995

- Hibbler, R.C, <u>Engineering Mechanics Statics and Dynamics</u>, 8th ed, Prentice Hall, Saddle River, NJ

- Hoelscher, Springer, Dobrovolny, <u>Graphics for Engineers</u>, John Wiley, 1968

- Jensen, Cecil, <u>Interpreting Engineering Drawings</u>, Glencoe, 2002

- Jensen & Helsel, <u>Engineering Drawing and Design</u>, Glencoe, 1990

- Lockhart & Johnson, <u>Engineering Design Communications</u>, Addison Wesley, 1999

- Olivo C., Payne, Olivo, T, <u>Basic Blueprint Reading and Sketching</u>, Delmar, 1988

- Planchard & Planchard, <u>Drawing and Detailing with SolidWorks</u>, SDC Pub., Mission, KS 2010

- Walker, James, <u>Machining Fundamentals</u>, Goodheart Wilcox, 1999

- 80/20 Product Manual, 80/20, Inc., Columbia City, IN, 2009

- Reid Tool Supply Product Manual, Reid Tool Supply Co., Muskegon, MI, 2007

- Simpson Strong Tie Product Manual, Simpson Strong Tie, CA, 2008

- Ticona Designing with Plastics - The Fundamentals, Summit, NJ, 2008

- SMC Corporation of America, Product Manuals, Indiana, 2011

- Gears Educational Design Systems, Product Manual, Hanover, MA, 2011

- Emhart - A Black and Decker Company, On-line catalog, Hartford, CT, 2009

☼ During the initial SolidWorks installation, you are requested to select either the ISO or ANSI drafting standard. ISO is typically a European drafting standard and uses First Angle Projection. The book is written using the ANSI (US) overall drafting standard and Third Angle Projection for drawings.

TABLE OF CONTENTS

Copy the model folders from the Video Instruction DVD to your local hard drive. Work from your local hard drive. View the 1.5 hour Video Instruction DVD for additional help.

Documents library
Chapter 7 Homework

- Clamp Term Project
- Drill Press Term Project
- Pulley Term Project
- Radial Engine Term Project
- Shock Term Project
- Butterfly Valve Term Project
- Welder Arm Term Project

Additional projects are included in the exercise section of Chapter 7. Copy the components from the Chapter 7 Homework folder located on the DVD. View all components. Create an ANSI Assembly document. Insert and create all needed components and mates to assemble the assembly and to simulate proper movement per the provided avi file located in the folders.

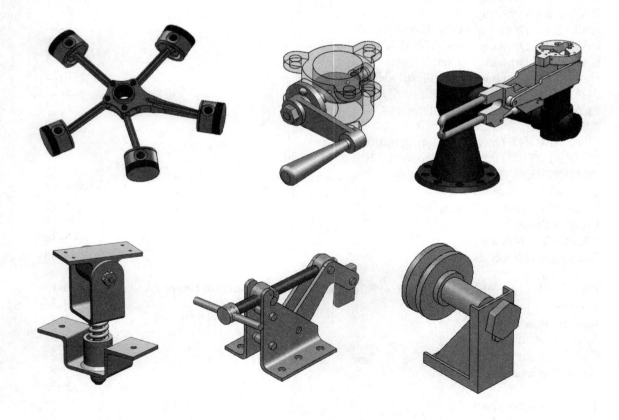

Overview of Chapters

Chapter 1: History of Engineering Graphics

Chapter 1 provides a broad discussion of the history of Engineering Graphics and the evolution from manual drawing/drafting along with an understanding of general sketching techniques, alphabet of lines, precedence of line types and Orthographic projection.

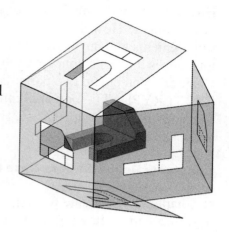

It also addresses the Glass Box method and the six principle orthographic views along with the difference between First Angle Projection and Third Angle Projection.

Chapter 2: Isometric Projection and Multi View Drawings

Chapter 2 provides a general introduction into Isometric Projection and sketching along with Additional Projections and arrangement of views.

It also covers advanced drawing views and an introduction to the evolution of from manual drafting to early CAD systems and finally to SolidWorks.

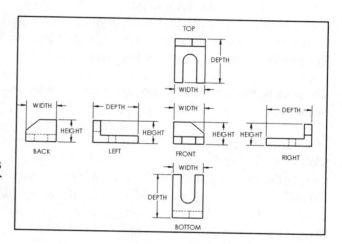

Chapter 3: Dimensioning Practices, Tolerancing and Fasteners

Chapter 3 provides a general introduction into dimensioning practices and systems, and the ASME ANSI Y14.5 standards along with fits, fasteners and general tolerancing.

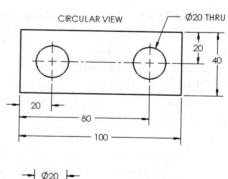

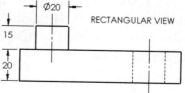

Chapter 4: Introduction to SolidWorks Part Modeling

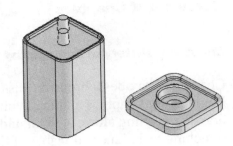

How do you start a model in SolidWorks? What is the design intent? How do you take a customer's requirements and covert them into a model? Chapter 4 introduces the basic concepts behind SolidWorks. In Chapter 4 you create two parts: BATTERY and BATTERYPLATE. You apply the following features: Extruded Boss, Extruded Base, Extruded Cut, Fillet, and Chamfer.

Chapter 5: Revolved Boss/Base Features

In Chapter 5 you create two parts: LENS and BULB. You apply the following features: Extruded Base, Extruded Boss, Extruded Cut, Revolved Base, Revolved Boss Thin, Revolved Thin Cut, Dome, Shell, Hole Wizard, and Circular Pattern along with the following Geometric relations: Equal, Coincident, Symmetric, Intersection, and Perpendicular.

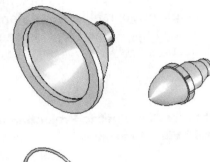

Chapter 6: Swept Boss/Base and Loft Boss/Base Features

In Chapter 6, you create four parts: O-RING, SWITCH, LENSCAP, and HOUSING. Chapter 6 covers the development of the Swept Base, Swept Boss, Loft Base, Loft Boss, Mirror, Draft, Shape, Rib, and Linear Pattern features and strengthens the use of the previously applied features.

Chapter 7: Assembly Modeling - Bottom-up

In Chapter 7, you learn about the Bottom-up assembly technique and create four assemblies: LENSANDBULB, CAPANDLENS, BATTERYANDPLATE, and the FLASHLIGHT assembly.

You insert the following Standard mate types: Coincident, Concentric, and Distance and use the following tools: Insert Component, Hide/Show, Suppress/Unsuppress, Mate, Move Component, Rotate Component, Exploded View, and Interference Detection.

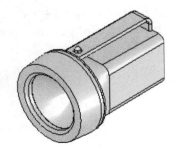

Chapter 8: Drawing Fundamentals

Chapter 8 covers the development of a customized drawing template. Develop a Company logo. Create a BATTERY drawing with five views. Develop and incorporate a Bill of Materials into the FLASHLIGHT assembly drawing.

Chapter 8 introduces Design Tables and configurations in the drawing. Create three configurations of the O-RING part. Display the three configurations in the O-RING-TABLE drawing.

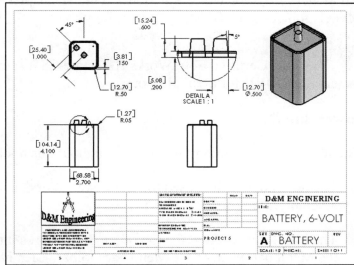

Chapter 9: Introduction to the Certified SolidWorks Associate Exam

Chapter 9 provides a basic introduction into the curriculum and exam categories for the Certified SolidWorks Associated CSWA Certification program. Review the exam procedure, process and required model knowledge needed to take and pass the exam.

- Review the five exam categories: *Drafting Competencies, Basic Part Creation and Modification, Intermediate Part Creation and Modification, Advanced Part Creation and Modification, and Assembly Creation and Modification*

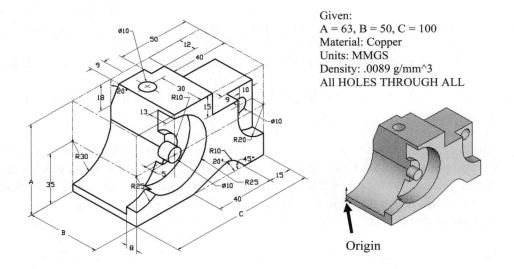

Given:
A = 63, B = 50, C = 100
Material: Copper
Units: MMGS
Density: .0089 g/mm^3
All HOLES THROUGH ALL

Origin

All model files for Chapter 9 are located in the Chapter 9 CSWA Models folder on the DVD.

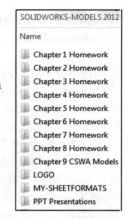

View the Certified SolidWorks Associate CSWA exam pdf file on the enclosed DVD for a sample exam.

About the Book

The following conventions are used throughout this book:

- The term document is used to refer a SolidWorks part, drawing, or assembly file.

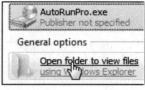

- The list of items across the top of the SolidWorks interface is the Menu bar menu or the Menu bar toolbar. Each item in the Menu bar has a pull-down menu. When you need to select a series of commands from these menus, the following format is used: Click **View**, check **Origins** from the Menu bar. The Origins are displayed in the Graphics window.

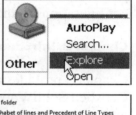

- The ANSI overall drafting standard and Third Angle projection is used as the default setting in this text. IPS (inch, pound, second) and MMGS (millimeter, gram, second) unit systems are used.

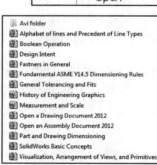

- The book is organized into various chapters. Each chapter is focused on a specific subject or feature. Additional pdf and ppt information and folders/models are provided on the enclosed 1.5 hour Video Instruction DVD.

- All *templates*, *logos* and needed *model documents* for this book are included on the enclosed DVD. Copy the information from the DVD to your local hard drive. Work from your local hard drive.

- View the enclosed 1.5 hour Video Instruction DVD for additional help on SolidWorks.

- Additional projects are included in the exercise section of Chapter 7. Copy the components from the Chapter 7 Homework folder located on the DVD. Create an ANSI Assembly document. Insert and create all needed components and mates to assemble the assembly and to simulate proper movement per the provided avi file located in the folders.

- Screen shots in the book were made using SolidWorks 2012 SP0 running Windows® 7.

The following command syntax is used throughout the text. Commands that require you to perform an action are displayed in **Bold** text.

Format:	Convention:	Example:
Bold	<ul><li>All commands actions.</li><li>Selected icon button.</li><li>Selected icon button.</li><li>Selected geometry: line, circle.</li><li>Value entries.</li></ul>	<ul><li>Click **Options** from the Menu bar toolbar.</li><li>Click **Corner Rectangle** ☐ from the Sketch toolbar.</li><li>Click **Sketch** ✐ from the Context toolbar.</li><li>Select the **centerpoint**.</li><li>Enter **3.0** for Radius.</li></ul>
Capitalized	<ul><li>Filenames.</li><li>First letter in a feature name.</li></ul>	<ul><li>Save the **FLATBAR** assembly.</li><li>Click the **Fillet** feature.</li></ul>

Windows Terminology in SolidWorks

The mouse buttons provide an integral role in executing SolidWorks commands. The mouse buttons execute commands, select geometry, display Shortcut menus and provide information feedback.

A summary of mouse button terminology is displayed below:

Item:	Description:
Click	Press and release the left mouse button.
Double-click	Double press and release the left mouse button.
Click inside	Press the left mouse button. Wait a second, and then press the left mouse button inside the text box. Use this technique to modify Feature names in the FeatureManager design tree.
Drag	Point to an object, press and hold the left mouse button down. Move the mouse pointer to a new location. Release the left mouse button.
Right-click	Press and release the right mouse button. A Shortcut menu is displayed. Use the left mouse button to select a menu command.
ToolTip	Position the mouse pointer over an Icon (button). The tool name is displayed below the mouse pointer.
Large ToolTip	Position the mouse pointer over an Icon (button). The tool name and a description of its functionality are displayed below the mouse pointer.
Mouse pointer feedback	Position the mouse pointer over various areas of the sketch, part, assembly or drawing. The cursor provides feedback depending on the geometry.

A mouse with a center wheel provides additional functionality in SolidWorks. Roll the center wheel downward to enlarge the model in the Graphics window. Hold the center wheel down. Drag the mouse in the Graphics window to rotate the model.

Visit SolidWorks website: http://www.solidworks.com/sw/support/hardware.html to view their supported operating systems and hardware requirements.

SolidWorks					
Operating Systems	**SolidWorks 2009**	**SolidWorks 2010**	**SolidWorks 2011**	**SolidWorks 2012**	**(SolidWorks 2013)**
Windows 7	✗	✓	✓	✓	✓
Windows Vista	✓	✓	✓	✓	✓
Windows XP	✓	✓	✓	✓	✗
Minimum Hardware	Configuring a SolidWorks Workstation				
RAM	2 GB or more				
Disk Space	5 GB or more				
Video Card	Certified cards and drivers				
Processor	Intel or AMD with SSE2 support. 64-bit operating system recommended				
Install Media	DVD Drive or Broadband Internet Connection				

The Instructors DVD contains PowerPoint presentations, Adobe files along with avi files, Term Projects, quizzes with the initial and final SolidWorks models.

The book is design to expose the new user to numerous tools and procedures. It may not always use the simplest and most direct process.

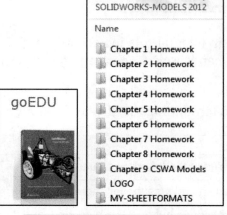

Documents library
SOLIDWORKS-MODELS 2012

Name

- Chapter 1 Homework
- Chapter 2 Homework
- Chapter 3 Homework
- Chapter 4 Homework
- Chapter 5 Homework
- Chapter 6 Homework
- Chapter 7 Homework
- Chapter 8 Homework
- Chapter 9 CSWA Models
- LOGO
- MY-SHEETFORMATS

goEDU

The book does not cover starting a SolidWorks session in detail for the first time. A default SolidWorks installation presents you with several options. For additional information for an Education Edition, visit the following sites: http://www.solidworks.com/goedu and http://www.solidworks.com/sw/education/6443_ENU_HTML.htm.

Installation Instructions

Education User License Agreement (EULA)

System Requirements

Product Description

Instructions to Access Instructors' Curriculum

Workgroup PDM Installation Instructions

Workgroup PDM Video Tutorials

- Avi folder
- Term Projects - Book
- 3D Modeling Features and Strategy
- Alphabet of lines and Precedent of Line T...
- Annotations in Drawings
- Assemblies and Mates in General
- Basic Sketching
- Boolean Operation
- Design Intent
- Drafting and Dimensioning Standards
- Fastners in General
- Flow Simulation Tutorial
- Fundamental ASME Y14.5 Dimensioning ...
- Gears using SolidWorks
- General GDT information
- General SolidWorks Tips
- General SolidWorks Tips
- General Tolerancing and Fits
- History of Engineering Graphics
- Materials in General
- Measurement and Scale
- Non-Standard Drawing View Types
- Open a Drawing Document 2012
- Open an Assembly Document 2011
- Open an Assembly Document 2012
- Part and Drawing Dimensioning
- Planes, Measurement tool, Equations, De...
- SolidWorks Basic Concepts
- SolidWorks Drawings Documents in Gen...
- SolidWorks Simulation FEA Overview
- SolidWorks Toolbox
- Split Line tool for a Static load analysis us...
- Surface Finish
- Sustainbility_Presentation
- Tolerance, Weld and Texture Symbols
- Visualization, Arrangement of Views, and...
- Weldments

Chapter 1

History of Engineering Graphics

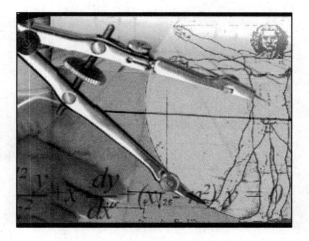

Below are the desired outcomes and usage competencies based on the completion of Chapter 1.

Desired Outcomes:	Usage Competencies:
• Appreciate the history of Engineering Graphics.	• Identify categories and disciplines related to Engineering Graphics.
• Grasp of the Cartesian Coordinate system.	• Apply 2D and 3D Cartesian Coordinate system: Absolute, Relative, Polar, Cylindrical and Spherical.
• Understand Geometric entities. • Comprehend Free Hand Sketches.	• Points, Circles, Arcs, Planes, etc. • Solid Primitives. • Generate basic 2D shapes and objects. • Create 2D and 3D freehand sketches.
• Recognize the Alphabet of Lines and Precedence of Line types.	• Create and understand correct line precedence.
• Be familiar with Orthographic Projection using the Glass Box method.	• Explain the difference between First and Third Angle Projection type. • Identify the six principal views using the Glass Box method.

Notes:

Chapter 1 - History of Engineering Graphics

Chapter Overview

Chapter 1 provides a broad discussion of the history of Engineering Graphics and the evolution from manual drawing/drafting along with an understanding of the Cartesian Coordinate system, Geometric entities, general sketching techniques, alphabet of lines, precedence of line types and Orthographic projection.

On the completion of this chapter, you will be able to:

- Appreciate the history of Engineering Graphics

- Understand 2D and 3D Cartesian Coordinate system:
  - Right-handed vs. Left-handed
  - Absolute, Relative, Polar, Cylindrical and Spherical

- Understand Geometric entities:

  - Point, Circle, Arc, Plane, etc.

- Recognize the Alphabet of lines

- Distinguish between the Precedence of line types

- Grasp Orthographic Projection using the Glass Box method

- Identify the six Principal views

- Explain the difference between First and Third Angle Projection type

History of Engineering Graphics

Engineering Graphics is the academic discipline of creating standardized technical drawings by architects, interior designers, drafters, design engineers, and related professionals.

Standards and conventions for layout, sheet size, line thickness, text size, symbols, view projections, descriptive geometry, dimensioning, tolerencing, abbreviations, and notation are used to create drawings that are ideally interpreted in only one way.

A technical drawing differs from a common drawing by how it is interpreted. A common drawing can hold many purposes and meanings, while a technical drawing is intended to concisely and clearly communicate all needed specifications to transform an idea into physical form for manufacturing, inspection or purchasing.

We are all aware of the amazing drawings and inventions of Leonardo da Vinci (1453-1528). It is assumed that he was the father of mechanical drafting. Leonardo was probably the greatest engineer the world has ever seen. Below are a few freehand sketches from his notebooks.

Example 1:

The first freehand sketch is of a crossbow. Note the detail and notes with the freehand sketch.

Example 2:

The second freehand sketch is of an early example of an exploded assembly view.

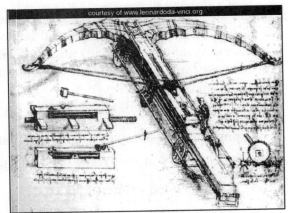

The only source for the detailed history of Leonardo's work is his own careful representations. His drawings were of an artist who was an inventor and a modern day engineer. His drawings were three-dimensional (3D) and they generally were without dimensional notations.

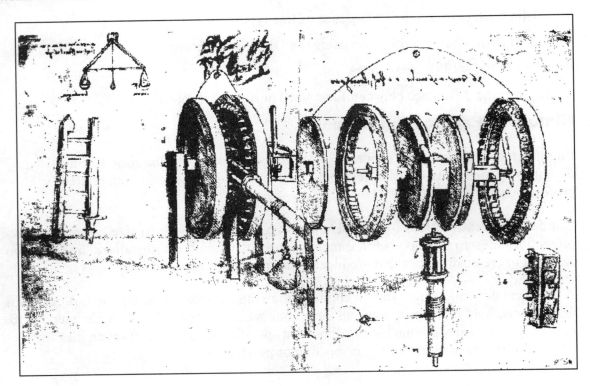

Craftsman created objects from his drawings, and each machine or device was one-of-a-kind creations. Assembly line manufacturing and interchangeable parts were not a concern.

Engineering graphics is a visual means to develop ideas and convey designs in a technical format for construction and manufacturing. Drafting is the systematic representation and dimensional specification and annotation of a design.

The basic mechanics of drafting is to place a piece of paper (or other material) on a smooth surface with right-angle corners and straight sides - typically a drafting table. A sliding straightedge known as a T-square, is then placed on one of the sides, allowing it to be slid across the side of the table, and over the surface of the paper.

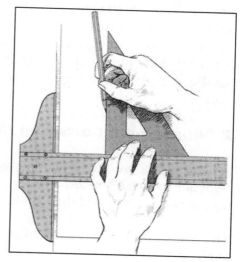

"Parallel lines" can be drawn simply by moving the T-square and running a pencil or technical pen along the T-square's edge, but more typically; the T-square is used as a tool to hold other devices such as set squares or triangles. In this case, the drafter places one or more triangles of known angles on the T-square - which is itself at right angles to the edge of the table - and can then draw lines at any chosen angle to others on the page.

Modern drafting tables (which have by now largely been replaced by CAD workstations) come equipped with a parallel rule that is supported on both sides of the table to slide over a large piece of paper. Because it is secured on both sides, lines drawn along the edge are guaranteed to be parallel.

In addition, the drafter uses several tools to draw curves and circles. Primary among these are the compasses, used for drawing simple arcs and circles, the French curve, typically made out of plastic, metal or wood composed of many different curves and a spline; rubber coated articulated metal that can be manually bent to most curves.

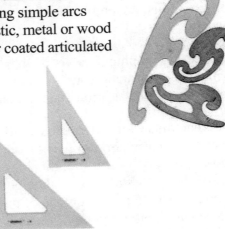

A drafting triangle always has one right angle 90°. This makes it possible to put a triangle against a T-square to draw vertical lines. A 30 60 90 triangle is used with a t-square or parallel straightedge to draw lines that are 30 60 or 90 degrees. A 45 90 triangle is used to draw lines with a T-square or parallel straightedge that are 45 or 90 degrees.

2 Dimensional Cartesian Coordinate system

A Cartesian coordinate system in two dimensions is commonly defined by two axes, at right angles to each other, forming a plane (an x,-y plane). The horizontal axis is normally labeled x, and the vertical axis is normally labeled y.

The axes are commonly defined as mutually orthogonal to each other (each at a right angle to the other). Early systems allowed "oblique" axes, that is, axes that did not meet at right angles, and such systems are occasionally used today, although mostly as theoretical exercises. All the points in a Cartesian coordinate system taken together form a so-called Cartesian plane. Equations that use the Cartesian coordinate system are called Cartesian equations.

The point of intersection, where the axes meet, is called the origin. The x and y axes define a plane that is referred to as the xy plane. Given each axis, choose a unit length, and mark off each unit along the axis, forming a grid. To specify a particular point on a two dimensional coordinate system, indicate the x unit first (abscissa), followed by the y unit (ordinate) in the form (x,-y), an ordered pair.

Example 1:

Example 1 displays an illustration of a Cartesian coordinate plane. Four points are marked and labeled with their coordinates: (2,3), (-3,1), (-1.5,-2.5), and the origin (0,0).

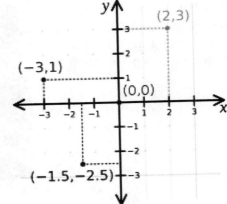

The intersection of the two axes creates four regions, called quadrants, indicated by the Roman numerals I, II, III and IV. Conventionally, the quadrants are labeled counter-clockwise starting from the upper right ("northeast") quadrant.

Example 2:

Example 2 displays an illustration of a Cartesian coordinate plane. Two points are marked and labeled with their coordinates: (3,5) and the origin (0,0) with four quadrants.

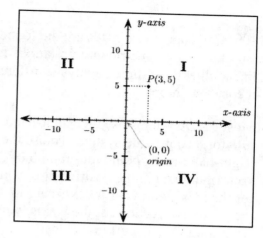

In the first quadrant, both coordinates are positive, in the second quadrant x-coordinates are negative and y-coordinates positive, in the third quadrant both coordinates are negative and in the fourth quadrant, x-coordinates are positive and y-coordinates negative.

3 Dimensional Cartesian Coordinate system

The three dimensional coordinate system provides the three physical dimensions of space: *height*, *width*, and *length*. The coordinates in a three dimensional system are of the form (x,y,z).

Once the x- and y-axes are specified, they determine the line along which the z-axis should lie, but there are two possible directions on this line. The two possible coordinate systems which result are called "Right-hand" and "Left-hand."

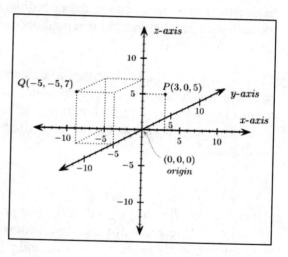

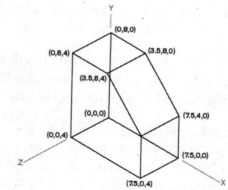

💡 Cartesian coordinates are the foundation of analytic geometry, and provide enlightening geometric interpretations for many other branches of mathematics, such as linear algebra, complex analysis, differential geometry, multivariate calculus, group theory, and more.

X- Always the thumb

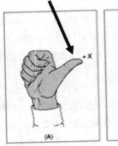  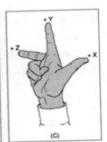

Most CAD systems use the Right-hand rule for a coordinate system. To use the Right-hand rule - point your thumb of your right hand in the positive direction for the x axis and your index finger in the positive direction for the y axis, your remaining fingers curl in the positive direction for the z axis as illustrated.

When the x,-y plane is aligned with the screen in a CAD system, the z axis is oriented horizontally (pointing towards you). In machining and many other applications, the z-axis is considered to be the vertical axis. In all cases, the coordinate axes are mutually perpendicular and oriented according to the Right-hand or Left-hand rule.

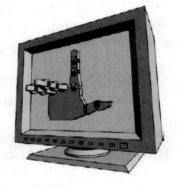

The Right-hand rule is also used to determine the direction of rotation. For rotation using the right-hand rule, point your thumb in the positive direction along the axis of rotation. Your fingers will curl in the positive direction for the rotation.

💡 Some CAD systems use a Left-hand rule. In this case, the curl of the fingers on your left hand provides the positive direction for the z axis. In this case, when the face of your computer monitor is the x,-y plane, and positive direction for the z axis would extend into the computer monitor, not towards you.

Models and drawings created in SolidWorks or a CAD system are defined and stored using sets of points in what is sometimes called World Space.

Each reference line is called a coordinate axis or just axis of the system, and the point where they meet is its origin. The coordinates can also be defined as the positions of the perpendicular projections of the point onto the two axes, expressed as a signed distance from the origin.

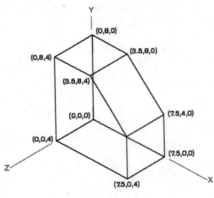

The origin in SolidWorks is displayed in blue in the center of the Graphics window. The origin represents the intersection of the three default reference planes: *Front Plane, Top Plane* and *Right Plane* illustrated in the FeatureManager. The positive x-axis is horizontal and points to the right of the origin in the Front view. The positive y-axis is vertical and point upward in the Front view. The FeatureManager contains a list of features, reference geometry, and settings utilized in the part.

Absolute Coordinates

Absolute coordinates are the coordinates used to store the location of points in your CAD system. These coordinates identify the location in terms of distance from the origin (0,0,0) in each of the three axis (x, y, z) directions of the Cartesian coordinate system.

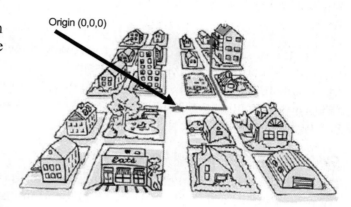

Origin (0,0,0)

As an example - someone provides directions to your house (or to a house in an area where the streets are laid out in nice rectangular blocks). A way to describe how to get to your house would be to inform the person how many blocks over and how many blocks up it is from two main streets (and how many floors up in the building, for 3D). The two main streets are like the x and y axes of the Cartesian coordinate system, with the intersection as the origin (0,0,0).

Relative Coordinates

Instead of having to specify each location from the origin (0,0,0), using relative coordinates allows you to specify a 3D location by providing the number of units from a previous location. In other words; the location is defined relative to your previous location. To understand relative coordinates, think about giving someone directions from his or her current position, not from two main streets. Using the same map as before but this time with the location of the house relative to the location of the person receiving directions.

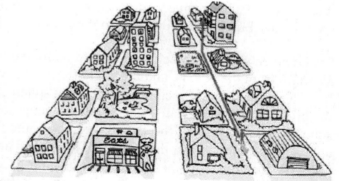

Polar Coordinates

Polar coordinates are used to locate an object by providing an angle (from the x axis) and a distance. Polar coordinates can either be absolute, providing the angle and distance from the origin (0,0,0), or they can be relative, providing the angle and distance from the current location.

Picture the same situation of having to provide directions. You could inform the person to walk at a specified angle from the crossing of the two main streets, and how far to walk. In the illustration, it shows the angle and direction for the shortcut across the empty lot using absolute polar coordinates. Polar coordinates can also be used to provide an angle and distance relative to a starting point.

Cylindrical and Spherical Coordinates

Cylindrical and spherical coordinates are similar to polar coordinates except that you specify a 3D location instead of one on a single flat plane (such as a map). Cylindrical coordinates specify a 3D location based on a radius, angle, and distance (usually in the z axis direction). It may be helpful to think about this as giving a location as though it were on the edge of a cylinder. The radius tells how far the point is from the center (or origin); the angle is the angle from the x axis along which the point is located; and the distance gives you the height where the point is located on the cylinder. Cylindrical coordinates are similar to polar coordinates, but they add distance in the z direction.

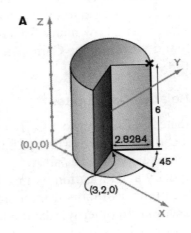

Spherical coordinates specify a 3D location by the radius, an angle from the X axis, and the angle from the x,y plane. It is helpful to think of locating a point on a sphere, where the origin of the coordinate system is at the center of the sphere. The radius gives the size of the sphere, the first angle gives a location on the equator. The second angle gives the location from the plane of the equator to the point on the sphere in line with the location specified on the equator.

Free Hand Sketching

Free hand sketching is a method of visualizing and conceptualizing your idea that allows you to communicate that idea with others. Sketches are NOT intended to be final engineering documents or drawings, but are a step in the process from an idea or thought to final design or to production.

Two types of drawings are generally associated with the four key stages of the engineering process: (1) Freehand sketches and (2) Detailed Engineering Drawings.

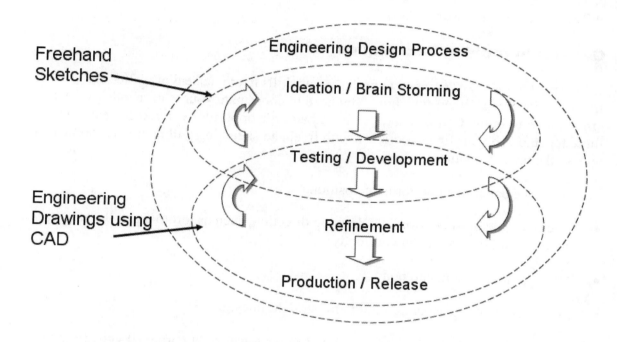

Freehand sketching is an important method to quickly document and to communicate your design ideas. Freehand sketching is a process of creating a rough, preliminary drawing to represent the main features of a design, whether it is a manufactured product, a chemical process, or a structure.

Sketches take many forms, and vary in level of detail. The designer or engineer determines the level of detail based on clarity and purpose of the sketch, as well as the intended audience. Sketches are important to record the fleeting thoughts associated with idea generation and brain storming in a group.

Freehand sketching is considered one of the most powerful methods to help develop visualization skills. The ability to sketch is helpful, not only to communicate with others, but also to work out details in ideas and to identify any potential problems.

Freehand sketching requires simple tools, a pencil, piece of paper, straight edge and can be accomplished almost anywhere. Creating freehand sketches does not require artistic ability, as some may assume.

General Sketching Techniques

Understand that it takes practice to perfect your skills in any endeavor, including freehand sketching. When sketching, you need to coordinate your eyes, hands (wrist and arm), and your brain. Chances are; you have had little opportunity in recent years to use these together, so your first experience with freehand sketching will be taxing. Some tips to ease the process include:

- Orient the paper in a comfortable position.

- Determine the most comfortable drawing direction, such as left to right, or drawing either toward or away from your body.

- Relax your hand, arm and body.

- Use the edge of the paper as a guide for straight lines.

- When using pencil, work from the top left to the lower right corner (if you are right-handed). This helps avoid smudging your work (your hand is resting on blank paper, rather than on your work).

- Remember that sketches are generally drawn without dimensions, since you are trying to represent the main features of your design concept.

- Use a wooden pencil with soft HB lead or a mechanical pencil in 5 mm or 7 mm.

☀ Today, you may not have a T-square available, but you can still sketch in your notebook and use good sketching techniques. You should also be prepared to sketch anywhere; even on the back of a napkin.

Geometric Entities

Points

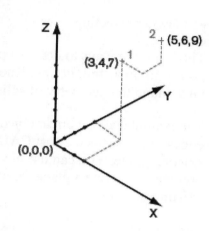

Points are geometrical constructs. Points are considered to have no width, height, or depth. Points are used to indicate locations in space. When you represent a point in a free hand sketch, the convention is to make a small cross or a bar if it is along a line, to indicate the location of the point.

In CAD drawings, a point is located by its coordinates and usually shown with some sort of marker like a cross, circle, or other representation. Many CAD systems allow you to choose the style and size of the mark that is used to represent points. Most CAD systems offer three ways to specify a point:

- End the coordinates for the point

- Select a point in the Graphics window

- Enter a point's location by its relationship to existing geometry (Example: a centerpoint, an endpoint of a line, or an intersection of two lines)

Picking a point from the screen is a quick way to enter points when the exact location is not important, but the accuracy of the CAD database makes it impossible to enter a location accurately in this way.

Lines

A straight line is defined as the shortest distance between two points. Geometrically, a line has length, but no other dimension such as width or thickness. Lines are used in drawings to represent the edge view of a surface, the limiting element of a contoured surface, or the edge formed where two surfaces on an object join. In CAD, 2D lines are typically stored by the coordinates (x,y) of their endpoints.

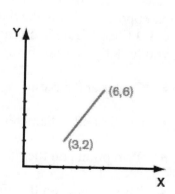

Planes

Planes are defined by:

- Two parallel lines

- Three points not lying in a straight line

- A point and a line

- Two intersecting lines

The last three ways to define a plane are all special cases of the more general case - three points not in a straight line. Knowing what can determine a plane can help you understand the geometry of solid objects - and use the geometry to work in CAD.

For example, a face on an object can is a plane that extends between the vertices and edges of the surface. Most CAD programs allow you to align new entities with an existing plane. You can use any face on the object - whether it is normal, inclined, or oblique - to define a plane for aligning a new entity. The plane can be specified using existing geometry.

Defining planes on the object or in 3D space is an important tool for working in 3D CAD. You will learn more about specifying planes to orient a user coordinate system to make it easy to create CAD geometry later in this text.

Circles

A circle is a set of points that are equidistant from a center point. The distance from the center to one of the points is the radius. The distance across the center to any two points on opposite sides is the diameter. The circumference of a circle contains 360° of arc. In a CAD file, a circle is often stored as a center point and radius. Most CAD systems allow you to define circles by specifying:

- Center and a radius

- Center and a diameter

- Two points on the diameter

- Three points on the circle

- Radius and two entities to which the circle is tangent

- Three entities to which the circle is tangent

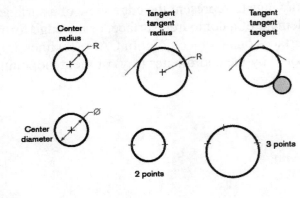

As with any points, the points defining a circle can be entered with absolute, relative, polar, cylindrical, or spherical coordinates; by picking points from the screen; or by specifying existing geometry.

Arcs

An arc is a portion of a circle. An arc can be defined by specifying:

- Center, radius, and angle measure (sometimes called the included angle or delta angle)

- Center, radius, and arc length

- Center, radius, and chord length

- Endpoints and arc length

- Endpoints and a radius

- Endpoints and one other point on the arc (3 points)

- Endpoints and a chord length

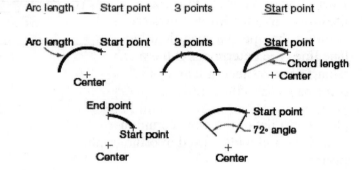

Solid Primitives

Many 3D objects can be visualized, sketched, and modeled in a CAD system by combining simple 3D shapes or primitives. Solid primitives are the building blocks for many solid objects. You should become familiar with these common primitive shapes and their geometry. The same primitives that helped you understand how to sketch objects can also help you create 3D models of them using your computer.

A common set of primitive solids that you can use to build more complex objects is illustrated: (a) box, (b) sphere, (c) cylinder, (d) cone, (e) torus, (f) wedge, and (g) pyramid.

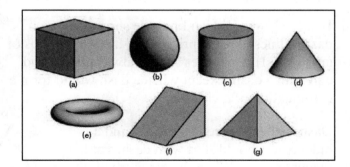

Look around the room and identify some solid primitives that make up the shapes you see. The ability to identify the primitive shapes can help you model features of the objects.

Alphabet of Lines

The lines used in drafting are referred to as the alphabet of lines.

Line types and conventions for mechanical drawings are covered in ANSI Standard Y14.2M. There are four distinct thicknesses of lines: Very Thick, Thick, Medium and Thin.

Every line on your drawing has a meaning. In other words, lines are symbols that mean a specific thing. The line type determines if the line is part of the object or conveys information about the object.

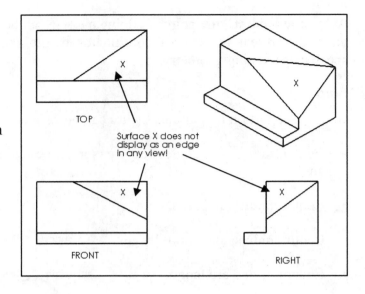

Below is a list of the most common line types and widths used in orthographic projection.

Visible lines: Visible lines (object or feature lines) are continuous lines used to represent the visible edges and contours (features) of an object. Since visible lines are the most important lines, they must stand out from all other secondary lines on the drawing. The line type is continuous and the line weight is thick (0.5 - 0.6 mm).

Hidden lines: Hidden lines are short-narrow dashed lines. They represent the hidden features of an object. Hidden lines should always begin and end with a dash, except when a dash would form a continuation of a visible line.

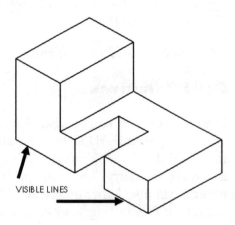

Dashes always meet at corners, and a hidden arc should start with dashes at the tangent points. When the arc is small, the length of the dash may be modified to maintain a uniform and neat appearance.

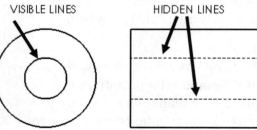

Excessive hidden lines are difficult to follow. Therefore, only lines or features that add to the clearness and the conciseness of the drawing should be displayed. Line weight is medium thick (0.35 - 0.45 mm).

Confusing and conflicting hidden lines should be eliminated. If hidden lines do not adequately define a part's configuration, a section should be taken that imaginarily cuts the part. Whenever possible, hidden lines are eliminated from the sectioned portion of a drawing. In SolidWorks, to hide a line, right click the line in a drawing view, click Hide.

Dimension lines: Dimension lines are thin lines used to show the extent and the direction of dimensions. Space for a single line of numerals is provided by a break in the dimension line.

If possible, dimension lines are aligned and grouped for uniform appearance and ease of reading. For example, parallel dimension lines should be spaced not less than (6 mm) apart, and no dimension line should be closer than (10 mm) to the outline of an object feature [(12 mm) is the preferred distance].

All dimension lines terminate with an arrowhead on mechanical engineering drawings, a slash or a dot in architecture drawings. The preferred ending is the arrowhead to an edge or a dot to a face. Line weight is thin (0.3 mm).

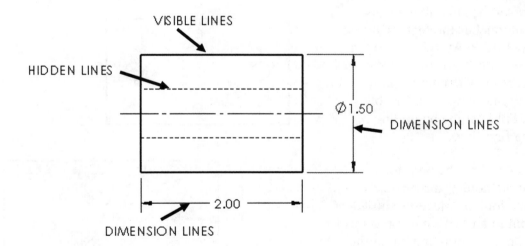

Extension lines: Extension lines are used to indicate the termination of a dimension. An extension line must not touch the feature from which it extends, but should start approximately (2 - 3mm) from the feature being dimensioned and extended the same amount beyond the arrow side of the last dimension line.

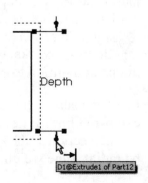

In SolidWorks, use the control points to create the needed extension line gap of ~1.5 - 2.5mms.

In SolidWorks, inserted dimensions in the drawing are displayed in gray. Imported dimensions from the part are displayed in black.

When extension lines cross other extension lines, dimension lines, leader lines, or object lines, they are usually not broken. When extension lines cross dimension lines close to an arrowhead, breaking the extension line is recommended for clarity. Line weight is thin (0.3 mm).

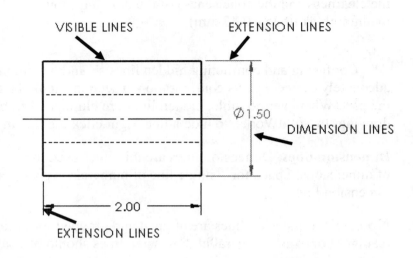

Leader lines: A leader line is a continuous straight line that extends at an angle from a note, a dimension, or other reference to a feature. An arrowhead touches the feature at that end of the leader. At the note end, a horizontal bar (6 mm) long terminates the leader approximately (3 mm) away from mid-height of the note's lettering, either at the beginning or end of the first line.

Leaders should not be bent to underline the note or dimension. Unless unavoidable, leaders should not be bent in any way except to form the horizontal terminating bar at the note end of the leader.

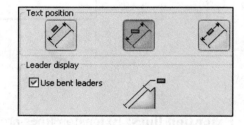

In SolidWorks, use the dimension option to control Leader display.

Leaders usually do not cross. Leaders or extension lines may cross an outline of a part or extension lines if necessary, but they usually remain continuous and unbroken at the point of intersection. When a leader is directed to a circle or a circular arc, its direction should be radial. Line weight is thin (0.3 mm).

Break lines: Break lines are applied to represent an imaginary cut in an object, so the interior of the object can be viewed or fitted to the sheet. Line weight is thick (0.5 – 0.6 mm).

In SolidWorks, Break lines are displayed as short dashes or continuous solid lines, straight, curved or zig zag.

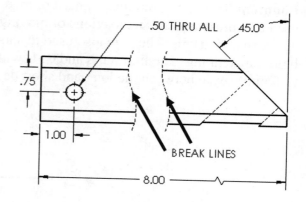

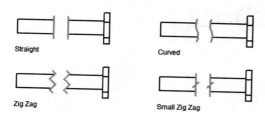

Centerlines: Centerlines are thin, long and short dashes, alternately and evenly spaced, with long dashes placed at each end of the line. The long dash is dependent on the size of the drawing and normally varies in length from (20 to 50 mm). Short dashes, depending on the length of the required centerline should be approximately (1.5 to 3.0 mm). Very short centerlines may be unbroken with dashes at both ends.

Centerlines are used to represent the axes of symmetrical parts of features, bolt circles, paths of motion, and pitch circles. They should extend about (3 mm) beyond the outline of symmetry, unless they are used as extension lines for dimensioning. Every circle, and some arcs, should have two centerlines that intersect at their center of the short dashes. Line weight is thin (0.3mm).

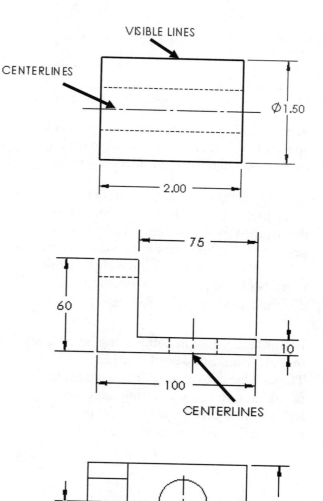

Phantom lines: Phantom lines consist of medium - thin, long and short dashes. They are used to represent alternate positions of moving parts, adjacent positions of related parts, and repeated details. They are also used to show the cast, or the rough shape, of a part before machining. The line starts and ends with the long dash of (15 mm) with about (1.5 mm) space between the long and short dashes. Line weight is usually (0.45 mm).

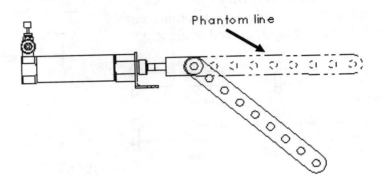

Phantom line

Section lines: Section lines are thin, uniformly spaced lines that indicate the exposed cut surfaces of an object in a sectional view.

Spacing should be approximately (3 mm) and at an angle of 45°. The section pattern is determined by the material being "cut" or sectioned. Section lines are commonly referred to as "cross-hatching." Line weight is thin (0.3 mm). Multiple parts in an assembly use different section angles for clarity.

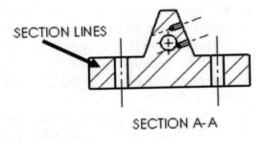

SECTION LINES

SECTION A-A

In this text, you will concentrate on creating 3D models using SolidWorks 2012. Three-dimensional modeling is an integral part of the design, manufacturing, and construction industry and contributes to increased productivity in all aspects of a project.

💡 Section lines can serve the purpose of identifying the kind of material the part is made from.

Below are few common section line types for various materials:

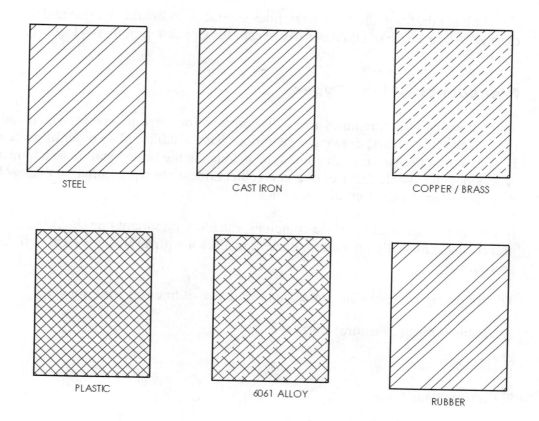

STEEL   CAST IRON   COPPER / BRASS

PLASTIC   6061 ALLOY   RUBBER

💡 A Section lined area is always completely bounded by a visible outline.

Cutting Plane lines: Cutting Plane lines show where an imaginary cut has been made through an object in order to view and understand the interior features. Line type is phantom. Line weight is very thick (0.6 - 0.8 mm).

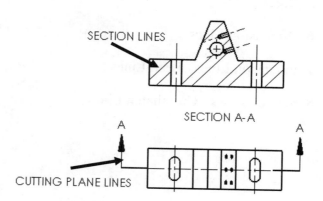

SECTION LINES

SECTION A-A

CUTTING PLANE LINES

Arrows are located at the ends of the cutting plane line and the direction indicates the line of sight into the object.

Line weight (also called hierarchy) refers to thickness.

In hand drafting, the contrast in lines should be in the line weight and not in the density. All lines are of equal density except for Construction lines - Light Thin so they can be erased.

Precedence of Line Types

When creating Orthographic views, it is common for one line type to overlap another line type. When this occurs, drawing conventions have established an order of precedence. For example - perhaps a visible line type belongs in the same location as a hidden line type, since the visible features of a part (object lines) are represented by thick solid lines, they take precedence over all other lines.

If a centerline and cutting plane coincides, the more important one should take precedence. Normally the cutting plane line, drawn with a thicker weight, will take precedence.

The following list gives the preferred precedence of lines on your drawing:

1. Visible (Object / Feature) Lines

2. Hidden Lines

3. Cutting Plane Lines

4. Centerlines

5. Break Lines

6. Dimension Lines

7. Extension Lines / Lead Lines

8. Section Lines / Crosshatch Lines

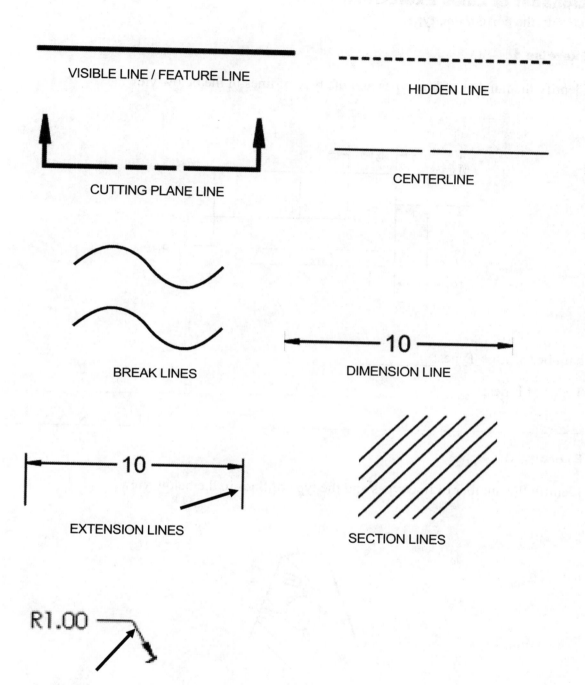

VISIBLE LINE / FEATURE LINE

HIDDEN LINE

CUTTING PLANE LINE

CENTERLINE

BREAK LINES

DIMENSION LINE

EXTENSION LINES

SECTION LINES

LEADER LINE - BENT

Alphabet of Lines Exercises:

Identify the correct line types:

Exercise 1:

Identify the number of line types and the type of lines in the below view.

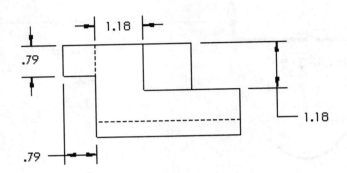

Number of Line Types:_____

Types of Lines: _____

_____

Exercise 2:

Identify the number of line types and the type of lines in the below view.

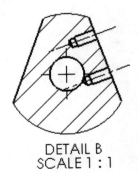

DETAIL B
SCALE 1 : 1

Number of Line Types:_____

Types of Lines: _____

_____

Exercise 3:

Identify the number of line types and the type of lines in the below view.

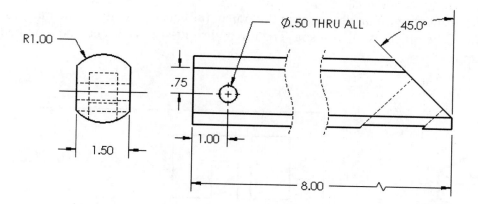

Number of Line Types:_____

Types of Lines: _____

_____

Orthographic Projection

Before an object is drawn or created, it is examined to determine which views will best furnish the information required to manufacture the object. The surface, which is to be displayed as the observer looks at the object, is called the Front view.

To obtain the front view of an object, turn the object (either physically or mentally) so that the front of the object is all you see. The top and right-side views can be obtained in a similar fashion.

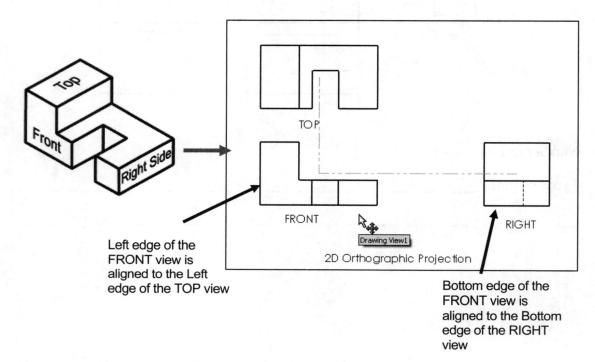

Left edge of the FRONT view is aligned to the Left edge of the TOP view

Bottom edge of the FRONT view is aligned to the Bottom edge of the RIGHT view

Orthographic projection is a common method of representing three-dimensional objects, usually by three two-dimensional drawings, in each of the object is viewed along parallel lines that are perpendicular to the plane of the drawing as illustrated. These lines remain parallel to the projection plane and are not convergent.

Orthographic projection provides the ability to represent the shape of an object using two or more views. These views together with dimensions and annotations are sufficient to manufacture the part.

The six principle views of an orthographic projection are illustrated. Each view is created by looking directly at the object in the indicated direction.

Glass Box and Six Principal Orthographic Views

The Glass box method is a traditional method of placing an object in an *imaginary glass box* to view the six principle views.

Imagine that the object you are going to draw is placed inside a glass box, so that the large flat surfaces of the object are parallel to the walls of the box.

From each point on the object, imagine a ray, or projector perpendicular to the wall of the box forming the view of the object on that wall or projection plane.

Then *unfold the sides* of the imaginary glass box to create the orthographic projection of the object.

There are two different types of Angle Projection: First and Third Angle Projection.

- First Angle Projection is used in Europe and Asia.

- Third Angle Projection is used in the United States.

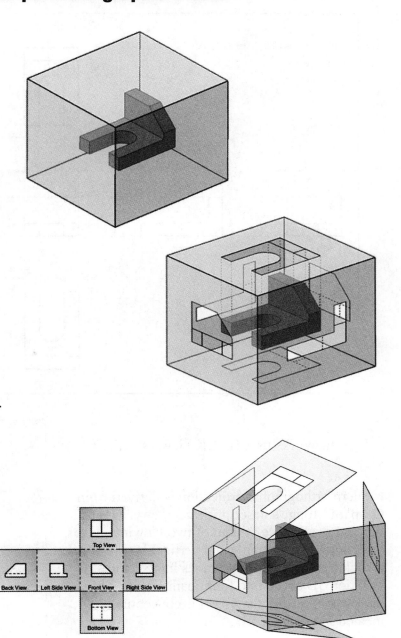

Third Angle Projection is used in the book. Imagine that the walls of the box are hinged and unfold the views outward around the front view. This will provide you with the standard arrangement of views.

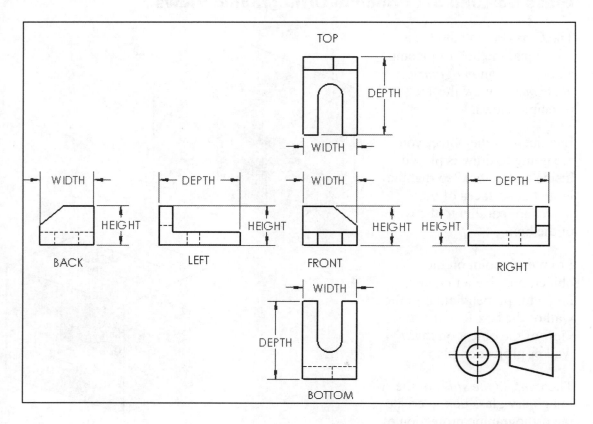

SolidWorks uses BACK view vs. REAR view.

Modern orthographic projection is derived from Gaspard Monge's descriptive geometry. Monge defined a reference system of two viewing planes, horizontal H ("ground") and vertical V ("backdrop"). These two planes intersect to partition 3D space into four quadrants. In **Third-Angle projection**, the object is conceptually located in quadrant III.

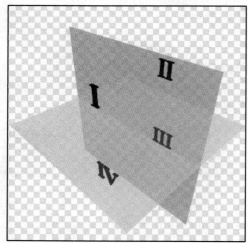

Both First Angle and Third Angle projections result in the same six views; the difference between them is the arrangement of these views around the box.

Below is an illustration of First Angle Projection.

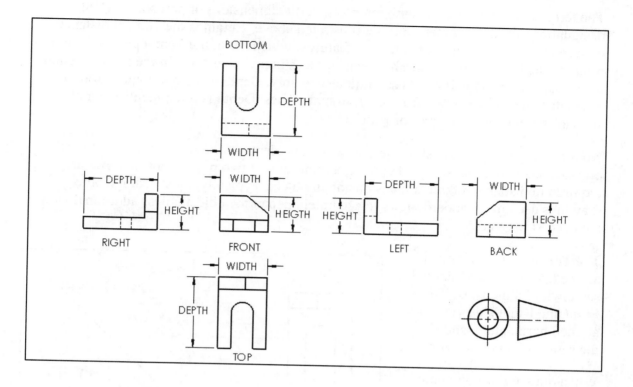

In First Angle projection, the object is conceptually located in quadrant I, i.e. it floats above and before the viewing planes, the planes are opaque, and each view is pushed through the object onto the plane furthest from it.

Both First Angle and Third Angle projections result in the same six views; the difference between them is the arrangement of these views around the box.

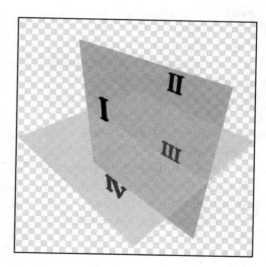

SolidWorks uses BACK view vs. REAR view.

In SolidWorks, when you create a new part or assembly, the three default Planes (Front, Right and Top) are aligned with specific views. The Plane you select for the Base sketch determines the orientation of the part or assembly.

Height, Width, and Depth Dimensions

The term height, width, and depth refer to specific dimensions or part sizes. ANSI designations for these dimensions are illustrated above. Height is the vertical distance between two or more lines or surfaces (features) which are in horizontal planes. Width refers to the horizontal distance between surfaces in profile planes. In the machine shop, the terms length and width are used interchangeably. Depth is the horizontal (front to back) distance between two features in frontal planes. Depth is often identified in the shop as the thickness of a part or feature.

No orthographic view can show height, width and depth in the same view. Each view only depicts two dimensions. Therefore, a minimum of two projections or views are required to display all three dimensions of an object. Typically, most orthographic drawings use three standard views to accurately depict the object unless additional views are needed for clarity.

The Top and Front views are aligned vertically and share the same width dimension. The Front and Right side views are aligned horizontally and share the same height dimension.

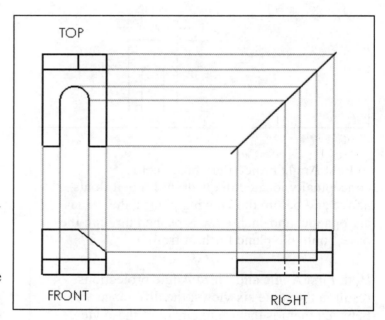

When drawing orthographic projections, spacing is usually equal between each of the views. The Front, Top, and Right views are most frequently used to depict or orthographic projection.

The Front view should show the **most features** or characteristics of the object. It usually contains the least number of hidden lines. All other views are based (projected) on the orientation chosen for the front view.

Transferring Dimensions

In SolidWorks, you can view the projection lines from the Front view placement.

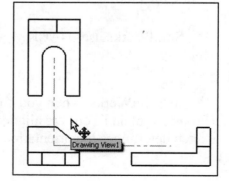

When transferring measurements between views, the width dimension can be projected from the Front view upward to the Top view or vice versa and that the height dimension can be projected directly across from the Front view to the Right view.

Depth dimensions are transferred from the Top view to the Right view or vice versa.

Height dimensions can be easily projected between two views using the grid on grid paper. Note: the grid is not displayed in the illustration to provide improved line and picture quality.

The miter line drawn at a 45° is used to transfer depth dimensions between the Top and Right view.

When constructing an Orthographic projection, you need to include enough views to completely describe the true shape of the part. You will address this later in the book.

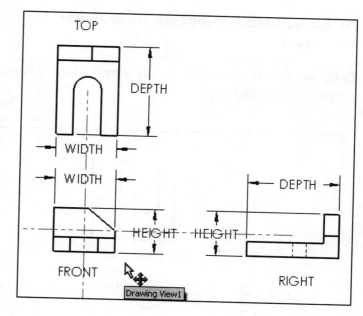

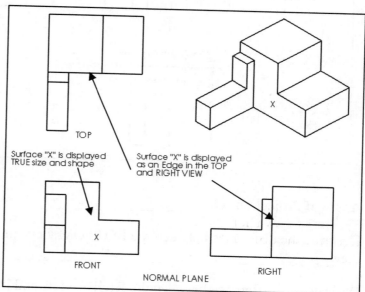

Orthographic Projection Exercises:
Exercise 1:

Label the four remaining Principle views with the appropriate view name. Identify the Angle of Projection type.

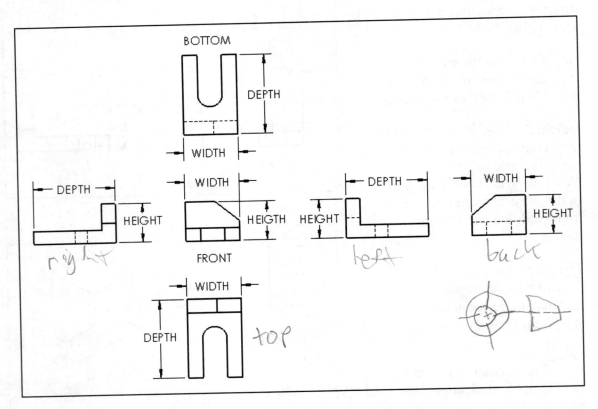

Angle of Projection type:_____ first _____

Describe the difference between the BOTTOM view and the TOP view_____

Describe the difference between the RIGHT view and the LEFT view. right has the slopty side _____

Which views have the least amount of Hidden Lines? frout, top _____

Exercise 2:

Identify the number of views required to completely describe the illustrated box.

1.) One view

(2.) Two views

3.) Three views

4.) Four views

5.) More than four views

Explain
Why? to get height, width, & depth

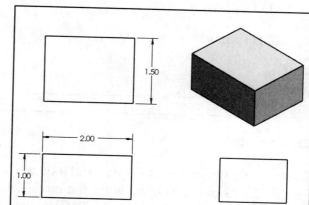

Exercise 3:

Identify the number of views required to completely describe the illustrated sphere.

(1.) One view

(2.) Two views

3.) Three views

4.) Four views

5.) More than four views

Explain
Why? 1 for measurement & 1 ortho to show its a sphere

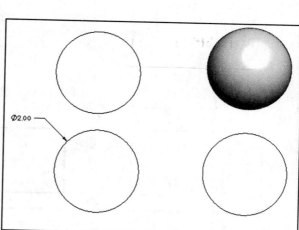

Exercise 4:

In Third Angle Projection, identify the view that displays
the most information about the illustrated model.

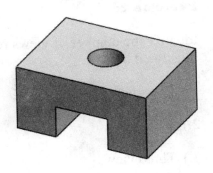

1.) FRONT view.

2.) TOP view

3.) BOTTOM view

4.) RIGHT view

Explain
Why? _shows all the features_

Exercise 5:

Third Angle Projection is displayed. Draw the Visible
Feature Lines of the TOP view for the model. Fill in the
missing lines in the FRONT view, RIGHT view and TOP view.

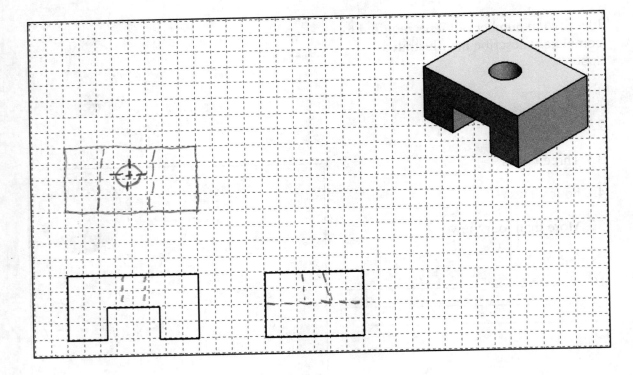

Exercise 6:

Third Angle Projection is displayed. Fill in the missing lines in the FRONT view, RIGHT view and TOP view.

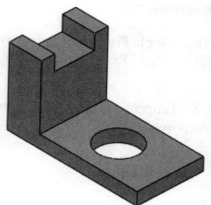

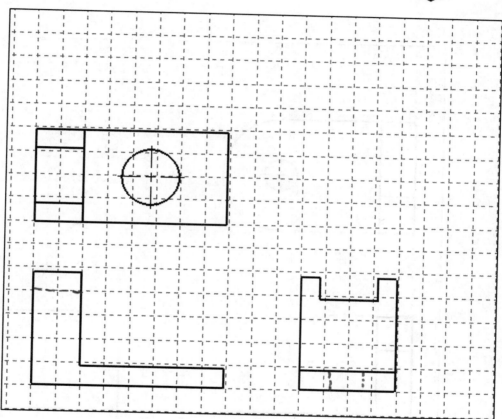

Exercise 7:

Third Angle Projection is displayed. Fill in the missing lines in the FRONT view, RIGHT view and TOP view.

Tangent Edges are displayed for educational purposes.

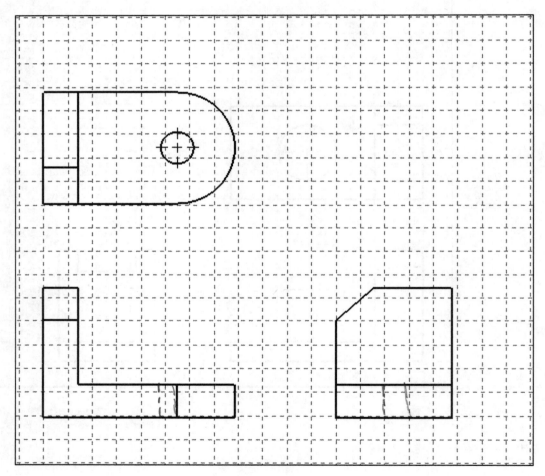

Planes (Normal, Inclined and Oblique)

Each type of plane (Normal, Inclined and Oblique) has unique characteristics when viewed in orthographic projection. To understand the three basic planes, each is illustrated.

Normal planes appear as an edge in two views and a true sized in the remaining view when using three views such as the Front, Top, and Right side views.

When viewing the six possible views in an orthographic projection, a normal plane appears as an edge in four views and a true sized plane in the other two views.

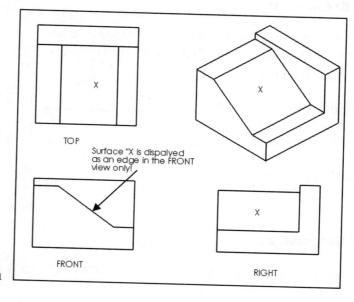

Inclined Planes appear as an edge view in one of the three views as illustrated. The inclined plane is displayed as a rectangular surface in the other two views. Note: The two rectangular surfaces appear "normal" they are foreshortened and no not display the true size or shape of the object.

Oblique Planes do not display as an edge view in any of the six principle orthographic views. They are not parallel or perpendicular to the projection planes. Oblique planes are displayed as a plane and have the same number of corners in each of the six views.

In SolidWorks, you can insert a sketch on any plane or face.

In SolidWorks, when you create a new part or assembly, the three default Planes (Front, Right and Top) are aligned with specific views. The Plane you select for the Base sketch determines the orientation of the part or assembly.

Plane Exercises:
Exercise 1:

Identify the surfaces with the appropriate letter that will appear in the FRONT view, TOP view and RIGHT view.

FRONT view surfaces:___D, A_____

TOP view surfaces:___C, A_____

RIGHT view surfaces:___B_____

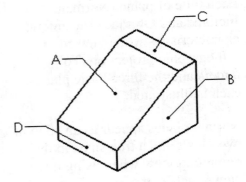

Exercise 2:

Estimate the size; draw the FRONT view, TOP view and RIGHT view of the illustrated part.

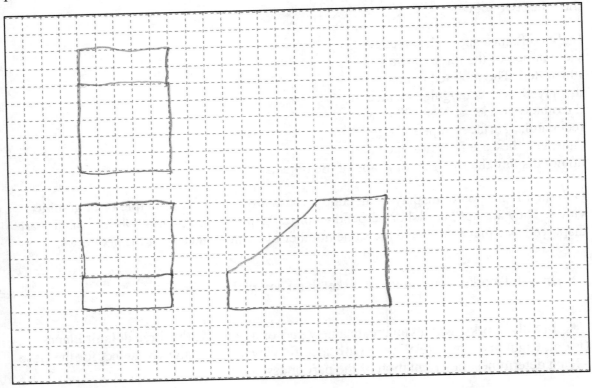

Exercise 3:

Identify the surfaces with the appropriate letter that will appear in the FRONT view, TOP view and RIGHT view.

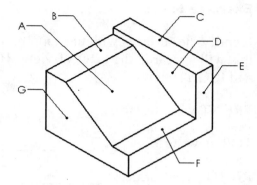

FRONT view surfaces: _G, D_

TOP view surfaces: _B, A, F, C_

RIGHT view surfaces: _F, E,_

Exercise 4:

Estimate the size; draw the FRONT view, TOP view and RIGHT view of the illustrated part.

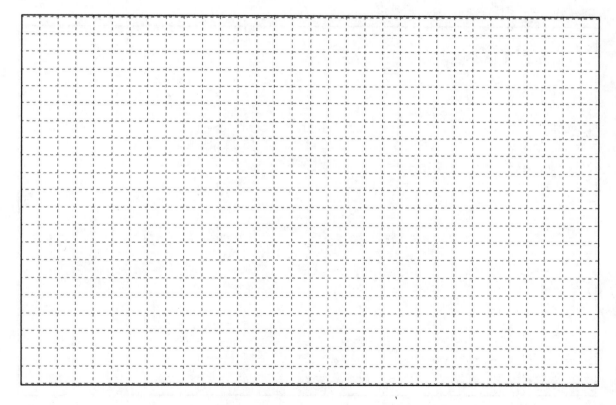

Exercise 5:

Identify the surfaces with the appropriate letter that will appear in the FRONT view, TOP view and RIGHT view.

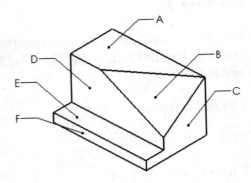

FRONT view surfaces:_____

TOP view surfaces:_____

RIGHT view surfaces:_____

Exercise 6:

Estimate the size; draw the FRONT view, TOP view and RIGHT view of the illustrated part.

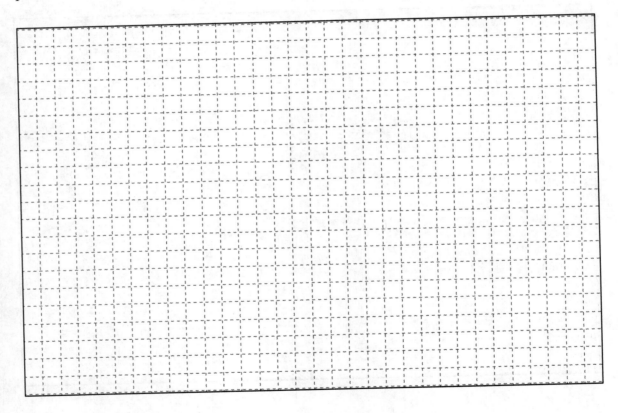

Exercise 7:

Identify the surfaces with the appropriate letter that will appear in the FRONT view, TOP view and RIGHT view.

FRONT view
surfaces:_____

TOP view surfaces:_____

RIGHT view surfaces:_____

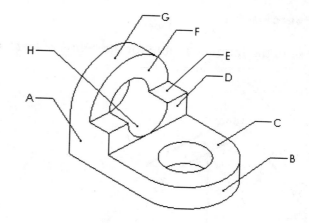

Exercise 8:

Identify the surfaces with the appropriate letter that will appear in the FRONT view, TOP view and RIGHT view.

FRONT view
surfaces:_____

TOP view surfaces:_____

RIGHT view surfaces:_____

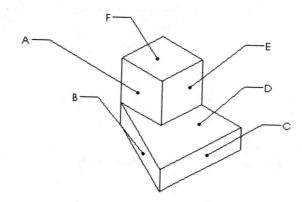

Exercise 9:

Identify the surfaces with the appropriate letter that will appear in the FRONT view, TOP view and RIGHT view.

FRONT view
surfaces:_____

TOP view surfaces:_____

RIGHT view
surfaces:_____

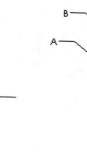
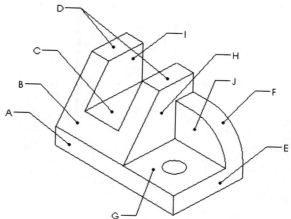

Exercise 10:

Fill in the following table for the below object.

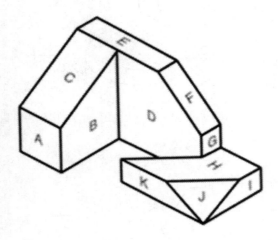

SURFACE	TOP	FRONT	RIGHT
A			
B			
C			
D			
E			
F			
G			
H			
I			
J			
K			

Exercise 11:

Fill in the following table for the below object.

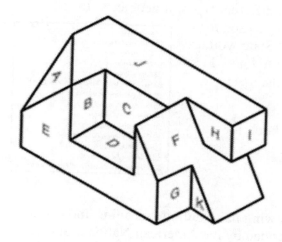

SURFACE	TOP	FRONT	RIGHT
A			
B			
C			
D			
E			
F			
G			
H			
I			
J			
K			

Chapter Summary

Chapter 1 provided a short discussion on the history of Engineering Graphics and the evolution of manual drawing/drafting. You were introduced to general sketching techniques and the 2D and 3D Cartesian Coordinate system. In engineering graphics there is a specific alphabet of lines that represent different types of geometry. In Orthographic projection, the Glass Box method was used to distinguish the six principle views. In the United State, you use Third Angle Projection. However, it is important to know First Angle Projection and other international standards.

View the files on the enclosed DVD in the book for additional information.

- Avi folder
- Alphabet of lines and Precedent of Line Types
- Boolean Operation
- Design Intent
- Fastners in General
- Fundamental ASME Y14.5 Dimensioning Rules
- General Tolerancing and Fits
- History of Engineering Graphics
- Measurement and Scale
- Open a Drawing Document 2012
- Open an Assembly Document 2012
- Part and Drawing Dimensioning
- SolidWorks Basic Concepts
- Visualization, Arrangement of Views, and Primitives

Chapter Terminology

Alphabet of Lines: Each line on a technical drawing has a definite meaning and is draw in a certain way. The line conventions recommended by the American National Standards Institute (ANSI) are presented in this text.

Axonometric Projection: A type of parallel projection, more specifically a type of orthographic projection, used to create a pictorial drawing of an object, where the object is rotated along one or more of its axes relative to the plane of projection.

Cartesian Coordinate system: Specifies each point uniquely in a plane by a pair of numerical coordinates, which are the signed distances from the point to two fixed perpendicular directed lines, measured in the same unit of length. Each reference line is called a coordinate axis or just axis of the system, and the point where they meet is its origin.

Depth: The horizontal (front to back) distance between two features in frontal planes. Depth is often identified in the shop as the thickness of a part or feature.

Engineering Graphics: Translates ideas from design layouts, specifications, rough sketches, and calculations of engineers & architects into working drawings, maps, plans, and illustrations which are used in making products.

First Angle Projection: In First Angle Projection the Top view is looking at the bottom of the part. First Angle Projection is used in Europe and most of the world. However America and Australia use a method known as Third Angle Projection.

French curve: A template made out of plastic, metal or wood composed of many different curves. It is used in manual drafting to draw smooth curves of varying radii.

Glass Box method: **A** traditional method of placing an object in an *imaginary glass box* to view the six principle views.

Grid: A system of fixed horizontal and vertical divisions.

Height: The vertical distance between two or more lines or surfaces (features) which are in horizontal planes.

Isometric Projection: A form of graphical projection, more specifically, a form of axonometric projection. It is a method of visually representing three-dimensional objects in two dimensions, in which the three coordinate axes appear equally foreshortened and the angles between any two of them are 120 °.

Precedence of Line types: When obtaining orthographic views, it is common for one type of line to overlap another type. When this occurs, drawing conventions have established an order of precedence.

Origin: The point of intersection, where the X,Y,Z axes meet, is called the origin.

Orthographic Projection: A means of representing a three-dimensional object in two dimensions. It is a form of parallel projection, where the view direction is orthogonal to the projection plane, resulting in every plane of the scene appearing in affine transformation on the viewing surface.

Right-Hand Rule: Is a common mnemonic for understanding notation conventions for vectors in 3 dimensions.

Scale: A relative term meaning "size" in relationship to some system of measurement.

T-Square: A technical drawing instrument, primarily a guide for drawing horizontal lines on a drafting table. It is used to guide the triangle that draws vertical lines. Its name comes from the general shape of the instrument where the horizontal member of the T slides on the side of the drafting table. Common lengths are 18", 24", 30", 36" and 42".

Third Angle Projection: In Third angle projection the Top View is looking at the Top of the part. First Angle Projection is used in Europe and most of the world. America and Australia use the Third Angle Projection method.

Units: Used in the measurement of physical quantities. Decimal inch dimensioning and Millimeter dimensioning are the two types of common units specified for engineering parts and drawings.

Width: The horizontal distance between surfaces in profile planes. In the machine shop, the terms length and width are used interchangeably.

Questions

1. Describe the Cartesian coordinate system.

2. Name the point of intersection, where the axes meet _____.

3. Explain the Right-hand rule in drafting.

4. Why is Freehand sketching important to understand?

5. Describe the different between First and Third Angle Projection type.

6. True or False. First Angle Projection type is used in the United States.

7. Explain the Precedent of Line types. Provide a few examples.

8. True or False. A Hidden Line has precedent over a Visible / Feature line.

9. True or False. The intersection of the two axes creates four regions, called quadrants, indicated by the Roman numerals I, II, III, and IV.

10. Explain the Glass Box method for Orthographic Projection.

11. True or False. Both First Angle and Third Angle Projection type result in the same six views; the difference between them is the arrangement of these views.

12. True or False. Section lines can serve the purpose of identifying the kind of material the part is made from.

13. True or False. All dimension lines terminate with an arrowhead on mechanical engineering drawings.

14. True or False. Break lines are applied to represent an imaginary cut in an object, so the interior of the object can be viewed or fitted to the sheet. Provide an example.

15. True or False. The Front view should show the most features or characteristics of the object. It usually contains the least number of hidden lines. All other views are based (projected) on the orientation chosen for the front view. Explain your answer.

Exercises

Exercise 1.1: Third Angle Projection type is displayed. Name the illustrated views in the below model.

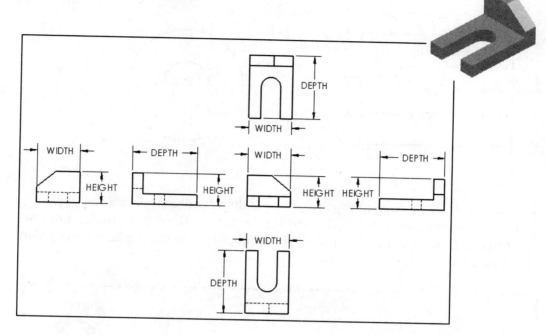

Exercise 1.2: First Angle Projection type is displayed. Name the illustrated views in the below model.

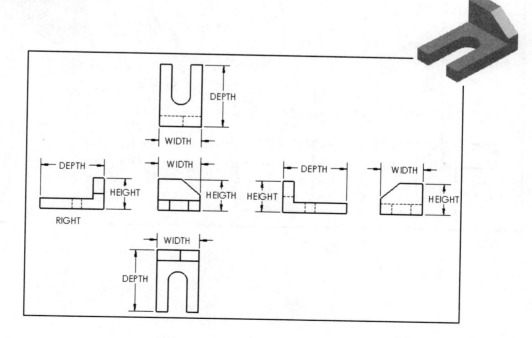

Exercise 1.3: Identify the various Line types in the below model.

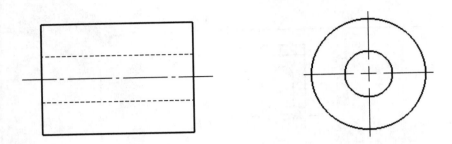

Exercise 1.4: Third Angle Projection type is displayed. Estimating the distance, draw the Visible Feature Lines of the TOP view for the model. Draw any Hidden lines or Centerlines if needed. Identify the view that displays the most information about the illustrated model.

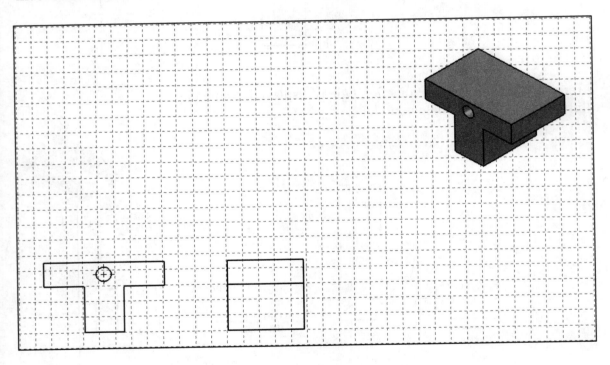

Exercise 1.5: Third Angle Projection type is displayed. Estimating the distance, draw the Visible Feature Lines of the Right view for the model. Draw any Hidden lines if needed. Identify the view that displays the most information about the illustrated model.

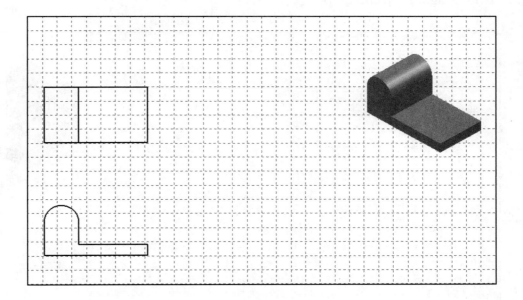

Exercise 1.6: Estimating the distance, draw the Front view, Top view and Right view. Draw the Visible Feature Lines for the model. Draw any Hidden lines if needed. Which view displays the most information about the illustrated model? Note: Third Angle Projection.

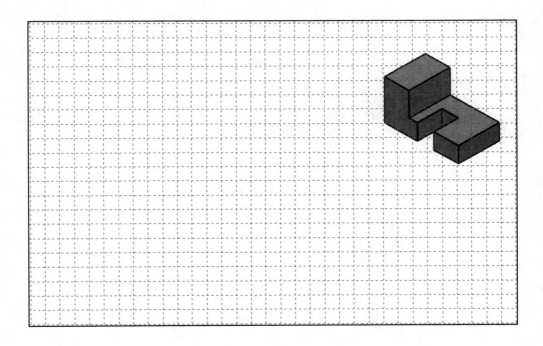

Exercise 1.7: Estimating the distance, draw the Front view, Top view and Right view. Draw the Visible Feature Lines for the model. Draw any Hidden lines or Centerlines if needed. Which view displays the most information about the illustrated model? Note: Third Angle Projection type.

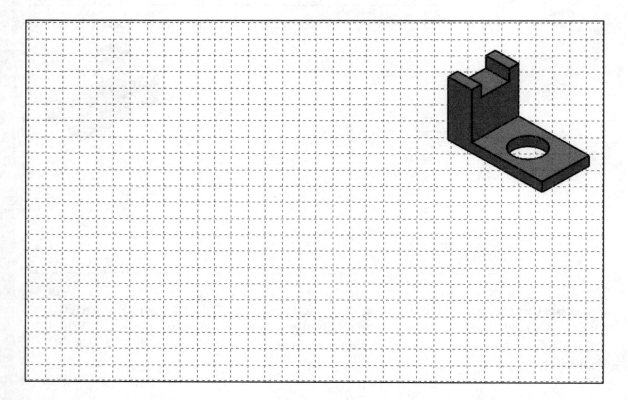

Exercise 1.8: Draw the Isometric view. Note: Third Angle Projection.

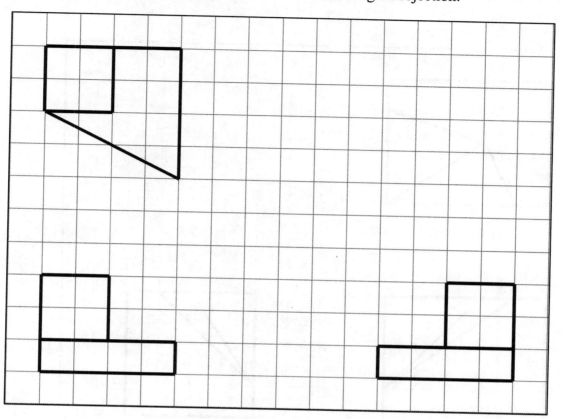

Exercise 1.9: Draw the Isometric view. Note: Third Angle Projection.

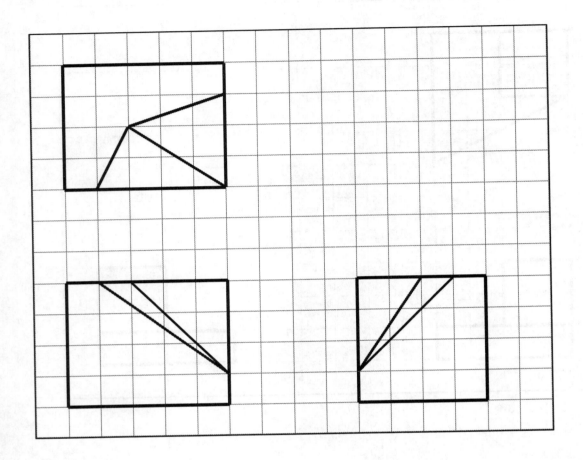

Exercise 1.10: Draw the Isometric view. Note: Third Angle Projection

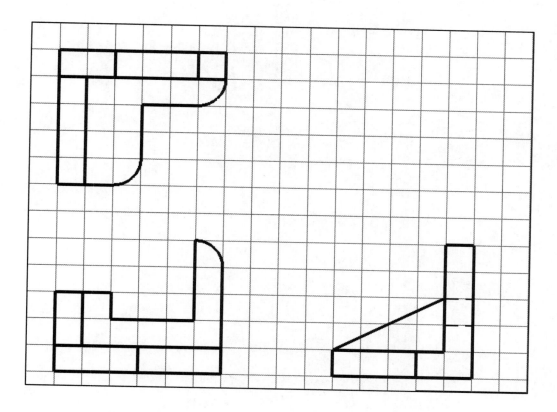

Exercise 1.11:

Identify the surfaces with the appropriate letter that will appear in the FRONT view, TOP view and RIGHT view.

FRONT view
surfaces:_____

TOP view surfaces:_____

RIGHT view
surfaces:_____

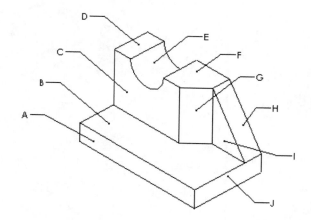

Notes:

Chapter 2

Isometric Projection and Multi View Drawings

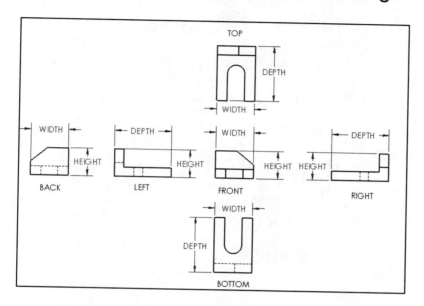

Below are the desired outcomes and usage competencies based on the completion of Chapter 2.

Desired Outcomes:	Usage Competencies:
• Understand Isometric Projection and 2D sketching.	• Identify the three main projection divisions in freehand engineering sketches and drawings: o Axonometric, Oblique, and Perspective
• Knowledge of additional Projection views and arrangement of drawing views.	• Create one and two view drawings.
• Comprehend the history and evolution of CAD and the development of SolidWorks. • Recognize Boolean operations and feature based modeling.	• Identify the development of in historic CAD systems and SolidWorks features, parameters and design intent of a sketch, part, assembly and drawing. • Apply the Boolean operation: Union, Difference and Intersection.

Notes:

Chapter 2 - Isometric Projection and Multi View Drawings

Chapter Overview

Chapter 2 provides a general introduction into Isometric Projection and Sketching along with Additional Projections and arrangement of views. It also covers advanced drawing views and an introduction from manual drafting to CAD.

On the completion of this chapter, you will be able to:

- Understand and explain Isometric Projection.

- Create an Isometric sketch.

- Identify the three main projection divisions in freehand engineering sketches and drawings:

  o Axonometric

  o Oblique

  o Perspective

- Comprehend the history and evolution of CAD.

- Recognize the following Boolean operations: Union, Difference, and Intersection.

- Understand the development of SolidWorks features, parameters and design intent of a sketch, part, assembly and drawing.

Isometric Projections

There are three main projection divisions commonly used in freehand engineering sketches and detailed engineering drawings; they are: 1.) Axonometric, with its divisions in Isometric, Dimetric and Trimetric, 2.) Oblique, and 3.) Perspective. Let's review the three main divisions.

Axonometric is a type of parallel projection, more specifically a type of Orthographic projection, used to create a pictorial drawing of an object, where the object is rotated along one or more of its axes relative to the plane of projection.

There are three main types of Axonometric projection: Isometric, Dimetric, and Trimetric projection depending on the exact angle at which the view deviates from the Orthogonal.

To display Isometric, Dimetric, or Trimetric of a 3D SolidWorks model, select the drop-down arrow from the View Orientation icon in the Heads-up view toolbar.

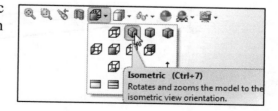

Axonometric drawings often appear distorted because they ignore the foreshortening effects of perspective (foreshortening means the way things appear to get smaller in both height and depth as they recede into the distance). Typically, Axonometric drawings use vertical lines for those lines representing height and sloping parallel edges for all other sides.

- *Isometric Projection.* Isometric projection is a method of visually representing three-dimensional objects in two dimensions, in which the three coordinate axes appear equally foreshortened and the angles between them are 120 °.

The term "Isometric" comes from the Greek for "equal measure", reflecting that the scale along each axis of the projection is the same (this is not true of some other forms of graphical projection).

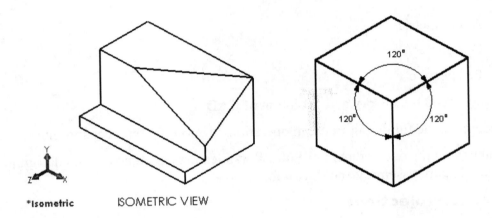

*Isometric   ISOMETRIC VIEW

- *Dimetric Projection.* A Dimetric projection is created using 3 axes but only two of the three axes have equal angles. The smaller these angles are, the less we see of the top surface. The angle is usually around 105°.

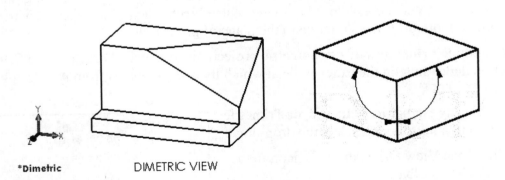

*Dimetric   DIMETRIC VIEW

- *Trimetric Projection.* A Trimetric projection is created using 3 axes where each of the angles between them is different (there are no equal angles). The scale along each of the three axes and the angles among them are determined separately as dictated by the angle of viewing. Approximations in trimetric drawings are common.

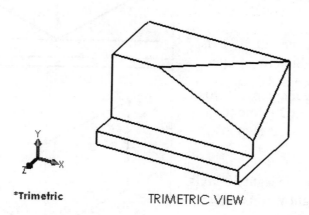

*Trimetric   TRIMETRIC VIEW

Isometric Sketching

Isometric sketches provide a 3D dimensional pictorial representation of an object. Isometric sketches helps in the visualization of an object.

The surface features or the axes of the object are drawn around three axes from a horizontal line; vertical axis, and 30° axis to the right, and a 30° axis to the left. All three axes intersect at a single point on the horizontal line.

All horizontal lines in an Isometric sketch are always drawn at 30° and parallel to each other, and are either to the left or to the right of the vertical.

For this reason, all shapes in an Isometric sketch are not true shapes, they are distorted shapes.

All vertical lines in an Isometric sketch are always drawn vertically, and they are always parallel to each other as illustrated in the following example.

View the additional presentations on the enclosed DVD for supplementary information.

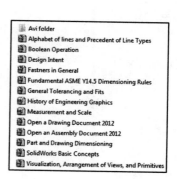

Example 1:

Exercise: Draw an Isometric sketch of a cube.

1. Draw a light horizontal axis (construction line) as illustrated on graph paper. Draw a light vertical axis. Draw a light 30° axis to the right. Draw a light 30° axis to the left.

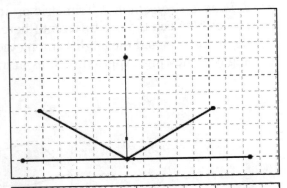

2. Measure the length along the left 30° axis, make a mark and draw a light vertical line.

3. Measure the height along the vertical axis, make a mark and draw a light 30° line to the left to intersect the vertical line drawn in step 2.

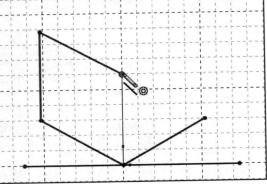

4. Measure the length along the right 30° axis, make a mark and draw a light vertical line.

5. From the height along the vertical axis, make a mark and draw a light 30° line to the right to intersect the vertical line drawn in step 4.

6. Draw a light 30° line to the right and a light 30° line to the left to complete the cube. Once the sketch is complete, darken the shape.

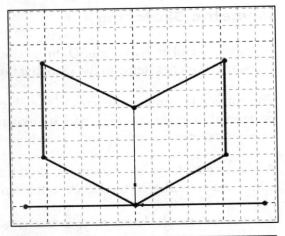

In an Isometric drawing, the object is viewed at an angle, which makes circles appear as ellipses.

Isometric Rule #1: Measurement can only be made on or parallel to the isometric axis.

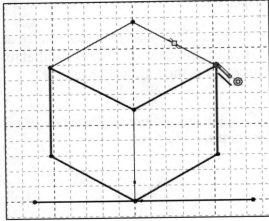

Circles drawn in Axonometric view

A circle drawn on a sloping surface in Axonometric projection will be drawn as an ellipse. An ellipse is a circle turned through an angle. All the examples shown above were box shapes without any curved surfaces. In order to draw curved surfaces we need to know how to draw an ellipse.

If you draw a circle and rotate it slowly, it will become an ellipse. As it is turned through 90° - it will eventually become a straight line. Rotate it 90° again, and it will eventually be back to a circle.

Example 1:

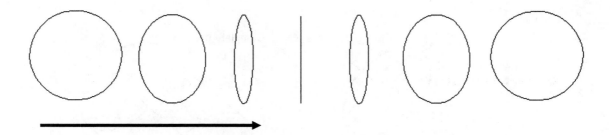

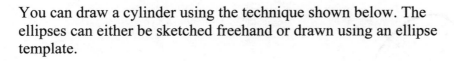

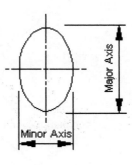

An ellipse has a major axis and a minor axis. The major axis is the axis about which the ellipse is being turned. The minor axis becomes smaller as the angle through which the ellipse is turned approaches 90°.

You can draw a cylinder using the technique shown below. The ellipses can either be sketched freehand or drawn using an ellipse template.

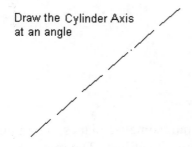

Draw the Cylinder Axis at an angle

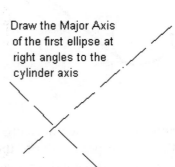

Draw the Major Axis of the first ellipse at right angles to the cylinder axis

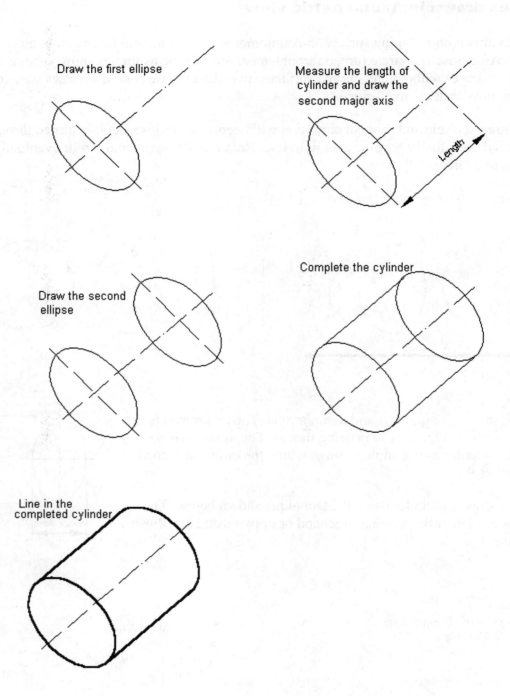

Draw the first ellipse

Measure the length of cylinder and draw the second major axis

Length

Draw the second ellipse

Complete the cylinder

Line in the completed cylinder

💡 Isometric Rule #2: When drawing ellipses on normal isometric planes, the minor axis of the ellipse is perpendicular to the plane containing the ellipse. The minor axis is perpendicular to the corresponding normal isometric plane.

Additional Projections

Oblique Projection: In Oblique projections; the front view is drawn true size, and the receding surfaces are drawn on an angle to give it a pictorial appearance. This form of projection has the advantage of showing one face (the front face) of the object without distortion. Generally, the face with the greatest detail; faces the front.

There are two types of Oblique projection used in engineering design.

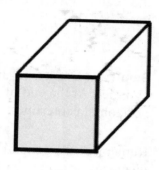

- *Cavalier*: In Cavalier Oblique drawings, all lines (including receding lines) are created to their true length or scale (1:1).

- *Cabinet*: In Cabinet Oblique drawings, the receding lines are shortened by one-half their true length or scale to compensate for distortion and to approximate more closely what the human eye would see. It is for this reason that Cabinet Oblique drawings are the most used form of Oblique drawings.

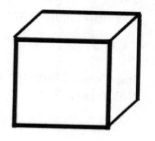

In Oblique drawings, the three axes of projection are vertical, horizontal, and receding. The front view (vertical & horizontal axis) is parallel to the frontal plane and the other two faces are oblique (receding). The direction of projection can be top-left, top-right, bottom-left, or bottom-right. The receding axis is typically drawn at 60°, 45° or 30°.

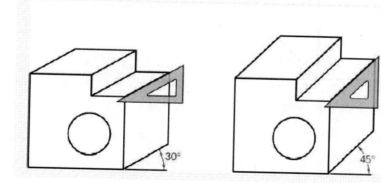
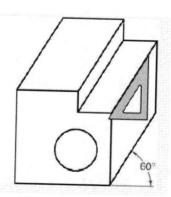

In the oblique pictorials coordinate system, only one axes is at an angle. The most commonly used angle is 45°.

Isometric Rule #1: A measurement can only be made on or parallel to the isometric axis. Therefore you cannot measure an isometric inclined or oblique line in an isometric drawing because they are not parallel to an isometric axis.

Example: Drawing cylinders in Oblique projection is quite simple if the stages outlined below are followed. In comparison with other ways of drawing cylinders (for example, perspective and isometric) using Oblique projection is relatively easy.

Step One: Draw vertical and horizontal centerlines to indicate the center of a circle, then use a compass to draw the circle itself.

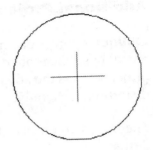

Step Two: Draw a 45° line to match the length on the cylinder. At the end of this line, draw vertical and horizontal centerlines.

Remember the general rule for Oblique is to half all distances projected backwards. If the cylinder is 100mm in length the distance back must be drawn to 50mm.

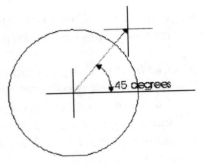

Step Three: Draw the second circle with a compass as illustrated.

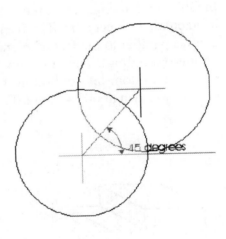

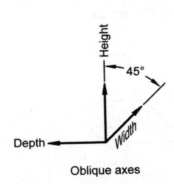

Oblique axes

Step Four: Draw two 45° lines - to join the front and back circles.

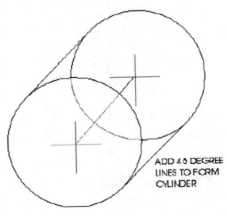

ADD 45 DEGREE LINES TO FORM CYLINDER

Step Five: Go over the outline of the cylinder with a fine pen or sharp pencil. Add shading - if required.

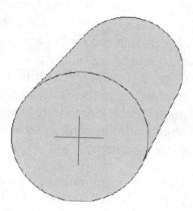

Perspective Projection: If you look along a straight road, the parallel sides of the road appear to meet at a point in the distance. This point is called the vanishing point and has been used to add realism. Suppose you want to draw a road that vanishes into the distance.

The rays from the points a given distance from the eye along the lines of the road, are projected to the eye. The angle formed by the rays decreases with increasing distance from the eye.

In SolidWorks, to display a Perspective view of the 3D model, click View, Display, Perspective from the Main toolbar.

A perspective drawing typically aims to reproduce how humans see the world: objects that are farther away seem smaller, etc. Depending on the type of perspective (1-pt, 2-pt, 3-pt), vanishing points are established in the drawing towards which lines recede, mimicking the effect of objects diminishing in size with distance from the viewer.

One vanishing point is typically used for roads, railroad tracks, or buildings viewed so that the front is directly facing the viewer as illustrated above.

Any objects that are made up of lines either directly parallel with the viewer's line of sight or directly perpendicular (the railroad slats) can be represented with one-point perspective.

The selection of the locations of the vanishing points, which is the first step in creating a perspective sketch, will affect the looks of the resulting images.

Two-point perspective can be used to draw the same objects as one-point perspective, rotated: looking at the corner of a house, or looking at two forked roads shrink into the distance, for example. One point represents one set of parallel lines, the other point represents the other. Looking at a house from the corner, one wall would recede towards one vanishing point, the other wall would recede towards the opposite vanishing point as illustrated.

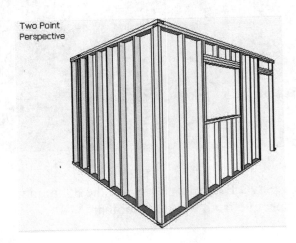

Two Point
Perspective

Three-point perspective is usually used for buildings seen from above (or below). In addition to the two vanishing points from before, one for each wall, there is now one for how those walls recede into the ground. This third vanishing point will be below the ground. Looking up at a tall building is another common example of the third vanishing point. This time the third vanishing point is high in space.

One-point, two-point, and three-point perspectives appear to embody different forms of calculated perspective. Despite conventional perspective drawing wisdom, perspective basically just means "position" or "viewpoint." of the viewer relative to the object.

Arrangement of Views

The main purpose of an engineering drawing is to provide the manufacturer with sufficient information needed to build, inspect or assemble the part or assembly according to the specifications of the designer. Since the selection and arrangement of views depends on the complexity of a part, only those views that are needed should be drawn.

The average part drawing which includes the Front view, Top view and Right view - are known as a three-view drawing. However, the designation of the views is not as important as the fact that the combination of views must give all the details of construction in clear, correct, and concise way.

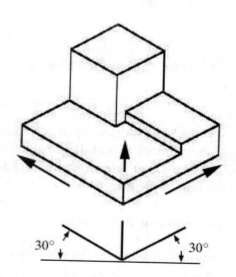

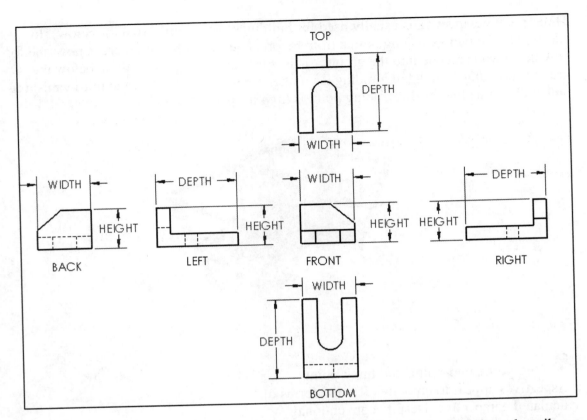

The designer usually selects as a Front view of the object that view which best describes the general shape of the part. This Front view may have no relationship to the actual front position of the part as it fits into an assembly.

The names and positions of the different views that may be used to describe an object are illustrated.

Third "3rd"Angle Projection type is displayed and used in this book.

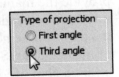

In SolidWorks, when you create a new part or assembly, the three default Planes (Front, Right and Top) are aligned with specific views. The Plane you select for the Base sketch determines the orientation of the part or assembly.

Two view drawing

Simple symmetrical flat objects and cylindrical parts, such as sleeves, shafts, rods, and studs require only two views to provide the full details of construction.

In the Front view, the centerline runs through the axis of the parts as a horizontal centerline. If the plug is in a vertical position, the centerline runs through the axis as vertical centerline.

The second view of the two-view drawing contains a horizontal and vertical centerline intersection at the center of the circles which make up the part in this view.

The selection of views for a two-view drawing rests largely with the designer/engineer.

Example 1:

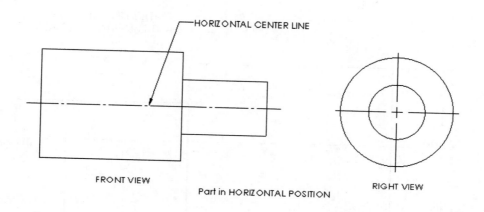

HORIZONTAL CENTER LINE

FRONT VIEW

Part in HORIZONTAL POSITION

RIGHT VIEW

Example 2:

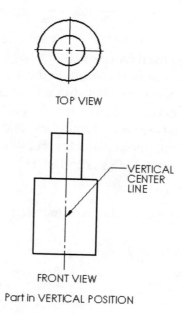

TOP VIEW

VERTICAL CENTER LINE

FRONT VIEW

Part in VERTICAL POSITION

One view drawing

Parts that are uniform in shape often require only one view to describe them adequately. This is particularly true of cylindrical objects where a one-view drawing saves time and simplifies the drawing.

When a one-view drawing of a cylindrical part is used, the dimension for the diameter (according to ANSI standards) must be preceded by the symbol Ø, as illustrated.

Example 1:

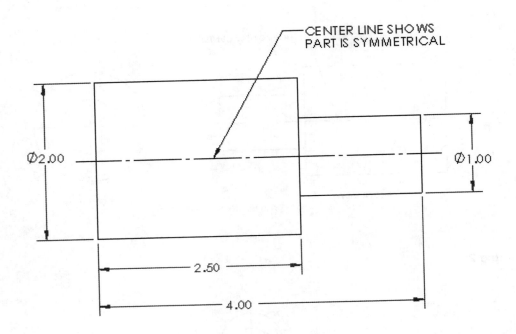

The one-view drawing is also used extensively for flat (Sheet metal) parts. With the addition of notes to supplement the dimensions on the view, the one view furnishes all the necessary information for accurately describing the part. In the first illustration, you have two view: Front view and Top view. In the section illustration, you replace the Top view with a Note: MATERIAL THICKNESS .125 INCH.

 Third Angle Projection type symbol is illustrated.

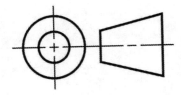

Example 1: No Note Annotation

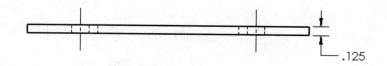

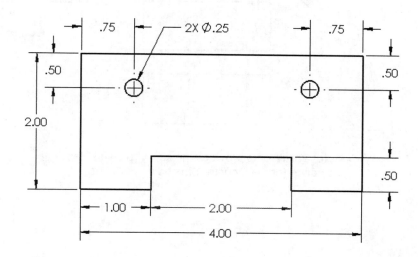

Example 2: Note Annotation to replace the TOP view

MATERIAL THICKNESS .125 INCH

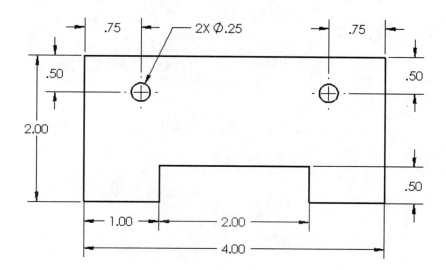

Example 3: Note Fastener Annotation

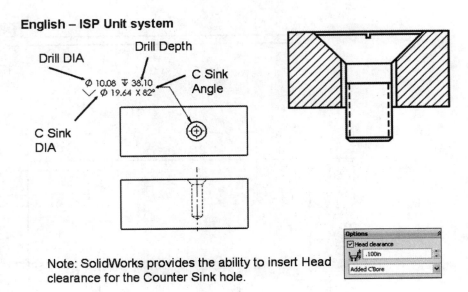

Note: SolidWorks provides the ability to insert Head clearance for the Counter Sink hole.

Exercises:
Exercise 1:

Draw an Isometric view. Approximate the size of the model.

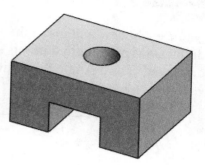

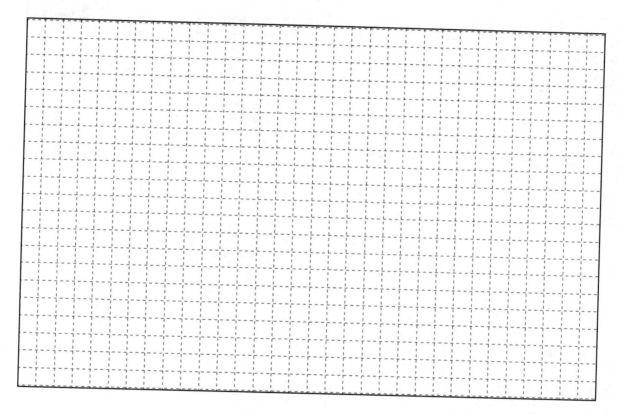

Exercise 2:

Name each view and insert the Width, Height, and Depth name. No dimensions are required in this exercise. Note: Centerlines are not displayed. Third Angle Projection is used.

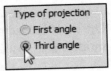

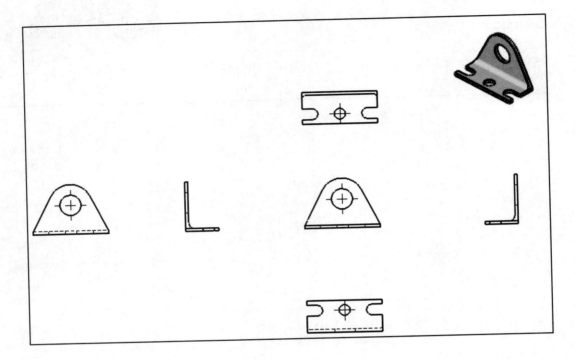

Drawing Views - Advanced

The standard views used in an orthographic projection are: Front view, Top view, Right view and Isometric view. Non-standard orthographic drawing views are used when the six principal views do not fully describe the part for manufacturing or inspection. Below are a few non-standard orthographic drawing views.

Section view

Section views are used to clarify the interior of a part that can't clearly be seen by hidden lines in a view.

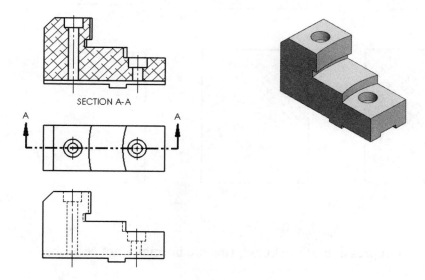

SECTION A-A

Think of an **imaginary** cutting (Plane) through the object and removing a portion. (*Imaginary)* is the key word!

A Section view is a child of the parent view. The Cutting Plane arrows used to create a Section view indicates the direction of sight. Section lines in the Section view are bounded by visible lines.

Section lines in the Section view can serve the purpose of identifying the kind of material the part is made from. Below are a few examples:

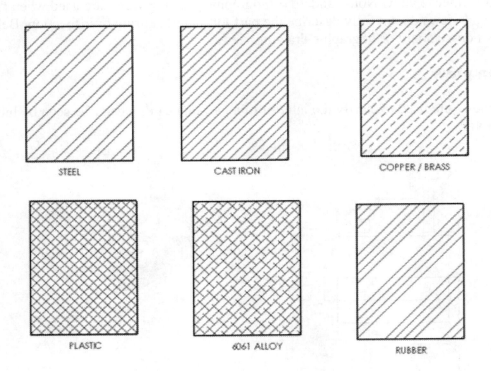

To avoid a false impression of thickness, ribs are normally not sectioned.

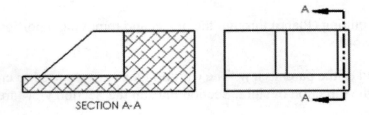

Detail View

The Detail view provides the ability to add a portion of a view, usually at an enlarged scale. A Detail view is a child of the parent view. Create a detail view in a drawing to display or highlight a portion of a view.

A Detail view may be of an Orthographic view, a non-planar (isometric) view, a Section view, a Crop view, an Exploded assembly view or another detail view.

Example 1:

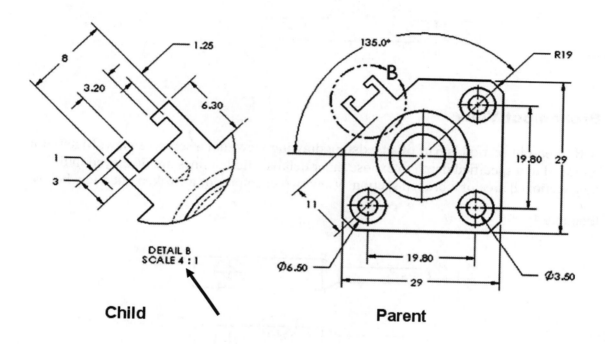

Child   **Parent**

If the Detail view has a different scale than the sheet, the scale needs to be supplied as an annotation as illustrated.

View the additional Power point presentations on the enclosed DVD for supplementary information.

Avi folder
Alphabet of lines and Precedent of Line Types
Boolean Operation
Design Intent
Fastners in General
Fundamental ASME Y14.5 Dimensioning Rules
General Tolerancing and Fits
History of Engineering Graphics
Measurement and Scale
Open a Drawing Document 2012
Open an Assembly Document 2012
Part and Drawing Dimensioning
SolidWorks Basic Concepts
Visualization, Arrangement of Views, and Primitives

Example 2:

Below is a Detail view of a Section view. The Detail view is a child view of the parent view (Detail view). The Section view cannot exist without the Detail view.

Parent view   Child view

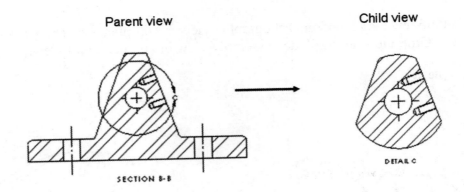

Broken out View

A Broken-out section is part of an existing drawing view, not a separate view. Material is removed to a specified depth to expose inner details. Hidden lines are displayed in the non-sectioned area of a broken section. View two examples of a Broken out View below.

Example 1:

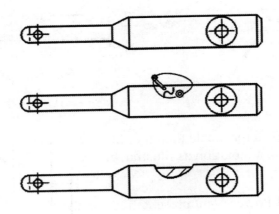

Example 2:

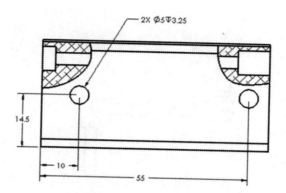

Break or Broken View

A Break view is part of an existing drawing view, not a separate view A Break view provides the ability to add a break line to a selected view. Create a Broken view to display the drawing view in a larger scale on a smaller drawing sheet size. Reference dimensions and model dimensions associated with the broken area reflect the actual model values.

Example 1:

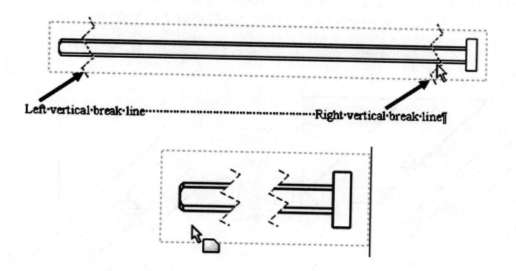

Example 2:

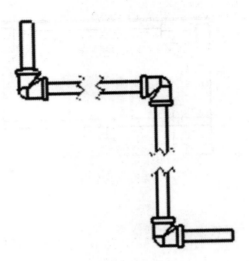

Crop View

A Crop view is a Child of the Parent view. A Crop view provides the ability to crop an existing drawing view. You can not create a Crop view on a Detail View, a view from which a Detail View has been created, or an Exploded view.

Create a Crop view to save time. Example: instead of creating a Section View and then a Detail View, then hiding the unnecessary Section View, create a Crop view to crop the Section View directly.

Example 1:

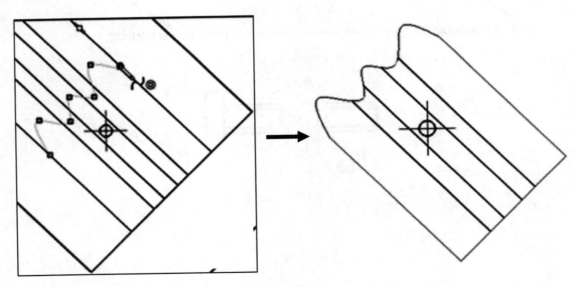

Auxiliary View

An Auxiliary view is a Child of the Parent view. An Auxiliary view provides the ability to display a plane parallel to an angled plane with true dimensions. A primary Auxiliary view is hinged to one of the six principal orthographic views.

Example 1:

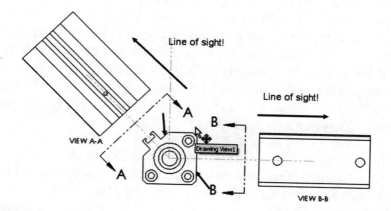

Exercises:

Exercise 1:

Label all of the name views below.

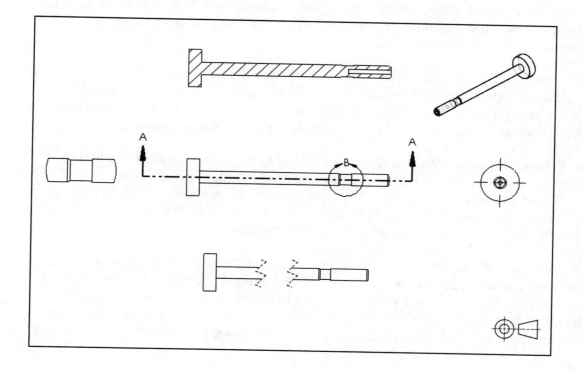

History of Computer Aided Design (CAD)

In 1963, Ivan Sutherland of MIT developed "Sketchpad", a graphical communication system, where with a light pen, Sutherland was able to select and modify geometry on a Cathode Ray System (CRT) and input values through a key pad. Geometric relationships were made between lines are arc and geometry could be moved and copied.

With aerospace and automotive technologies becoming more complex and IBM mainframe computers commercially available in the late 1960's and early 1970's, companies such as MacDonald-Douglas, Lockheed, General Motors, and Ford were utilizing their own internal CAD systems to design, manipulate and store models and drawings. Digital Equipment Corporation (DEC) and Prime Computer introduced computer hardware platforms that made CAD data storage and development more affordable. Ford's Product Design Graphics System (PDGS) developed into one of the largest integrated CAD systems in the 1980's.

By 1980, Cambridge Interact Systems (UK) introduced CIS Medusa, that was bought and distributed by Prime Computer and ran on a proprietary workstation and used Prime mini computers for data storage. Mid size companies, such as AMP and Carrier, were now using CAD in their engineering departments. Other CAD software companies also introduced new technology. Computervision utilized both proprietary hardware and SUN workstations and become a leader in 2D drafting technology.

But in the early 80's, 3D CAD used Boolean algorithms for solid geometry that were a challenge for engineers to manipulate. Other major CAD players were Integraph, GE Calma, SDRC, and IBM (Dassault Systèmes). Dassault Systèmes, with its roots in the aerospace industry, expanded development in CAD surface modeling software technology with Boeing and Ford.

In the late 80's, Parametric Technology Corporation (PTC) introduced CAD software to the market with the ability to manipulate a 3D solid model, running on a UNIX workstation platform. By changing dimensions directly on the 3D model, driven by dimensions and parameters, the model updated and was termed, parametric.

By the early 90's, the Personal Computer (PC) was becoming incorporated in the engineer's daily activities for writing reports and generating spreadsheets. In 1993, SolidWorks founder, Jon Hirschtick recruited a team of engineers to build a company and develop an affordable, 3D CAD software application that was easy to use and ran on an intuitive Windows platform, without expensive hardware and software to operate.

In 1995, SolidWorks was introduced to the market as the first 3D feature based, parametric solid modeler running on a PC. The company's rapidly growing customer base and continuous product innovation quickly established it as a strong competitor in the CAD market. The market noticed, and global product lifecycle technology giant Dassault Systèmes S.A. acquired SolidWorks for $310 million in stock in June of 1997.

SolidWorks went on to run as an independent company, incorporating finite element analysis (FEA) which has advanced dynamics, nonlinear, fatigue, thermal, steady state and turbulent fluid flow (CFD) and electromagnetic analysis capabilities, as well as design optimization. SolidWorks open software architecture as resulted in over 700 partner applications such as Computer Aided Manufacturing (CAM), robot simulation software, and process management. Today, SolidWorks software has the most worldwide users in production - more than 1,000,000 users at over 120,000 locations in more than 100 countries.

Note: There are many university researches and commercial companies that have contributed to the history of computer aided design. We developed this section on the history of CAD based on the institutions and companies that we worked for and worked with over our careers and as it relates to the founders of SolidWorks.

Boolean operations

To understand the difference between parametric solid modeling and Boolean based solid modeling you will first review Boolean operations. In the 1980s, one of the key advancements in CAD was the development of the Constructive Solid Geometry (CSG) method. Constructive Solid Geometry describes the solid model as combinations of basic three-dimensional shapes or better known as primitives. Primitives are typically simple shape: cuboids, cylinders, prisms, pyramids, spheres, and cones.

Two primitive solid objects can be combined into one using a procedure known as the Boolean operations. There are three basic Boolean operations:

- Boolean Union

- Boolean Difference

- Boolean Intersection

Boolean Operation	Result
Boolean Union - The merger of two separate objects into one. A + B	
Boolean Difference - The subtraction of one object from another. A - B	
Boolean Intersection - The portion common to both objects. A ∩ B	

Even today, Boolean operations assist the SolidWorks designer in creating a model with more complex geometry by combining two bodies together with a Boolean intersection.

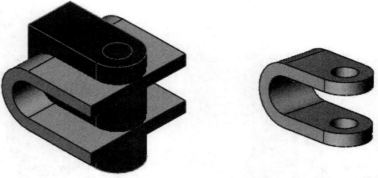

In the Help menu, the SolidWorks Tutorial, Multibody Parts provides Boolean model examples.

Mold Design
Molded Product Design - Advanced
Mouse Gestures ☼
Multibody Parts
Pattern Features
Revolves and Sweeps

What is SolidWorks?

SolidWorks is a design automation software package used to produce parts, assemblies and drawings. SolidWorks is a Windows native 3D solid modeling CAD program. SolidWorks provides easy to use, highest quality design software for engineers and designers who create 3D models and 2D drawings ranging from individual parts to assemblies with thousands of parts.

The SolidWorks Corporation, headquartered in Waltham, Massachusetts, USA develops and markets innovative design solutions for the Microsoft Windows platform. Additional information on SolidWorks and its family of products can be obtained at their URL, www.SolidWorks.com.

In SolidWorks, you create 3D parts, assemblies and 2D drawings. The part, assembly, and drawing documents are related.

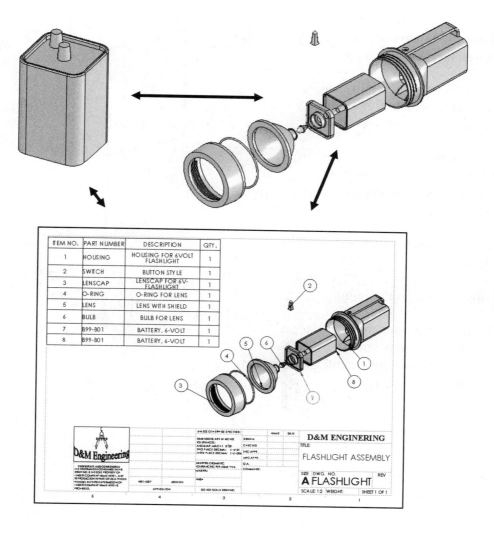

Features are the building blocks of parts. Use features to create parts, such as: Extruded Boss/Base and Extruded Cut. Extruded features begin with a 2D sketch created on a Sketch plane.

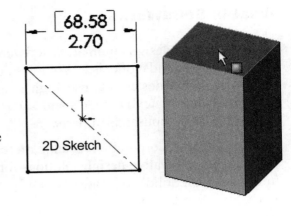

The 2D sketch is a profile or cross section. Sketch tools such as: lines, arcs and circles are used to create the 2D sketch. Sketch the general shape of the profile. Add Geometric relationships and dimensions to control the exact size of the geometry.

Create features by selecting edges or faces of existing features, such as a Fillet. The Fillet feature rounds sharp corners.

Dimensions drive features. Change a dimension, and you change the size of the part.

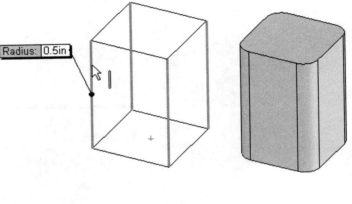

Apply Geometric relationships: Vertical, Horizontal, Parallel, etc. to maintain Design intent.

Create a hole that penetrates through a part. SolidWorks maintains relationships through the change.

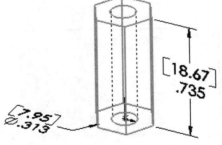

The step-by-step approach used in this text allows you to create parts, assemblies and drawings by doing, not just by reading.

The book provides the knowledge to modify all parts and components in a document. Change is an integral part of design.

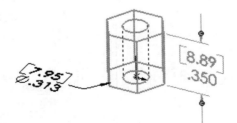

Design Intent

What is design intent? All designs are created for a purpose. Design intent is the intellectual arrangements of features and dimensions of a design. Design intent governs the relationship between sketches in a feature, features in a part and parts in an assembly.

The SolidWorks definition of design intent is the process in which the model is developed to accept future modifications. Models behave differently when design changes occur.

Design for change! Utilize geometry for symmetry, reuse common features, and reuse common parts. Build change into the following areas that you create:

- Sketch

- Feature

- Part

- Assembly

- Drawing

When editing or repairing geometric relations, it is considered best practice to edit the relation vs. deleting it.

Design Intent in a sketch

Build design intent in a sketch as the profile is created. A profile is determined from the Sketch Entities. Example: Rectangle, Circle, Arc, Point, Slot, etc. Apply symmetry into a profile through a sketch centerline, mirror entity and position about the reference planes and Origin.

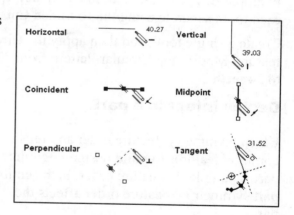

Build design intent as you sketch with automatic Geometric relations. Document the decisions made during the up-front design process. This is very valuable when you modify the design later.

A rectangle (Center Rectangle Sketch tool) contains Horizontal, Vertical and Perpendicular automatic Geometric relations.

Apply design intent using added Geometric relations if needed. Example: Horizontal, Vertical, Collinear, Perpendicular, Parallel, Equal, etc.

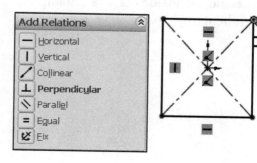

Example A: Apply design intent to create a square profile. Sketch a rectangle. Apply the Center Rectangle Sketch tool. Note: No construction reference centerline or Midpoint relation is required with the Center Rectangle tool. Insert dimensions to fully define the sketch.

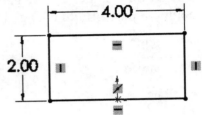

Example B: Develop a rectangular profile. Apply the Corner Rectangle Sketch tool. The bottom horizontal midpoint of the rectangular profile is located at the Origin. Add a Midpoint relation between the horizontal edge of the rectangle and the Origin. Insert two dimensions to fully define the rectangle as illustrated.

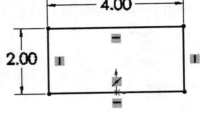

Design intent in a feature

Build design intent into a feature by addressing symmetry, feature selection, and the order of feature creation.

Example A: The Extruded Base feature remains symmetric about the Front Plane. Utilize the Mid Plane End Condition option in Direction 1. Modify the depth, and the feature remains symmetric about the Front Plane.

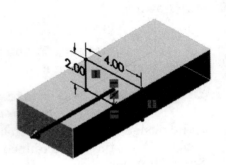

Example B: Create 34 teeth in the model. Do you create each tooth separate using the Extruded Cut feature? No.

Create a single tooth and then apply the Circular Pattern feature. Modify the Circular Pattern from 32 to 24 teeth.

Design intent in a part

Utilize symmetry, feature order and reusing common features to build design intent into a part. Example A: Feature order. Is the entire part symmetric? Feature order affects the part.

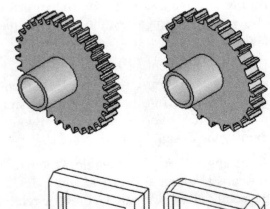

Apply the Shell feature before the Fillet feature and the inside corners remain perpendicular.

Design intent in an assembly

Utilizing symmetry, reusing common parts and using the Mate relation between parts builds the design intent into an assembly.

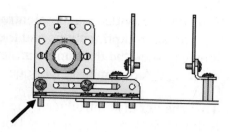

Example A: Reuse geometry in an assembly. The assembly contains a linear pattern of holes. Insert one screw into the first hole. Utilize the Component Pattern feature to copy the machine screw to the other holes.

Design intent in a drawing

Utilize dimensions, tolerance and notes in parts and assemblies to build the design intent into a drawing.

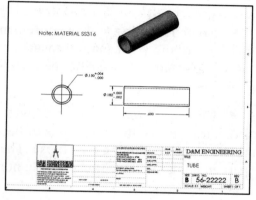

Example A: Tolerance and material in the drawing. Insert an outside diameter tolerance +.000/-.002 into the TUBE part. The tolerance propagates to the drawing.

Define the Custom Property Material in the Part. The Material Custom Property propagates to your drawing.

Create a sketch on any of the default planes: Front, Top, Right or a created plane.

Additional information on design process and design intent is available in SolidWorks Help.

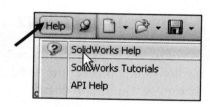

The book is design to expose the new user to many tools, techniques and procedures. It does not always use the most direct tool or process.

Every license of SolidWorks contains a copy of SolidWorks SustainabilityXpress. SustainabilityXpress calculates environmental impact on a model in four key areas: *Carbon Footprint, Energy Consumption, Air Acidification and Water Eutrophication*. Material and Manufacturing process region and Transportation Usage region are used as input variables.

Chapter Summary

Chapter 2 provided a general introduction into isometric projection and sketching along with additional projections and the arrangement of standard views and advanced views. You explored the three main projection divisions in freehand engineering sketches and drawings: Axonometric, Oblique, and Perspective.

This chapter also introduced you to the history of CAD and the development of DS SolidWorks Corp. From early Boolean CAD software, you explored Union, Difference, and Intersection operations which are modeling techniques still used today. You were also introduced to the fundamentals of SolidWorks, its feature based modeling, driven by parameters that incorporates your design intent into a sketch, part, assembly and drawing.

Isometric Rule #1: A measurement can only be made on or parallel to the isometric axis. Therefore you cannot measure an isometric inclined or oblique line in an isometric drawing because they are not parallel to an isometric axis.

Isometric Rule #2: When drawing ellipses on normal isometric planes, the minor axis of the ellipse is perpendicular to the plane containing the ellipse. The minor axis is perpendicular to the corresponding normal isometric plane.

View the Power Point files on the enclosed DVD in the book for additional information.

Avi folder
Alphabet of lines and Precedent of Line Types
Boolean Operation
Design Intent
Fastners in General
Fundamental ASME Y14.5 Dimensioning Rules
General Tolerancing and Fits
History of Engineering Graphics
Measurement and Scale
Open a Drawing Document 2012
Open an Assembly Document 2012
Part and Drawing Dimensioning
SolidWorks Basic Concepts
Visualization, Arrangement of Views, and Primitives

Chapter Terminology

Axonometric Projection: A type of parallel projection, more specifically a type of orthographic projection, used to create a pictorial drawing of an object, where the object is rotated along one or more of its axes relative to the plane of projection.

CAD: The use of computer technology for the design of objects, real or virtual. CAD often involves more than just shapes.

Cartesian Coordinate system: Specifies each point uniquely in a plane by a pair of numerical coordinates, which are the signed distances from the point to two fixed perpendicular directed lines, measured in the same unit of length. Each reference line is called a coordinate axis or just axis of the system, and the point where they meet is its origin.

Depth: The horizontal (front to back) distance between two features in frontal planes. Depth is often identified in the shop as the thickness of a part or feature.

Engineering Graphics: Translates ideas from design layouts, specifications, rough sketches, and calculations of engineers & architects into working drawings, maps, plans, and illustrations which are used in making products.

First Angle Projection: In First Angle Projection the Top view is looking at the bottom of the part. First Angle Projection is used in Europe and most of the world. However America and Australia use a method known as Third Angle Projection.

Foreshortening: The way things appear to get smaller in both height and depth as they recede into the distance.

Grid: A system of fixed horizontal and vertical divisions.

Height: The vertical distance between two or more lines or surfaces (features) which are in horizontal planes.

Isometric Projection: A form of graphical projection, more specifically, a form of axonometric projection. It is a method of visually representing three-dimensional objects in two dimensions, in which the three coordinate axes appear equally foreshortened and the angles between any two of them are 120 °.

Oblique projection: A simple type of graphical projection used for producing pictorial, two-dimensional images of three-dimensional objects.

Origin: The point of intersection, where the X,Y,Z axes meet, is called the origin.

Orthographic Projection: A means of representing a three-dimensional object in two dimensions. It is a form of parallel projection, where the view direction is orthogonal to the projection plane, resulting in every plane of the scene appearing in affine transformation on the viewing surface.

Perspective Projection: The two most characteristic features of perspective are that objects are drawn: smaller as their distance from the observer increases and Foreshortened: the size of an object's dimensions along the line of sight are relatively shorter than dimensions across the line of sight.

Right-Hand Rule: Is a common mnemonic for understanding notation conventions for vectors in 3 dimensions.

Scale: A relative term meaning "size" in relationship to some system of measurement.

Third Angle Projection: In Third angle projection the Top View is looking at the Top of the part. First Angle Projection is used in Europe and most of the world. However America and Australia use a method known as Third Angle Projection.

Units: Used in the measurement of physical quantities. Decimal inch dimensioning and Millimeter dimensioning are the two types of common units specified for engineering parts and drawings.

Width: The horizontal distance between surfaces in profile planes. In the machine shop, the terms length and width are used interchangeably.

Questions

1. Name the three main projection divisions commonly used in freehand engineering sketches and detailed engineering drawings: _____ , _____ and _____

2. Name the projection divisions within Axonometric projection: _____ , _____ , and _____ .

3. True or False: In oblique projections; the front view is drawn true size, and the receding surfaces are drawn on an angle to give it a pictorial appearance.

4. Name the two types of Oblique projection used in engineering design: _____ , _____ .

5. Describe Perspective Projection. Provide an example.

6. True or False: Parts that are uniform in shape often require only one view to describe them adequately.

7. True or False: The designer usually selects as a Front view of the object that view which best describes the general shape of the part. This Front view may have no relationship to the actual front position of the part as it fits into an assembly.

8. True or False: When a one-view drawing of a cylindrical part is used, the dimension for the diameter (according to ANSI standards) must be preceded by the symbol Ø.

9. Draw a Third Angle Projection Symbol.

10. Draw a First Angle Projection Symbol.

11. Describe the different between First and Third Angle Projection.

12. True or False. First Angle Projection is used in the United States.

13. True or False. Section lines can serve the purpose of identifying the kind of material the part is made from.

14. True or False. All dimension lines terminate with an arrowhead on mechanical engineering drawings.

15. True or False. Break lines are applied to represent an imaginary cut in an object, so the interior of the object can be viewed or fitted to the sheet. Provide an example.

Exercises

Exercise 2.1: Hand draw the Isometric view for the illustrated model below.

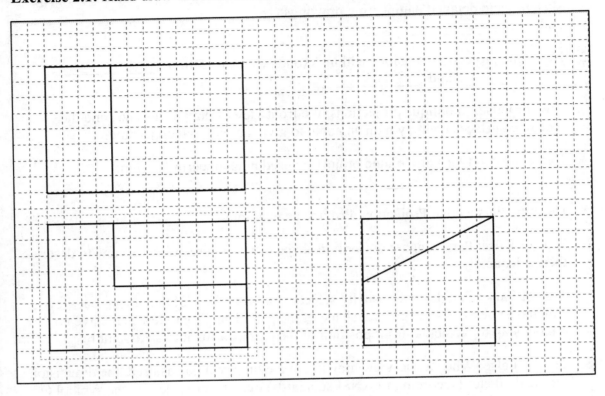

Exercise 2.2: Hand draw the Isometric view for the following models. Approximate the size of the model.

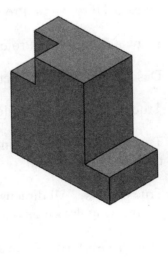

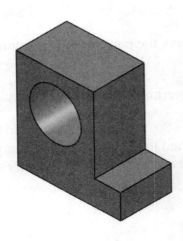

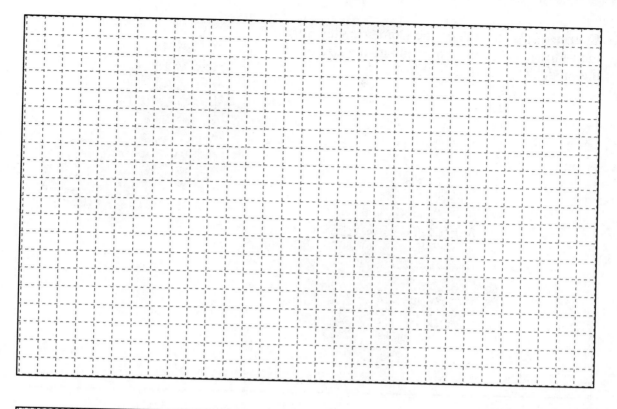

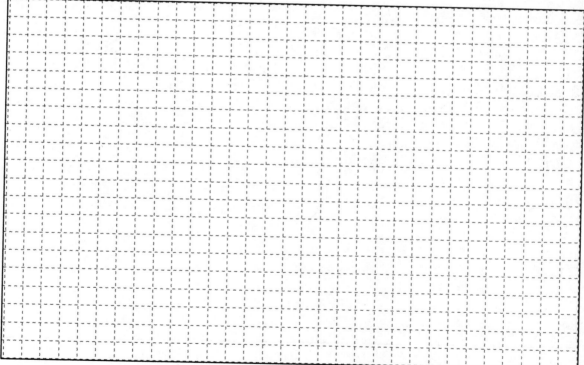

Exercise 2.3: Hand draw the Isometric view for the following models. Approximate the size of the model.

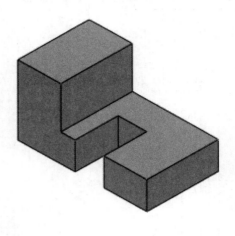

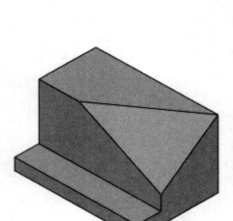

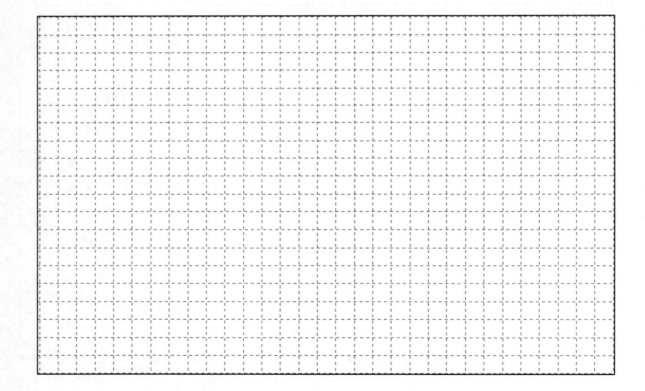

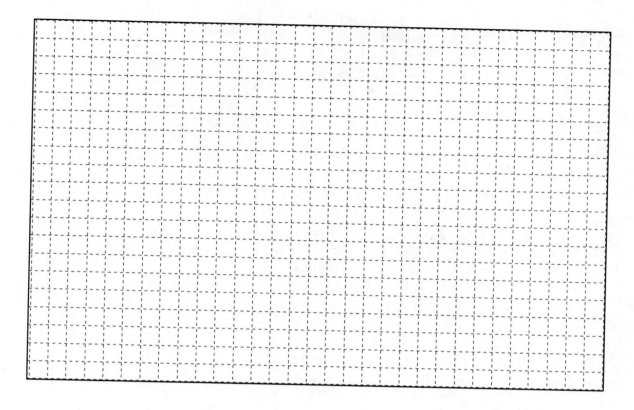

Notes:

Chapter 3

Dimensioning Practices, Tolerancing, and Fasteners

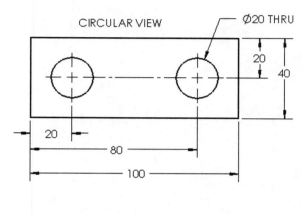

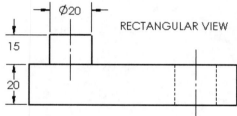

Below are the desired outcomes and usage competencies based on the completion of Chapter 3.

Desired Outcomes:	Usage Competencies:
• Knowledge of dimensioning systems and the ASME Y14.5 - 1994 standard.	• Dimension the following simple shapes: hole, cylinder, angle, point or center, arc, and chamfer.

• Awareness of the IPS Unit System, the MMGS Unit System and the Dual dimensioning system.	• Apply dual dimensioning.
	  o IPS Unit System
	  o MMGS Units System

• Understand Tolerancing for a drawing. • Comprehend Fasteners and hole dimensioning. • Recognize Fit types.	• Apply dimension and drawing Tolerances. • Read and understand general Fastener and hole annotations. • Apply Fit types.

Notes:

Chapter 3 - Dimensioning Practices, Tolerancing and Fasteners

Chapter Overview

Chapter 3 provides a general introduction into dimensioning systems and the ANSI Y14.5 2009 standards along with fasteners and general tolerancing.

On the completion of this chapter, you will be able to:

- Understand and apply various dimensioning systems.

- Correctly dimension the following simple shapes: hole, cylinder, angle, point or center, arc, and chamfer.

- Knowledge of dimensioning systems and the ANSI Y14.5 standards.

- Apply dual dimensioning.

- Understand and apply part and drawing Tolerance.

- Read and understand Fastener notation.

- Recognize single, double or tripe threads.

- Distinguish between Right and Left handed threads.

- Recognize annotations for a simple hole, Counterbore and Countersink in a drawing.

- Identify Fit types.

Size and Location Dimensions

To ensure some measure of uniformity in industrial drawings, the American National Standards Institute (ANSI) has established drafting standards; these standards are called the language of drafting and are in general use throughout the United States.

ANSI was originally formed in 1918, when five engineering societies and three government agencies founded the American Engineering Standards Committee (AESC). In 1928, the AESC became the American Standards Association (ASA). In 1966, the ASA was reorganized and became the United States of America Standards Institute (USASI). The present name was adopted in 1969 and the standards are published by the American Society of Mechanical Engineers (ASME).

While these drafting standards or practices may vary in some respects between industries, the principles are basically the same. The practices recommended by ANSI for dimensioning and for marking notes are followed in this book.

Dimensioning Systems

A dimension is a numerical value that is being assigned to the size, shape or location of the feature being described. There are basically three types of dimensioning systems use in creating parts and drawings:

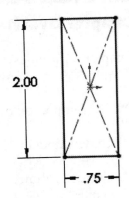

U.S. - ANSI standard for U.S. dimensioning use the decimal inch value. When the decimal inch system is used, a zero is not used to the left of the decimal point for values less than one inch, and trailing zeros are used.

The U.S. unit system is also known as the Inch, Pound, Second (IPS) unit system.

Metric - ANSI standards for the use of metric dimensioning required all the dimensions to be expressed in millimeters (mm). The (mm) is not needed on each dimension, but it is used when a dimension is used in a notation. No trailing zeros are used.

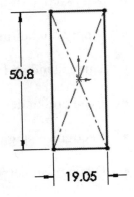

The Metric or International System of Units (S.I.) unit system in drafting is also known as the Millimeter, Gram Second (MMGS) unit system.

Dual Dimensioning - Working drawings are usually drawn with all U.S. or all metric dimensions. Sometimes the object manufactured requires using both the U.S. and metric measuring system. In this illustration, the secondary units (mm) are displayed in parenthesis. The Primary units are inches.

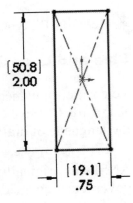

Do not over dimension a drawing view (sometimes called double dimension).

View the additional Power point presentations on the enclosed DVD for supplementary information.

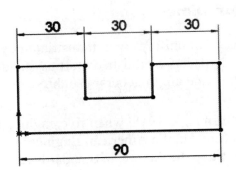

Standards for Dimensioning

All drawings should be dimensioned completely so that a minimum of computation is necessary, and the part can be built without scaling the drawing. However, there should not be a duplication of dimensions unless such dimensions make the drawing clearer and easier to read. These dimensions are called reference dimensions and are enclosed in parentheses.

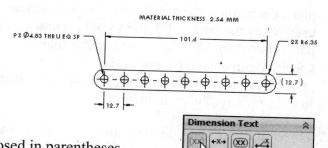

Part Dimensions / Construction Dimensions

Dimensions used in creating a part are sometimes called construction dimensions or part dimensions. These dimensions serve two purposes: (1) indicate size, and (2) provide exact locations. For example, to drill a through all hole in a part, the designer must know the diameter of the hole, and the exact location of the center of the hole relative to the origin. Note: Gaps are required between the dimension extension line and the feature line. Do not cross Leader lines and limit the leader line length.

Example 1:

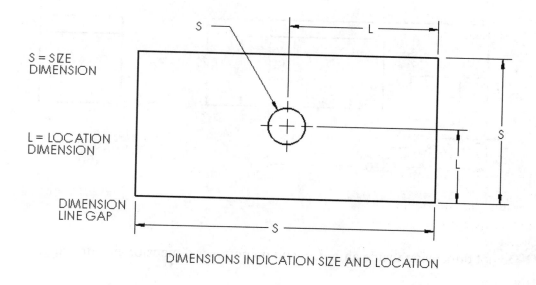

DIMENSIONS INDICATION SIZE AND LOCATION

Dimensions should not be duplicated or the same information given in two different ways. If a reference dimension is used, the size value is placed within parentheses (X).

Always position the diameter dimension up and off the model.

Two Place Decimal Dimensions

Dimensions may appear on drawings as two-place decimals (IPS) unit system. This is widely used when the range of dimensional accuracy of a part is between 0.01" larger or smaller than a specified dimension (nominal size). Where possible, two-place decimal dimensions are given in even hundredths of an inch.

Type	Unit	Decimals
Basic Units		
Length	inches	.12
Dual Dimension Length	millimeters	.12
		.123
Angle	degrees	.1234
		.12345
Mass/Section Properties		.123456
Length	inches	.1234567

Three and four place decimal dimensions continue to be used for more precise dimensions requiring machining accuracies in thousandths or ten-thousandths of an inch.

Size Dimensions

Every solid or part has three size dimensions: width or length, height, and depth. In the case of the Glass box method, two of the dimensions are usually placed on the Principal view and the third dimension is located on one of the other views.

Example 1:

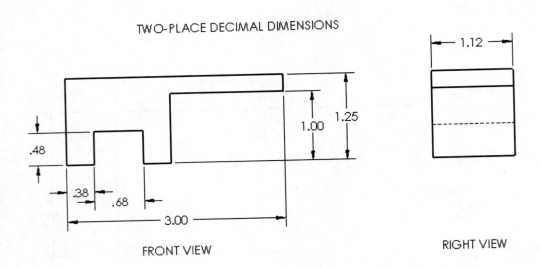

TWO-PLACE DECIMAL DIMENSIONS

FRONT VIEW   RIGHT VIEW

☀ Do not dimension inside an object and do not over dimension in a drawing.

☀ Always locate the dimensions off of the view if possible. Only place dimensions on the inside of the view if they add clarity, simplicity, and ease of reading.

☀ There should be a visible gap ~1.5mm between the object (feature) line and the beginning of each extension line.

Continuous Dimensions

Set of dimension lines and dimensions should be located on drawings close enough so they may be read easily without any possibility of confusing one dimension with another. If a series of dimensions is required, the dimensions should be placed in a line as continuous dimensions (chain dimensioning or point-to-point dimensioning) as illustrated below. This method is preferred over the staggering of dimensions, because of ease in reading, appearance, and simplified dimensioning. Note: Tolerance stack-up can be an issue with this method!

Example 1:

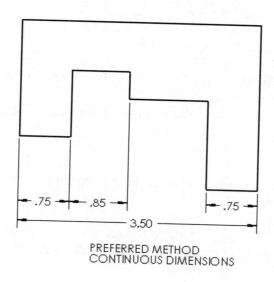

PREFERRED METHOD
CONTINUOUS DIMENSIONS

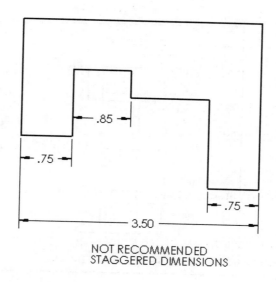

NOT RECOMMENDED
STAGGERED DIMENSIONS

 Spacing between dimension lines should be uniform throughout the drawing.

Other Dimension Placements

Dimensions should be placed in such a way as to enhance the communication of your drawing. There are a few key rules that you should. They are:

- Place dimensions between views.

- Group dimensions whenever possible.

- Locate dimensions in the view where the shape is best shown.

- Dimension a hole in a circular view.

- Dimension a cylinder in a rectangular view.

- Dimension a hole by its diameter.

- Dimension a slot in a view where the contour of the slot is visible.

- Dimension an arc by its radius.

- Place a smaller dimension inside a larger dimension on a drawing view to avoid dimension line crossing.

- ANSI standard states, "Dimensioning to hidden lines should be avoided wherever possible". However, sometimes it is necessary if additional views are needed to fully define the model as illustrated below.

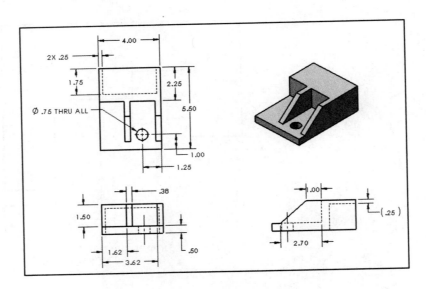

Example 1:

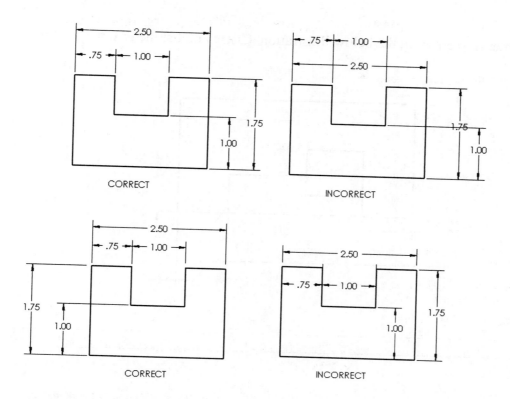

Arrowheads are drawn between extension lines if space is available. If space is limited, see the preferred arrowhead and dimension location order below.

Example 1:

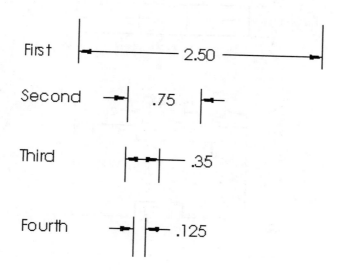

Dimension Exercises:
Exercise 1:

Identify the dimension errors in the below illustration. Circle and list the errors.

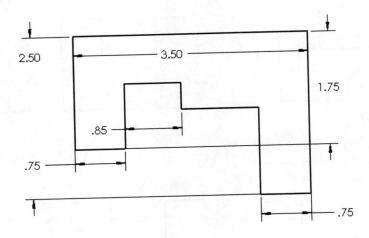

Errors:_____

Exercise 2:

Identify the dimension errors in the below illustration. Circle and list the errors.

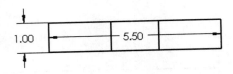

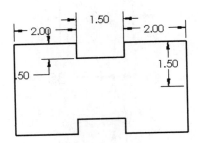

Errors:_____

Exercise 3:

Identify the duplicate dimensions and cross out the ones that you feel should be omitted. Explain why. Are there any dimensioning mistakes in this drawing? Explain.

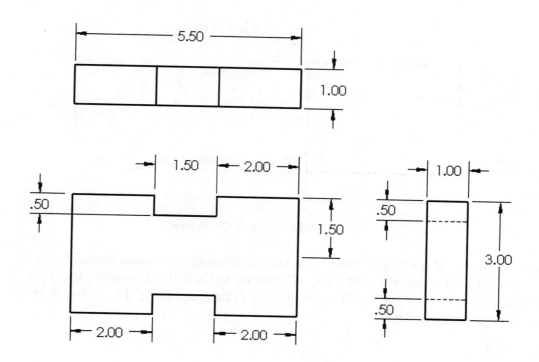

Explain:

_____

_____

_____

Dimensioning Cylinders

The length and diameter of cylinders are usually place in the view which shows the cylinder as a rectangle as illustrated below.

Example 1:

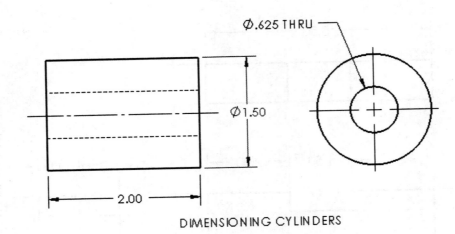

DIMENSIONING CYLINDERS

Note: Many round parts with cylindrical surface symmetrical about the axis, can be represented with a one-view drawing. A diameter is identified according to ANSI standards by using the symbol Ø preceding the dimension. Note: The below model is displayed in millimeters.

Example 2:

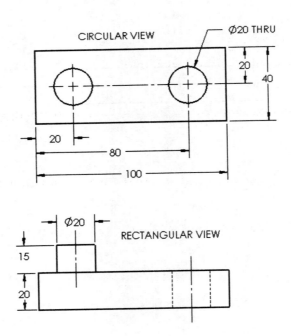

A cylinder is dimensioned by giving its diameter and length in the rectangular view, and is located in the circular view.

Holes are dimensioned by giving their diameter and location in the circular view.

Example 3:

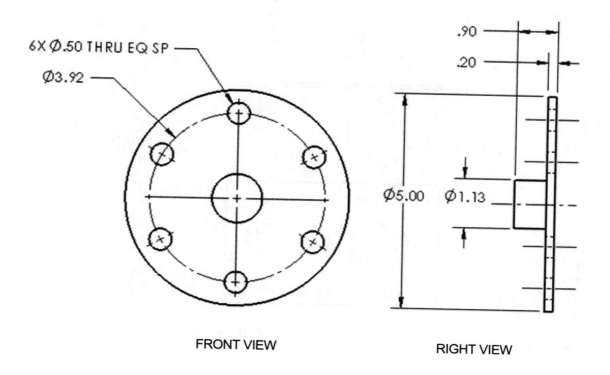

FRONT VIEW   RIGHT VIEW

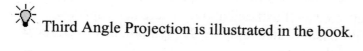 Third Angle Projection is illustrated in the book.

Your choice of dimensions will directly influence the method used to manufacture the part.

Always position the diameter dimension up and off the model.

Insert the needed dimension to a single view (best view). Do not over dimension.

Dimension lines should not cross, if avoidable.

Dimensioning a Simple Hole

The diameters of holes which are to be formed by drilling, reaming, or punching should have the diameter, preferable on a leader, followed by a note indicating the operation to be performed and the number of holes to be produced, as illustrated. Note: The diameter of a hole is *place in the view which shows the hole as a circle* as illustrated below.

Example 1:

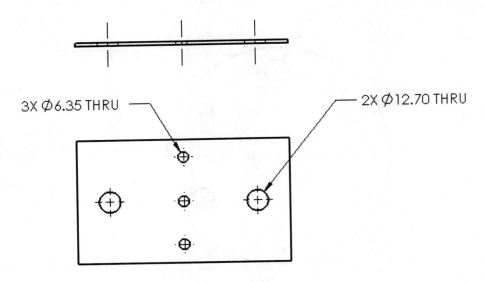

DIMENSIONING A SIMPLE HOLE

Repetitive features or dimensions can be specified by using the symbol "X" along with the number of times the feature is repeated as illustrated above. There is no space between the number of times the feature is repeated and the "X" symbol; however, there is a space between the symbol "X" and the dimension.

If a hole goes completely through the feature and it is not clearly shown on the drawing, the abbreviation "THRU" in all upper case follows the dimension. All notes should be in UPPER CASE LETTERS.

The Front View should be the most descriptive view in the drawing document.

Dimensioning Angles

The design of a part may require some lines to be drawn at an angle. The amount of the divergence (the amount the lines move away from each other) is indicated by an angle measured in degrees or fractional parts of a degree. The degree is indicated by a symbol °placed after the numerical value of the angle.

Example 1:

The dimension line for an angle should be an arc whose ends terminate in arrowheads.

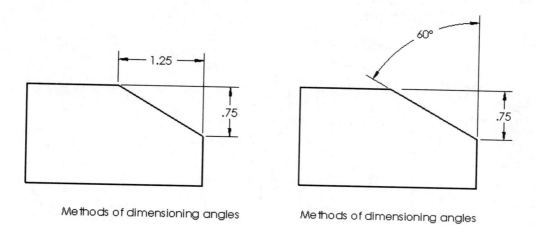

Methods of dimensioning angles   Methods of dimensioning angles

The numeral indication the number of degrees in the angle is read in a horizontal position, except where the angle is large enough to permit the numerals to be placed along the arc.

Example 2:

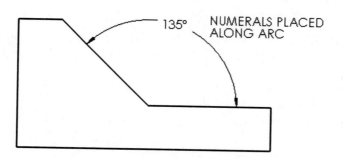

Dimensioning a Point or a Center

A point or a center of an arc or circle is generally measured from two finished surfaces. The method of location the center is preferred to making an angular measurement. As illustrated, the center of the circle and arc may be found easily be scribing the vertical and horizontal center lines from the machined surfaces.

Example 1:

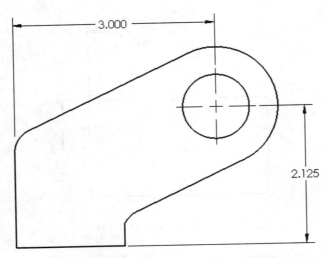

Dimensioning the center of a circle

Dimensioning equally spaced holes on a Circle

If a number of holes are to be equally spaced on a circle, then the exact location of the first hole is given by location dimension. To locate the remaining holes, the location dimension is followed by 1.) diameter of the holes, 2.) number of holes, and 3.) notation EQUALLY SPACED or "EQ SP" as illustrated.

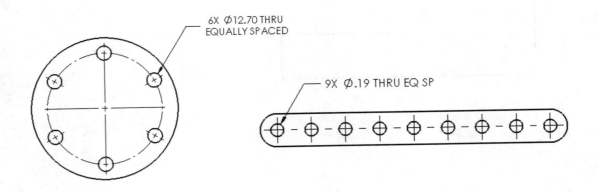

Dimensioning Holes not on a circle

Holes are often dimensioned in relation to one another and to a finished surface. Dimensions are usually given, in such cases, in the view which the shape of the holes, that is, square, round, or elongated. The preferred method of placing these dimensions is illustrated below.

Example 1:

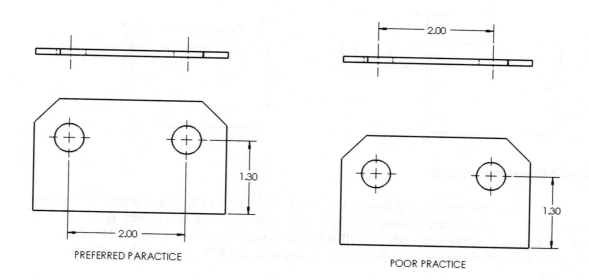

PREFERRED PARACTICE

POOR PRACTICE

Dimensioning Arcs

An arc is always dimensioned by its radius. ANSI standards require a radius dimension to be preceded by the letter (symbol) R as illustrated.

Example 1:

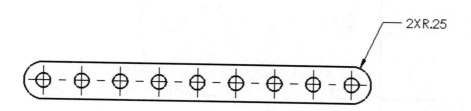

2XR.25

The radial dimension line should have only one arrowhead, and it should touch the arc.

Dimensioning Chamfers

There are two ways to dimension a chamfer feature as illustrated below.

Example 1:

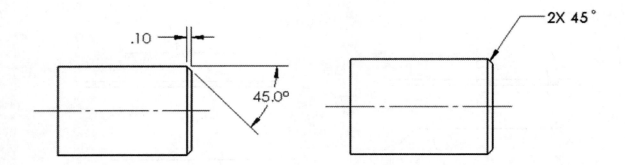

Dual Dimensioning

Working drawings are usually drawn with all U.S. or all metric dimensions. Sometimes the object is manufactured using both U.S. and metric measuring system. Dual dimensioning may be necessary. As illustrated, the primary unit system is IPS (Inch, Pounds, Second) and the secondary unit system is in MMGS (Millimeters, Grams, Second).

Example 1:

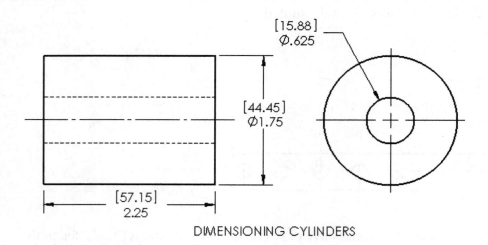

DIMENSIONING CYLINDERS

Dimension Exercises:
Exercise 1:

Identify the dimension errors in the below illustration. Circle and list the errors.

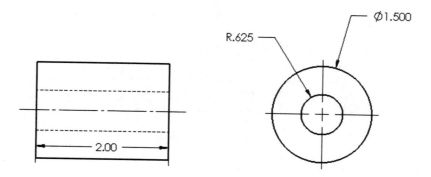

Errors:_____

_____

Exercise 2:

Identify the dimension errors in the below illustration. Circle and list the errors.

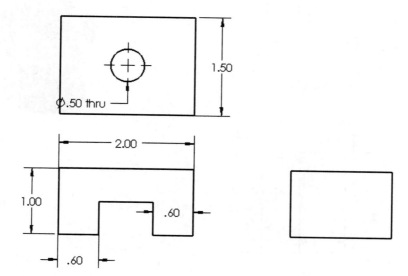

Errors:_____

_____

Exercise 3:

Identify the dimension errors in the below illustration. Circle and list the errors.

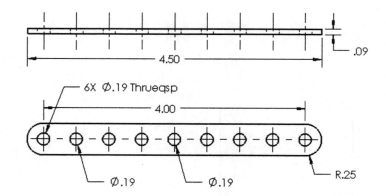

Explain:

_____

_____

Exercise 4:

Identify the dimension errors in the below illustration. Circle and list the errors.

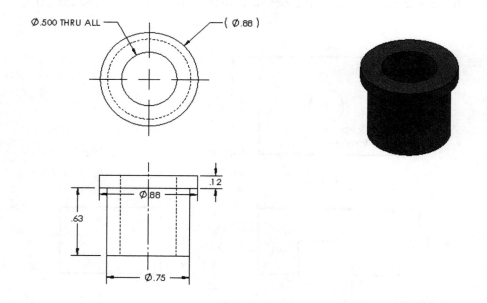

Explain:

_____

_____

Precision and Tolerance

In a manufacturing environment, quality and cost are two of the main considerations for an engineer or designer. Engineering drawings with local and general notes and dimensions often serve as purchasing documents, construction, inspection, and legal contracts to ensure the proper function and design of the product. When dimensioning a drawing, it is essential to reflect on the precision required for the model.

Precision is the degree of accuracy required during manufacturing. However, it is unfeasible to produce any dimension to an absolute, accurate measurement. Some discrepancy must be provided or allowed in the manufacturing process.

Specifying higher precision on a drawing may ensure better quality of a product, but doing so can increase the cost of the part and make it cost prohibited in being competitive with similar products.

For example, consider a design that contains cast components. A cast part usually has two types of surfaces: 1.) mating surfaces, and 2.) non-mating surfaces.

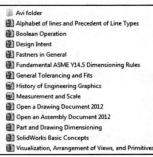

Mating surfaces work together with other surfaces, typically machined to a specified finish. Mating surfaces typical require higher precision on all corresponding dimensions.

Non-mating surfaces are usually left in the original rough-cast form. They have no significant connection with other surfaces. The dimensions on a drawing must clearly indicate which surfaces are to be finished and provide the degree of precision needed for the finishing.

The method of specifying the degree of precision is called Tolerancing. Tolerance in simple terms is the amount of _size variation_ permitted and provides a useful means to achieve the precision necessary in a design. Tolerancing makes certain interchangeability in manufacturing. Parts can be manufactured by different companies in various locations while maintaining the proper functionality of the intended design.

In tolerancing; each dimension is permitted to vary within a specified amount. By assigning as large a tolerance as possible, without interfering with the functionality or intended design of a part, the production costs can be reduced and the product can be competitive in the real world. The smaller the tolerance range specified, the more expensive it is to manufacture. There is always a trade off in design.

View the additional Power point presentations on the enclosed DVD for supplementary information.

Avi folder
Alphabet of lines and Precedent of Line Types
Boolean Operation
Design Intent
Fastners in General
Fundamental ASME Y14.5 Dimensioning Rules
General Tolerancing and Fits
History of Engineering Graphics
Measurement and Scale
Open a Drawing Document 2012
Open an Assembly Document 2012
Part and Drawing Dimensioning
SolidWorks Basic Concepts
Visualization, Arrangement of Views, and Primitives

Tolerance for a drawing

The two most common Tolerance Standard agencies are: American National Standards Institute (ANSI) / (ASME) and the International Standards Organization (ISO). This book covers the ANSI (US) standards.

In this section - we will discuss Dimensional Tolerances vs. Geometric Tolerances.

General Tolerance - Title Block:

General tolerances are typically provided in the Title Block. General tolerances are applied to the dimensions in which tolerances are not given in the drawing.

As a part is designed, the engineer should consider: 1.) function either as a separate unit or as a component relation to other components in an assembly, 2.) manufacturing operations, 3.) material, 4.) quantity (run size), 5.) sustainability, and 6.) cost.

The dimensions displayed on a drawing (obtained from the part) indicate the accuracy limits for manufacturing. The limits are called tolerances and are normal displayed in decimal notation. Tolerances can be specified in various unit systems. ANSI, specifications are normally specified either in English (IPS) or Metric (MMGS).

Tolerances on decimal dimensions, which are expressed in terms of one, two, three, or more decimal places. This information can be documented on a drawing is several ways. One of the common methods of specifying a tolerance that applies to all dimensions is to use a general note in the Title block as illustrated.

Example 1 & 2:

UNLESS OTHERWISE SPECIFIED:
DIMENSIONS ARE IN INCHES TOLERANCES: ANGULAR: ± 1 ° ONE PLANE DECIMAL   ± .1 TWO PLACE DECIMAL   ± .01 THREE PLACE DECIMAL ± .005

UNLESS OTHERWISE SPECIFIED:
DIMENSIONS ARE IN MILLIMETERS TOLERANCES: ANGULAR: MACH± 0 °30' ONE PLACE DECIMAL   ±0.5 TWO PLACE DECIMAL ±0.15

Local Tolerance - Dimension:

A Local Tolerance note indicates a special situation which is not covered by the General Title box. A Local Tolerance is located on the drawing (NOT in the Title box) with the dimension.

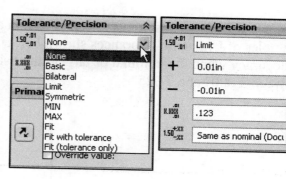

The three most common Tolerance types are: *Limit, Bilateral, and Unilateral.*

Limit Tolerance is when a dimension has a high (upper) and low (Lower) limits stated. In a limit tolerance, the higher value is placed on top, and the lower value is placed on the bottom as illustrated.

$$\phi^{1.001}_{.999} \quad \text{or} \quad \phi.999 - 1.001$$

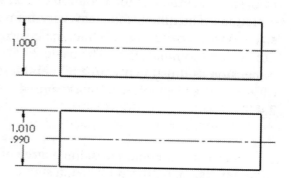

Limits are the maximum and minimum size that a part can obtain and still pass inspection and function in the intended assembly. When both limits are placed on a single line, the lower limit precedes the higher limit. The tolerance for the dimension illustrated above is the total amount of variation permitted or .002.

In the angle example - the dimension may vary between 60 °and 59°45'.

Note: Each degree is one three hundred and sixtieth of a circle (1/360). The degree (°) may be divided into smaller units called minutes ('). There are 60 minutes in each degree. Each minute may be divided into smaller units called seconds ("). There are 60 seconds in each minute. To simplify the dimensioning of angles, symbols are used to indicate degrees, minutes and second as illustrated below.

Name	**Symbol**
Degrees	°
Minutes	'
Seconds	"

Unilateral Tolerance is the variation of size in a single direction - either (+) or (-). The examples of Unilateral tolerances shown below indicate that the first part meets standards of accuracy when the nominal or target dimension varies in one direction only and is between 3.000" and 3.025".

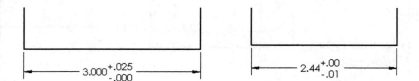

Bilateral Tolerance is the variation of size in both directions. The dimensions may vary from a larger size (+) to a smaller size (-) than the basic dimension (nominal size). The basic 2.44" dimension as illustrated with a bilateral tolerance of +-.01" is acceptable within a range of 2.45" and 2.43".

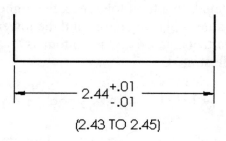

Specify a tolerance with the degree of accuracy that is required for the design to work properly and is cost effective.

You can also create a note on the drawing referring to a specific dimension or specify general tolerances in the Title block.

Formatting Inch Tolerances

The basic dimension and the plus and minus values should have the same number of decimal places. Below are examples or *Unilateral* and *Bilateral* tolerances.

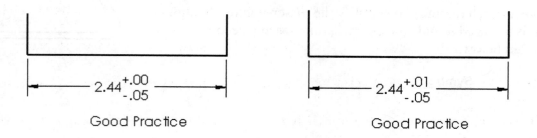

Metric Dimension Specifications

For Metric dimension specification, the book uses the Metric International System of Units (SI). The millimeter is the common unit of measurement used on engineering drawings made to the metric system.

In industry, a general note would be displayed in the Title block section of the drawing to invoke the metric system. A general note is: "UNLESS OTHERWISE SPECIFIED: DIMENSIONS ARE IN MILLIMETERS."

Three conventions are used when specifying dimensions in metric units. They are:

> UNLESS OTHERWISE SPECIFIED:
>
> DIMENSIONS ARE IN MILLIMETERS
> TOLERANCES:
> ANGULAR: MACH± 0°30'
> ONE PLACE DECIMAL   ±0.5
> TWO PLACE DECIMAL   ±0.15

1.) When a metric dimension is a whole number, the decimal point and zero are omitted.

2.) When a metric dimension is less than 1 millimeter, a zero precedes the decimal point. Example - 0.2 has a zero to the left of the decimal point.

3.) When a metric dimension is not a whole number, a decimal point with the portion of a millimeter (10ths or 100ths) is specified.

General Nomenclature

The followings are general important definitions in tolerancing as defined in the ANSI/ASME Y 14.5M standard.

- **Actual size** - The measured dimension. A shaft of nominal diameter 10 mm may be measured to be an actual size of 9.975 mm.
- **Allowance** - The minimum clearance space or maximum interference intended between two mating parts under the Maximum Material Condition (MMC).
- **Basic dimension / Basic size** - The theoretical size from which limits of size are derived. It is the size from which limits are determined for the size or location of a feature in a design.
- **Fit** - The general term used to signify the range of tightness in the design of mating parts.
- **Least Material Condition (LMC)** - The size of the part when it consists of the least material.
- **Maximum Material Condition (MMC)** - The size of the part when it consists of the most material.
- **Nominal size** - Size used for general identification – not exact size.
- **Tolerance** - The total permissible variation of a size. The tolerance is the difference between the limits.

Fit - Hole Tolerance

In the figure below, what is the minimum clearance (Allowance)? Minimum clearance is the minimum amount of space which exists between the hole and the shaft.

Minimum Clearance (Allowance) = $(0.49d_{hole})$ - $(0.51D_{shaft})$ = -0.02in

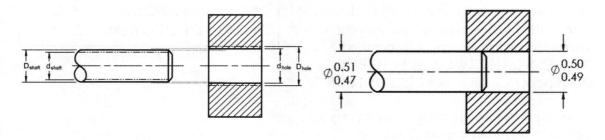

In the figure above, what is the maximum clearance (Allowance)? Maximum clearance is the difference between the largest hole diameter D_{hole} and the smallest shaft diameter d_{shaft}.

Maximum Clearance (Allowance) = $(0.50D_{hole})$ - $(0.47d_{shaft})$ = 0.03

Fit between Mating Parts

Fit is the general term used to signify the range of tightness in the design of mating parts. In ANSI/ASME Y 14.5M, four general types of fits are designated for mating parts:

1. **Clearance Fit**

2. **Interference Fit**

3. **Transition Fit**

4. **Line Fit**

Clearance Fit: The difference between the hole and shaft sizes before assembly is positive. Clearance fits have limits of size prearranged such that a clearance always results when the mating parts are assembled. Clearance fits are intended for accurate assembly of parts and bearings .The parts can be assembled by hand because the hole is always larger than the shaft. Min. Clearance > 0. Two examples: Lock and Key, Door and Door frame.

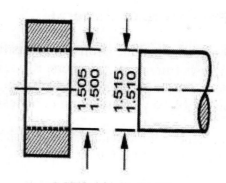

INTERFERENCE FIT

Interference Fit: The arithmetic difference between the hole and shaft sizes before assembly is negative. Interference fits have limits of size prearranged that an interference always results when mating parts are assembled. The hole is always smaller than the shaft .Interference fits are for permanent assemblies of parts which require rigidity and alignment, such as dowel pins and bearings in casting. Max. Clearance ≤ 0. Two examples: Hinge pin and pin in a bicycle chain.

Transition Fit: May provide either clearance or interference, depending on the actual value of the tolerance of individual parts. Transition fits are a compromise between the clearance and Interference fits .They are used for applications where accurate location is important, but either a small amount of clearance or interference is permissible. Max. Clearance > 0, Min. Clearance < 0.

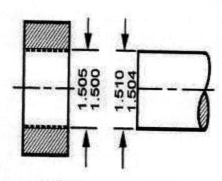

TRANSITION FIT

Line Fit: The condition in which the limits of size so that a clearance or surface contact may result between the mating parts. A space or a contact (hole diameter = shaft diameter). Max. Clearance > 0. Min. Clearance = 0.

Why is this information important? By specifying the correct allowances and tolerances, mating components in an assembly can be completely interchangeable.

Sometimes the desired fit may require very small allowances and tolerances, and the production cost may become too high and cost prohibited. In these cases, either manual or computer-controlled selective assembly is often used. The manufactured parts are then graded as small, medium and large based on the actual sizes. In this way, very satisfactory fits are achieved at a much lower cost than to manufacturing all parts to very accurate dimensions.

Fasteners in General

Fasteners include: Bolts and Nuts (threaded), Set screws (threaded), Washers, Keys, Pins to name a few. Fasteners are not a permanent means of assembly such as welding or adhesives.

Fasteners and threaded features should be specified on your engineering drawing.

- Threaded features: Threads are specified in a thread note. In SolidWorks, apply the Hole Wizard feature.

- General Fasteners: Purchasing information must be given to allow the fastener to be ordered correctly.

Representing External (Male) Threads

Screw threads are used widely (1) to fasten two or more parts together in position, (2) to transmit power such as a feed screw on a machine, and (3) to move a scale on an instrument used for precision measurements.

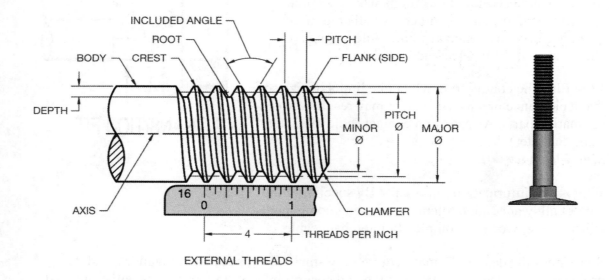

*Profile of UNIFIED and the American National Threads.

An external thread is an edge of a uniform section in the form of a helix on the **external** surface of a cylinder or cone; **"A"** suffix.

Cutting External (Male) Threads

Start with a shaft the same size as the major diameter. An external thread is cut using a die or a lathe as illustrated below.

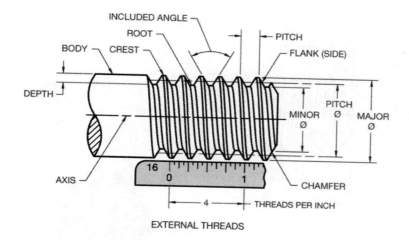

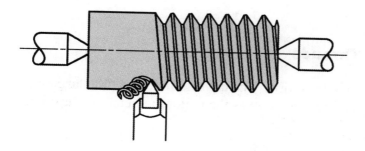

A chamfer on the end of the screw thread makes it easier to engage the nut.

Representing Internal (Female) Threads

An internal thread is a ridge of a uniform section in the form of a helix on the **internal** surface of a cylinder or cone; **"B"** suffix.

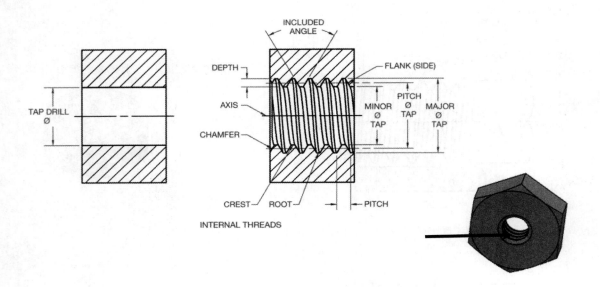

Cutting Internal (Female) Threads

In general, a tap drill hole is cut with a twist drill. The tap drill hole is a little larger than the minor diameter. Start with a shaft the same size as the major diameter as illustrated below.

Minor Diameter: The smallest diameter (fractional diameter or number) of a screw thread.

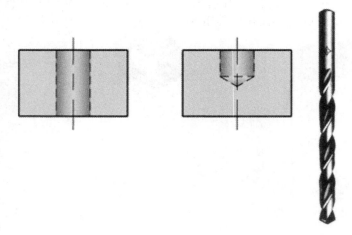

Then the threads are cut using a tap. Major tap types:

- **Taper** tap

- **Plug** tap

- **Bottoming** tap

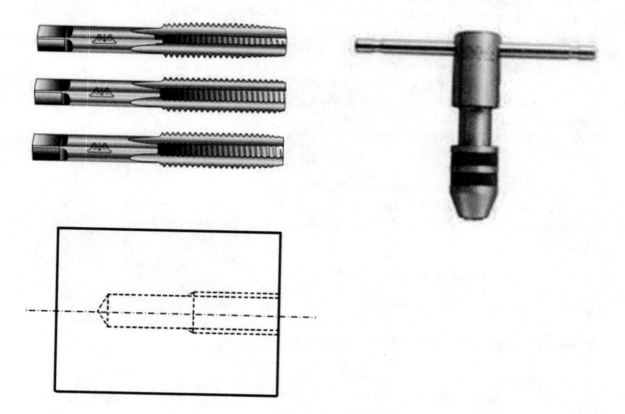

The process of cutting threads using a tap is called tapping, whereas the process using a die is called threading. Both tools can be used to clean up a thread, which is called chasing.

View the additional Power point presentations on the enclosed DVD for supplementary information.

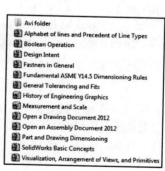

- Avi folder
- Alphabet of lines and Precedent of Line Types
- Boolean Operation
- Design Intent
- Fastners in General
- Fundamental ASME Y14.5 Dimensioning Rules
- General Tolerancing and Fits
- History of Engineering Graphics
- Measurement and Scale
- Open a Drawing Document 2012
- Open an Assembly Document 2012
- Part and Drawing Dimensioning
- SolidWorks Basic Concepts
- Visualization, Arrangement of Views, and Primitives

American National Standard and Unified Screw threads

The basic profile is the theoretical profile of the thread. An essential principle is that the actual profiles of both the nut and bolt threads must never cross or transgress the theoretical profile. So bolt threads will always be equal to, or smaller than, the dimensions of the basic profile. Nut threads will always be equal to, or greater than, the basic profile. To ensure this in practice, tolerances and allowances are applied to the basic profile.

The most common screw thread form is the symmetrical V-Profile with an included angle of 60 degrees. This form is prevalent in the Unified National Screw Thread Series (UN, UNC, UNF, UNRC, UNRF) form as well as the ISO/Metric.

A thread may be either right-hand or left-hand. A right-hand thread on an external member advances into an internal thread when turned clockwise.

A left-hand thread advances when turned counterclockwise. (Bike pedal, older propane tanks, etc).

Single vs. Double or Triple Threads

If a single helical groove is cut or formed on a cylinder, it is called a single-thread screw. Should the helix angle be increased sufficiently for a second thread to be cut between the grooves of the first thread, a double thread will be formed on the screw. Double, triple, and even quadruple threads are used whenever a rapid advance is desired, as on valves.

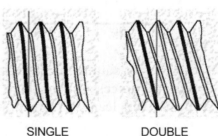

RIGHT HANDED   LEFT HANDED

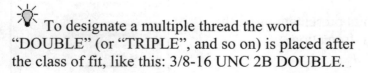

SINGLE   DOUBLE

🔆 To designate a multiple thread the word "DOUBLE" (or "TRIPLE", and so on) is placed after the class of fit, like this: 3/8-16 UNC 2B DOUBLE.

Pitch and Major Diameter

Pitch and major diameter designates a thread. Lead is the distance advanced parallel to the axis when the screw is turned one revolution.

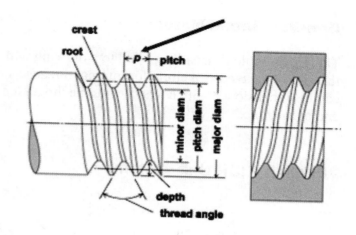

For a single thread, *lead is equal to the pitch*; for a double thread, lead is twice the pitch. For a straight thread, the pitch diameter is the diameter of an imaginary coaxial cylinder that would cut the thread forms at a height where the width of the thread and groove would be equal.

Thread Class of Fit

Classes of fit are tolerance standards; they set a plus or minus figure that is applied to the pitch diameter of bolts or nuts. The classes of fit used with almost all bolts sized in inches are specified by the ANSI/ASME Unified Screw Thread standards (which differ from the previous American National standards). There are three major Thread classes of fits, they are:

Class 1: The loosest fit. Used on parts which require assembly with a minimum of binding. Only found on bolts ¼ inch in diameter and larger.

Class 2: By far the most common class of fit. General purpose threads for bolts, nuts, and screws used in mass production.

Class 3: The closest fit. Used in precision assemblies where a close fit is desired to withstand stress and vibration.

Thread class identifies a range of thread tightness or looseness.

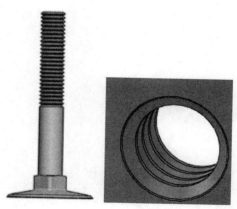

Classes for **External (male)** threads have an "**A**" suffix, for example, "2A" and classes for **Internal threads** have a "**B**" suffix.

General Thread Notes

The Thread note is usually applied to a drawing with a leader in the view where the thread is displayed as a circle for internal threads as illustrated below. External threads can be dimensioned with a leader with the thread length given as a dimension or at the end of the note.

English – ISP Unit system

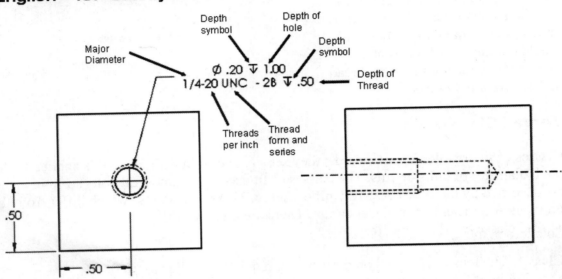

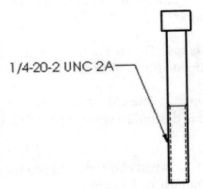

Pitch = 1/20: UNC = Unified National Series – Course: Thread Class 2: - B: = Internal. If not stated, the thread is always **Right-handed, Single**.

Dimensioning a Counterbore Hole

A Counterbore hole is a cylindrical flat bottom hole that has been machined to a larger diameter for a specified depth, so that a bolt or pin will fit into the recessed hole. The Counterbore provides a flat surface for the bolt or pin to seat against. In SolidWorks, use the Hole Wizard to insert the hole callout for a Counterbore.

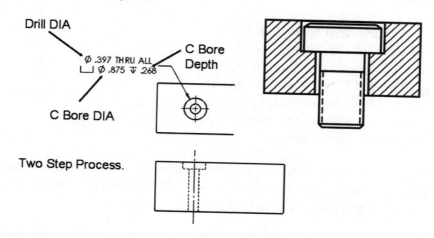

Counterbore holes are dimensioned by giving 1.) the diameter of the drill, 2.) the depth of the drill, 3.) the diameter of the counterbore , 4.) the depth of the counterbore, and 4.) the number of holes. Counterbore holes are displaced with the abbreviation C'BORE, C BORE or the symbol ⌴ .

💡 The difference between a C'BORE and a SPOTFACE is that the machining operation occurs on a curved surface.

Dimensioning a Countersunk Hole

The Countersunk hole, as illustrated below, is a cone-shaped recess machined in a part to receive a cone-shaped flat head screw or bolt.

A Countersunk hole is dimensioned by giving 1.) the diameter of the hole, 2.) the depth of the hole 3.) the diameter of the Countersunk, 4.) the angle at which the hole is to be Countersunk, 5.) the Counterbore diameter, 6.) the depth of the Counterbore, and 7.) the number of holes to be Countersunk.

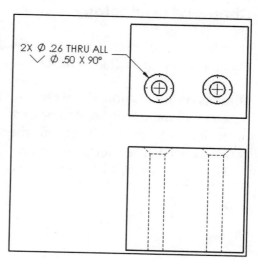

Adding a Counterbore for head clearance to a Countersink is optional. In SolidWorks, the Head Clearance option is located in the Hole Wizard Property Manager. The symbol for a Countersunk hole on a drawing annotation is CSK or ∨ .

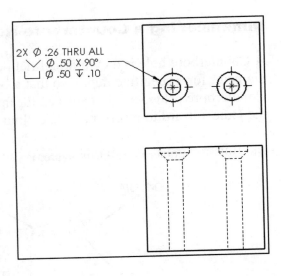

Chapter Summary

In Chapter 3 you reviewed basic dimensioning practices and were introduced to general tolerancing terminology according the ASME ANSI Y14.5 standard. You reviewed various dimensioning systems, fits and were presented with right and wrong way to dimension simple shapes, lines, angles, circles and arcs.

Dimensioning a drawing is a means to communicate the requirements to manufacture a part. It requires special annotations for fasteners, threads, countersunk holes, counterbored holes and other types of holes.

Tolerances determine the maximum and minimum variation that a dimension on a part is manufactured to. By understanding the required tolerance can save both time and money to create a part from your drawing.

Although SolidWorks automatically generates most annotations for a part, it is up to the designer to determine if all the required information is available to manufacture the part. The annotations must be presented according a dimensioning standard. There is no partial credit in the machine shop.

 View the pdf files on the enclosed DVD in the book for additional information.

Chapter Terminology

Depth: The horizontal (front to back) distance between two features in frontal planes. Depth is often identified in the shop as the thickness of a part or feature.

Dimensioning Standard - Metric - ASME standards for the use of metric dimensioning required all the dimensions to be expressed in millimeters (mm). The (mm) is not needed on each dimension, but it is used when a dimension is used in a notation. No trailing zeros are used. The Metric or International System of Units (S.I.) unit system in drafting is also known as the Millimeter, Gram Second (MMGS) unit system.

Dimensioning Standard - U.S. - ASME standard for U.S. dimensioning use the decimal inch value. When the decimal inch system is used, a zero is not used to the left of the decimal point for values less than one inch, and trailing zeros are used. The U.S. unit system is also known as the Inch, Pound, Second (IPS) unit system.

Engineering Graphics: Translates ideas from design layouts, specifications, rough sketches, and calculations of engineers & architects into working drawings, maps, plans, and illustrations which are used in making products.

Fasteners: Includes: Bolts and nuts (threaded), Set screws (threaded), Washers, Keys, Pins to name a few. Fasteners are not a permanent means of assembly such as welding or adhesives.

First Angle Projection: In First Angle Projection the Top view is looking at the bottom of the part. First Angle Projection is used in Europe and most of the world. However America and Australia use a method known as Third Angle Projection.

Height: The vertical distance between two or more lines or surfaces (features) which are in horizontal planes.

Origin: The point of intersection, where the X,Y,Z axes meet, is called the origin.

Part dimensions: Used in creating a part are sometimes called construction dimensions.

Thread Class or Fit: Classes of fit are tolerance standards; they set a plus or minus figure that is applied to the pitch diameter of bolts or nuts. The classes of fit used with almost all bolts sized in inches are specified by the ANSI/ASME Unified Screw Thread standards (which differ from the previous American National standards).

Thread Lead: The distance advanced parallel to the axis when the screw is turned one revolution. For a single thread, lead is equal to the pitch; for a double thread, lead is twice the pitch.

Third Angle Projection: In Third angle projection the Top View is looking at the Top of the part. First Angle Projection is used in Europe and most of the world. However America and Australia use a method known as Third Angle Projection.

Tolerance: The permissible range of variation in a dimension of an object. Tolerance may be specified as a factor or percentage of the nominal value, a maximum deviation from a nominal value, an explicit range of allowed values, be specified by a note or published standard with this information, or be implied by the numeric accuracy of the nominal value.

Questions

1. True or False: Dimensions should not be duplicated or the same information given in two different ways. If a reference dimension is used, the size value is placed within parentheses (X).

2. The U.S. unit system is also known as the (IPS) unit system. What does IPS stand for?

3. The diameter of a hole is place in the view in which shows the hole as a _____.

4. The length and diameter of cylinder are usually place in the view which shows the cylinder as a _____.

5. Dimension a hole by its _____.

6. True or False: Dimensioning to hidden lines should be avoided wherever possible.

7. If a hole goes completely through the feature and it is not clearly shown on the drawing, the abbreviation _____ follows the dimension.

8. True or False: A dimension is said to have a *Unilateral* (single) tolerance when the total tolerance is in one direction only, either (+) or (-).

9. The degree (°) may be divided into smaller units called _____. There are 60 _____ in each degree. Each minute may be divided into smaller units called _____.

10. Classes for an **External (male thread)** have a _____ suffix.

11. Classes for an **Internal (female thread)** have a _____ suffix.

12. There are three major Thread classes of fits, they are: _____, _____, _____. Explain the differences.

13. Identify the pitch of the following Thread note: 3/8-16 UNC 2B DOUBLE_____.

14. Identify the symbol of a Counterbore and Countersunk: _____, _____.

Exercises

Exercise 3.1: Estimate the dimensions in a whole number. Dimension the below illustration. Note: There is more than one way to dimension an angle.

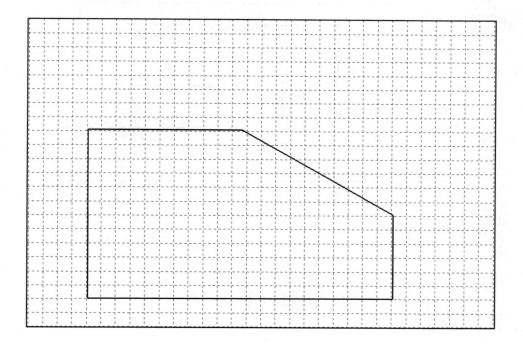

Exercise 3.2: Estimate the dimensions in a whole number. Dimension the illustration. Note: There is more than one way to dimension an angle.

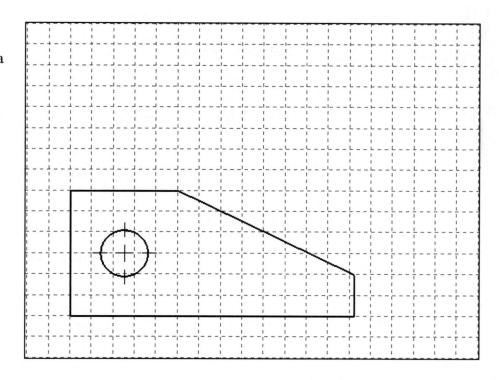

Exercise 3.3: Identify the dimension errors in the below illustration. Circle and list the errors.

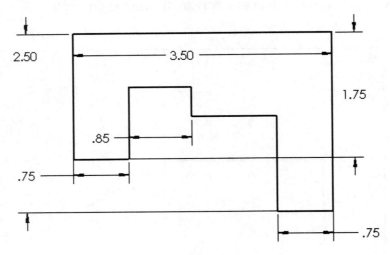

Errors:_____

_____

Exercise 3.4: Arrowheads are drawn between extension lines if space is available. If space is limited, the preferred Arrowhead and dimension location order is? List the preferred order.

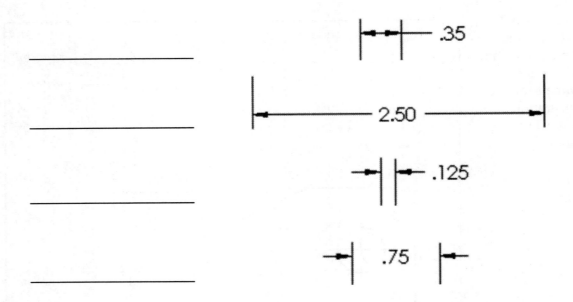

Exercise 3.5: Identify the dimension errors in the below illustration. Circle and list the errors.

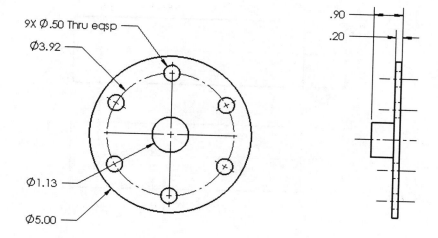

Errors:_____

_____

Exercise 3.6: Identify the dimension errors in the below illustration. Circle and list the errors.

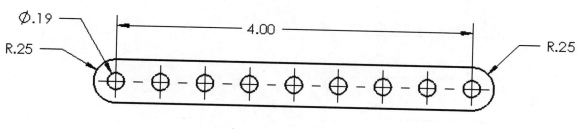

material thickness .10

Errors:_____

_____

Exercise 3.7: Place a *limit* tolerance of 002 on the below model.

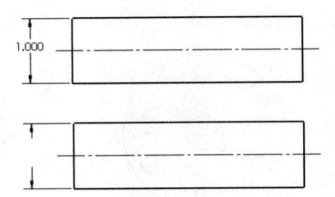

Exercise 3.8: Name three of the most common Tolerance Types.

1._____

2._____

3. _____ __

Exercise 3.9: Identify the following symbols.

_____ $\vee$

_____ $\sqcup$

_____ $\emptyset$

_____ $\underline{\vee}$

Exercise 3.10: Describe the following hole callouts (symbols and meanings) in detail.

⌀ .2500 THRU ALL
⌴ ⌀ .5000 ▽ .1250

⌀ .3970 THRU ALL
∨ ⌀ .7731 X 82°
⌴ ⌀ .7731 ▽ .0402

Exercise 3.11: True / False - The loosest fit is a Class 1 fit. A Class 1 fit is used on parts which require assembly with a minimum of binding.

Exercise 3.12: The two most common Tolerance Standard agencies are: American National Standards Institute (ANSI) / (ASME) and the International Standards Organization (ISO). In the ANSI (US) standard: This is a two part question.

True or False:
T   F   The higher limit is placed below the lower limit.
T   F   When both limits are placed on one line, the lower limit precedes the higher limit.

Exercise 3.13: There are basically two types of dimensioning systems use in creating parts and drawings - **U.S.** and **Metric**.

True or False: The U.S. system uses the decimal inch value. When the decimal inch system is used, a zero is not used to the left of the decimal point for values less than one inch and trailing zeros are not used.

True or False: The Metric system normally is expressed in millimeters. When the millimeter system is used, the number is rounded to the nearest whole number. Trailing zeros are used.

Exercise 3.14: Identify the illustrated Thread Note.
Remember units!

.250-20 UNC-2A-LH

1.) Pitch of the Thread:_____

2.) Major Thread Diameter:_____

3.) Internal or External Threads:_____

Chapter 4

Introduction to SolidWorks Part Modeling

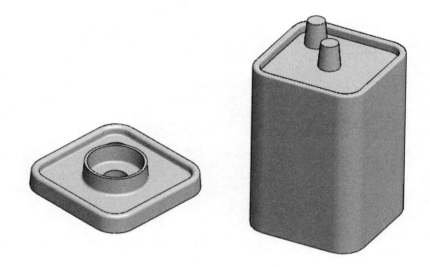

Below are the desired outcomes and usage competencies based on the completion of Chapter 4.

Desired Outcomes:	Usage Competencies:
• A comprehensive understanding of the SolidWorks 2012 User Interface.	• Ability to establish a SolidWorks session. Use the SolidWorks User Interface: CommandManager, Toolbars, Task Pane, Search, Confirmation Corner and more.
• Address File Management with file folders.	• Aptitude to create file folders for various Projects and Templates.

• Create two Part Templates:
  ○ PART-IN-ANSI
  ○ PART-MM-ISO
	• Skill to address System Options and Document Properties.

• Create two Parts:
  ○ BATTERY
  ○ BATTERYPLATE
	• Specific knowledge and understanding of 2D sketching and the following 3D features; Extruded Boss/Base, Extruded Cut, Fillet and Instant3D.

Notes:

Chapter 4 - Introduction to SolidWorks Part Modeling

Chapter Overview

Provide a comprehensive understanding of the SolidWorks default User Interface and CommandManager: *Menu bar toolbar, Menu bar menu, Drop-down menu, Right-click Pop-up menus, Context toolbars / menus, Fly-out tool button, System feedback icons, Confirmation Corner, Heads-up View toolbar and an understanding of System Options, Document Properties, Part templates, File management and more.*

Create two Part Templates utilized for the parts in this chapter:

- PART-IN-ANSI

- PART-MM-ISO

A Template is the foundation for a SolidWorks document. Templates are part, drawing, and assembly documents that include user-defined parameters and are the basis for new documents.

Create two parts for the FLASHLIGHT assembly:

- BATTERY

- BATTERYPLATE

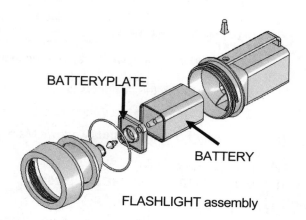

FLASHLIGHT assembly

On the completion of this chapter, you will be able to:

- Start a SolidWorks session.

- Understand and navigate through the SolidWorks (UI) and CommandManager.

- Apply and understand System Options and Document Properties.

- Create a Part Template.

- Open, Save and Close Part documents and Templates.

- Create 2D sketch profiles on the correct Sketch plane.

- Utilize the following Sketch tools: Smart Dimension, Line, Centerline, Circle, Center Rectangle, Convert Entities, Offset Entities and Mirror Entities.

- Apply and edit sketch dimensions.

- Establish Geometric relations, dimensions, and determine the status of the sketch.
  - Under defined, Fully defined and Over defined
- Utilize the Instant3D tool to create an Extruded Boss/Base feature.
- Utilize the Save As, Delete, Edit Feature and Modify tools.

 Screen shots and illustrations in the book display the SolidWorks user default setup.

File Management

File management organizes parts, assemblies, drawings, and templates. Why do you require file management? Answer: A top level assembly has hundreds or even thousands of documents that requires organization. Utilize folders to organize projects, vendor components, templates, and libraries. Create the first folder named SOLIDWORKS-MODELS 2012. Create two sub-folders named MY-TEMPLATES and PROJECTS.

Activity: File Management

Create a new folder in Windows. Note: The procedure will be different depending on your operating system.

1) Click **Start** from the Windows taskbar.

2) Click **Documents** in the Windows dialog box.

3) Click **New Folder** from the Windows Main menu.

Enter the new folder name.

4) Enter **SOLIDWORKS-MODELS 2012**.

 Screen shots in the text were made using SolidWorks 2012 SP0 running on Windows® 7. Illustrations may vary depending on your SolidWorks version and operating system.

Create the first sub-folder. Note: The procedure maybe different depending on your Operating System.

5) Double-click the **SOLIDWORKS-MODELS 2012** folder.

6) Click **New folder**.

7) Enter **MY-TEMPLATES** for the folder name.

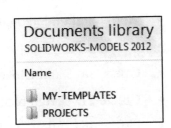

Create the second sub-folder.

8) Click **New folder**.

9) Enter **PROJECTS** for the second sub-folder name.

Return to the SOLIDWORKS-MODELS 2012 folder.

10) Click the **SOLIDWORKS-MODELS 2012** folder.

Utilize the MY-TEMPLATES folder and the PROJECTS folder throughout the text.

Start a SolidWorks 2012 Session

The SolidWorks application is located in the Programs folder. SolidWorks displays the Tip of the Day box. Read the Tip of the Day every day to obtain additional information on SolidWorks.

Create a new part. Click File, New from the Menu bar menu or click New ☐ from the Menu bar toolbar. There are two options for new documents: *Novice* and *Advanced*. Select the Advanced option. Select the Part document.

Activity: Start a SolidWorks 2012 Session

Start a SolidWorks 2012 session.

11)   Click **Start** on the Windows Taskbar.

12)   Click **All Programs**.

13)   Click the **SolidWorks 2012** folder.

14)   Click **SolidWorks 2012** application. The SolidWorks program window opens. Note: Do not open a document at this time.

 If available, double-click the SolidWorks 2012 icon on the Windows Desktop to start a SolidWorks session.

Read the Tip of the Day dialog box.

15)   If you do not see this screen, click the SolidWorks

Resources 🏠 icon on the right side of the Graphics window located in the Task Pane.

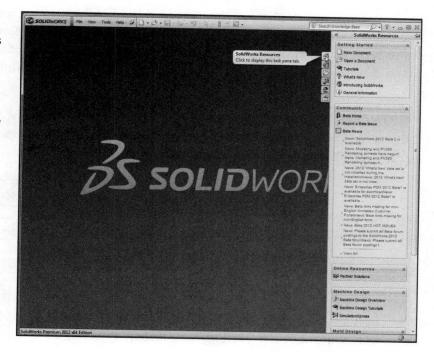

Activity: Understanding the SolidWorks UI and CommandManager

Menu bar toolbar

SolidWorks 2012 (UI) is design to make maximum use of the Graphics window area. The default Menu bar toolbar contains a set of the most frequently used tool buttons from the Standard toolbar. The available tools are: **New** ⬜ - Creates a new document, **Open** 🗁 - Opens an existing document, **Save** 💾 - Saves an active document, **Print** 🖨 - Prints an active document, **Undo** ↺ - Reverses the last action, **Select** ⬉ ▾ - Selects Sketch entities, components and more, **Rebuild** 🖇 - Rebuilds the active part, assembly or drawing, **File Properties** 🗐 - Shows the summary information on the active document, **Options** ▤ - Changes system options and Add-Ins for SolidWorks.

Menu bar menu / Menu bar toolbar

Click SolidWorks in the Menu bar toolbar to display the Menu bar menu. SolidWorks provides a Context-sensitive menu structure. The menu titles remain the same for all three types of documents, but the menu items change depending on which type of document is active.

Example: The Insert menu includes features in part documents, mates in assembly documents, and drawing views in drawing documents. The display of the menu is also dependent on the workflow customization that you have selected. The default menu items for an active document are: *File, Edit, View, Insert, Tools, Window, Help* and *Pin*.

☼ The Pin 📌 option displays the Menu bar toolbar and the Menu bar menu as illustrated. Throughout the book, the Menu bar menu and the Menu bar toolbar is referred as the Menu bar.

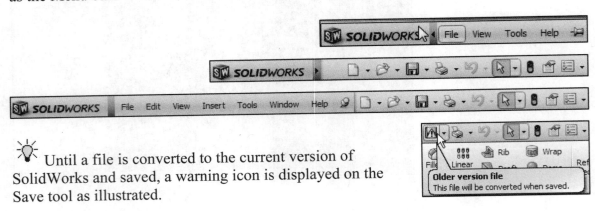

☼ Until a file is converted to the current version of SolidWorks and saved, a warning icon is displayed on the Save tool as illustrated.

Drop-down menu

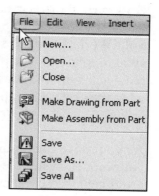

SolidWorks takes advantage of the familiar Microsoft® Windows® user interface. Communicate with SolidWorks either through the; *Drop-down menu, Pop-up menu, Shortcut toolbar, Fly-out toolbar* or the *CommandManager*.

A command is an instruction that informs SolidWorks to perform a task. To close a SolidWorks drop-down menu, press the Esc key. You can also click any other part of the SolidWorks Graphics window, or click another drop-down menu

Right-click

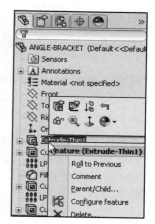

Right-click in the: *Graphics window, FeatureManager*, or *Sketch* to display a Context-sensitive toolbar. If you are in the middle of a command, this toolbar displays a list of options specifically related to that command.

Press the **s** key to view/access previous command tools in the Graphics window.

Consolidated toolbar

Similar commands are grouped in the CommandManager. Example: Variations of the Rectangle sketch tool are grouped in a single fly-out button as illustrated.

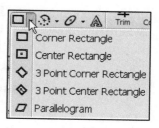

If you select the Consolidated toolbar button without expanding:

- For some commands such as Sketch, the most commonly used command is performed. This command is the first listed and the command shown on the button.

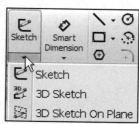

- For commands such as rectangle, where you may want to repeatedly create the same variant of the rectangle, the last used command is performed. This is the highlighted command when the Consolidated toolbar is expanded.

System Feedback

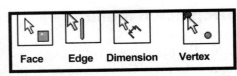

SolidWorks provides system feedback by attaching a symbol to the mouse pointer cursor. The system feedback symbol indicates what you are selecting or what the system is expecting you to select.

As you move the mouse pointer across your model, system feedback is provided to you in the form of symbols, riding next to the cursor arrow as illustrated.

Confirmation Corner

When numerous SolidWorks commands are active, a symbol or a set of symbols are displayed in the upper right hand corner of the Graphics window. This area is called the Confirmation Corner.

When a sketch is active, the confirmation corner box displays two symbols. The first symbol is the sketch tool icon. The second symbol is a large red X. These two symbols supply a visual reminder that you are in an active sketch. Click the sketch symbol icon to exit the sketch and to saves any changes that you made.

When other commands are active, the confirmation corner box provides a green check mark and a large red X. Use the green check mark to execute the current command. Use the large red X to cancel the command.

A goal of this book is to expose the new SolidWorks user to various design tools and features.

During the initial SolidWorks installation, you were requested to select either the ISO or ANSI drafting standard. ISO is typically a European drafting standard and uses First Angle Projection. The book is written using the ANSI (US) overall drafting standard and Third Angle Projection for drawings.

When you create a new part or assembly, the three default Planes (Front, Right and Top) are aligned with specific views. The Plane you select for the Base sketch determines the orientation of the part or assembly.

Heads-up View toolbar

SolidWorks provides the user with numerous view options from the Standard Views, View and Heads-up View toolbar. The Heads-up View toolbar is a transparent toolbar that is displayed in the Graphics window when a document is active.

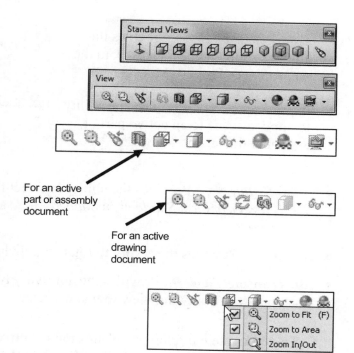

You can hide, move or modify the Heads-up View toolbar. To modify the toolbar: right-click on a tool and select or deselect the tools that you want to display. The following views are available: Note: Views are document dependent.

- *Zoom to Fit* : Zooms the model to fit the Graphics window.

- *Zoom to Area* : Zooms to the areas you select with a bounding box.

- *Previous View* : Displays the previous view.

- *Section View* : Displays a cutaway of a part or assembly, using one or more cross section planes.

- *View Orientation* : Provides the ability to select a view orientation or the number of viewports. The available options are: *Top, Isometric, Trimetric, Dimetric, Left, Front, Right, Back, Bottom, Single view, Two view - Horizontal, Two view - Vertical, Four view.*

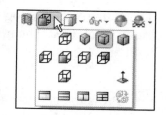

- *Display Style* : Provides the ability to display the style for the active view. The available options are: *Wireframe, Hidden Lines Visible, Hidden Lines Removed, Shaded, Shaded With Edges.*

- *Hide/Show Items* : Provides the ability to select items to hide or show in the Graphics window. Note: The available items are document dependent.

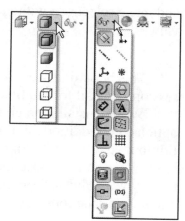

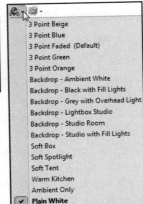

- *Edit Appearance* : Provides the ability to apply appearances from the Appearances PropertyManager.

- *Apply Scene* : Provides the ability to apply a scene to an active part or assembly document. View the available options.

- *View Setting* : Provides the ability to select the following: *RealView Graphics*, *Shadows in Shaded Mode* and *Perspective*.

- *Rotate* : Provides the ability to rotate a drawing view.

- *3D Drawing View* : Provides the ability to dynamically manipulate the drawing view to make a selection.

To deactivate the reference planes for an active document, click **View**, uncheck **Planes** from the Menu bar. To deactivate the grid, click **Options** , **Document Properties** tab. Click **Grid/Snaps**, uncheck the **Display grid** box.

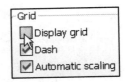

Modify the Heads-up View toolbar. Press the **space** key. The Orientation dialog box is display. Click the **New View** tool. The Name View dialog box is displayed. Enter a new **named** view. Click **OK**. The new view is displayed in the Heads-up View toolbar.

Press the **g** key to activate the Magnifying glass tool. Use the Magnifying glass tool to inspect a model and make selections without changing the overall view.

This book does not cover starting a SolidWorks session in detail for the first time. A default SolidWorks installation presents you with several options. For additional information for an Education Edition, visit the following sites: http://www.solidworks.com/goedu and http://www.solidworks.com/sw/docs/EDU_2011_Inst allation_Instructions.pdf.

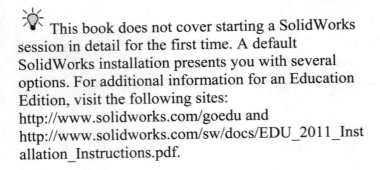

CommandManager

The SolidWorks CommandManager is a *Context-sensitive toolbar* that automatically updates based on the toolbar you want to access. By default, it has toolbars embedded in it based on your active document type. When you click a tab below the CommandManager, it updates to display that toolbar. Example, if you click the Sketch tab, the Sketch toolbar is displayed. The default Part tabs are: *Features, Sketch, Evaluate, DimXpert* and Office *Products*.

Below is an illustrated CommandManager for a default Part document.

 If you have SolidWorks, SolidWorks Professional, or SolidWorks Premium, the Office Products tab appears on the CommandManager as illustrated.

 Select the Add-In directly from the Office Products tab.

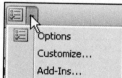

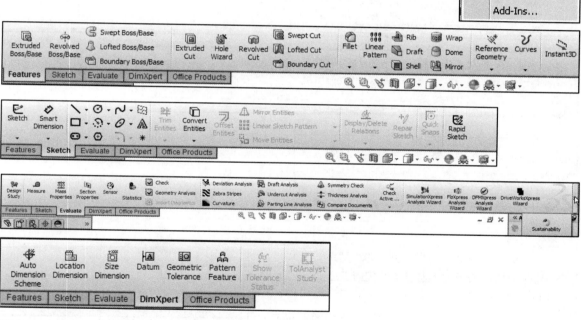

To customize the CommandManager, right-click on a tab and select Customize CommandManager.

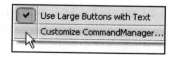

Below is an illustrated CommandManager for the default Drawing document. The default Drawing tabs are: *View Layout*, *Annotation*, *Sketch*, *Evaluate* and *Office Products*.

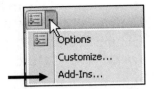

Double-clicking the CommandManager when it is docked will make it float. Double-clicking the CommandManager when it is floating will return it to its last position in the Graphics window.

Select the Add-In directly from the Office Products tab.

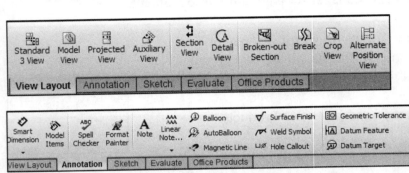

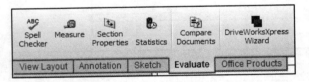

To add a custom tab to your CommandManager, right-click on a tab and click Customize CommandManager from the drop-down menu. The Customize dialog box is displayed.

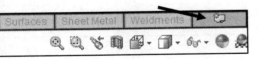

You can also select to add a blank tab as illustrated and populate it with custom tools from the Customize dialog box.

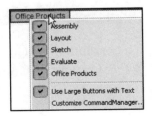

Below is an illustrated CommandManager for the default Assembly document. The default Assembly tabs are: *Assembly, Layout, Sketch, Evaluate* and *Office Products*.

If you have SolidWorks, SolidWorks Professional, or SolidWorks Premium, the Office Products tab appears on the CommandManager

 Select the Add-In directly from the Office Products tab.

 Instant3D and Rapid Sketch tool is active by default.

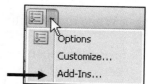

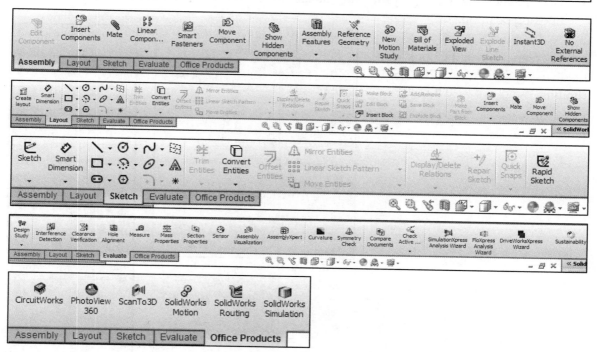

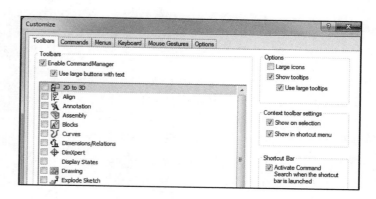

By default, the illustrated options are selected in the Customize box for the CommandManager.

Right-click on an existing tabs, and click Customize CommandManager to view your options.

Drag or double-click the CommandManager and it becomes a separate floating window. Once it is floating, you can drag the CommandManager anywhere on or outside the SolidWorks window.

To dock the CommandManager when it is floating, perform one of the following actions:

- While dragging the CommandManager in the SolidWorks window, move the pointer over a docking icon - ⬆ Dock above , ◀ Dock left , ▶ Dock right and click the needed command.

- Double-click the floating CommandManager to revert the CommandManager to the last docking position.

Screen shots in the book were made using SolidWorks 2012 SP0 running Windows® 7 Ultimate.

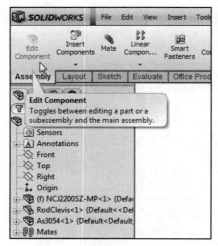

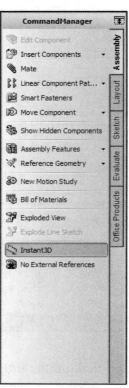

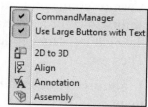

To save space in the CommandManager, right-click in the CommandManager and un-check the Use Large Buttons with Text box. This eliminates the text associated with the tool.

If you want to add a custom tab to your CommandManager, right-click on a tab and select the toolbar you want to insert. You can also select to add a blank tab as illustrated and populate it with custom tools from the Customize dialog box.

FeatureManager Design Tree

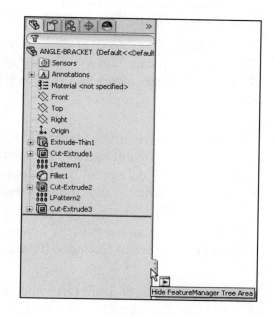

The FeatureManager design tree is located on the left side of the SolidWorks Graphics window. The FeatureManager provides a summarize view of the active part, assembly, or drawing document. The tree displays the details on how the part, assembly or drawing document was created.

Understand the FeatureManager design tree to troubleshoot your model. The FeatureManager is used extensively throughout this book.

The FeatureManager consist of five default tabs:

- *FeatureManager design tree*

- *PropertyManager*

- *ConfigurationManager*

- *DimXpertManager*

- *DisplayManager*

Select the Hide FeatureManager Tree Area

arrows as illustrated to enlarge the Graphics window for modeling.

When you create a new part or assembly, the three default Planes (Front, Right and Top) are aligned with specific views. The Plane you select for the Base sketch determines the orientation of the part or assembly.

Various commands provide the
ability to control what is displayed in
the FeatureManager design tree.
They are:

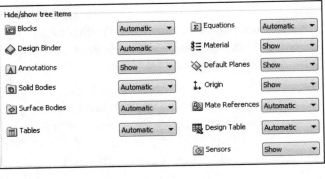

1. Show or Hide FeatureManager
items.

Click **Options** from the
Menu bar. Click **FeatureManager**
from the System Options tab.
Customize your FeatureManager
from the Hide/Show Tree Items dialog box.

2. Filter the FeatureManager design tree. Enter information in the
filter field. You can filter by: *Type of features, Feature names,
Sketches, Folders, Mates, User-defined tags* and *Custom
properties*.

Tags are keywords you can add to a
SolidWorks document to make them easier to filter
and to search. The Tags icon is located in the
bottom right corner of the Graphics window.

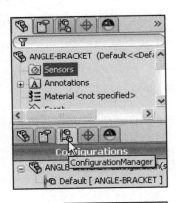

To collapse all items in the FeatureManager,
right-click and select **Collapse items**, or press the **Shift +C**
keys.

The FeatureManager design tree and the Graphics window are
dynamically linked. Select sketches, features, drawing views,
and construction geometry in either pane.

Split the FeatureManager design tree and either display two
FeatureManager instances, or combine the FeatureManager
design tree with the ConfigurationManager or
PropertyManager.

Move between the FeatureManager design tree,
PropertyManager, ConfigurationManager, and
DimXpertManager by selecting the tabs at the top of the menu.

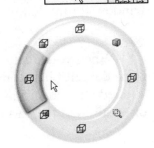

Right-click and drag in the Graphics area to display the
Mouse Gesture wheel. You can customize the default commands
for a sketch, part, assembly or drawing.

The ConfigurationManager is located to the right of the FeatureManager. Use the ConfigurationManager to create, select and view multiple configurations of parts and assemblies.

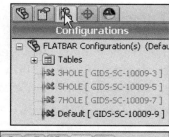

The icons in the ConfigurationManager denote whether the configuration was created manually or with a design table.

The DimXpertManager tab provides the ability to insert dimensions and tolerances manually or automatically. The DimXpertManager provides the following selections: *Auto Dimension Scheme* , *Show Tolerance Status* , *Copy Scheme* and *TolAnalyst Study* .

 TolAnalyst is available in SolidWorks Premium.

Fly-out FeatureManager

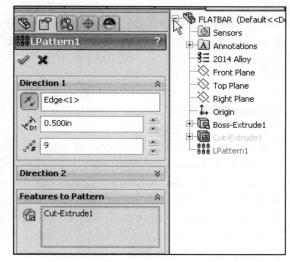

The fly-out FeatureManager design tree provides the ability to view and select items in the PropertyManager and the FeatureManager design tree at the same time.

Throughout the book, you will select commands and command options from the drop-down menu, fly-out FeatureManager, Context toolbar or from a SolidWorks toolbar.

Another method for accessing a command is to use the accelerator key. Accelerator keys are special key strokes which activate the drop-down menu options. Some commands in the menu bar and items in the drop-down menus have an underlined character.

Press the Alt key followed by the corresponding key to the underlined character activates that command or option.

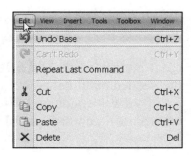

Illustrations may vary slightly depending on your SolidWorks version.

Task Pane

The Task Pane is displayed when a SolidWorks session starts. The Task Pane can be displayed in the following states: *visible or hidden*, *expanded or collapsed*, *pinned or unpinned*, *docked or floating*. The Task Pane contains the following default tabs: *SolidWorks Resources* ⌂, *Design Library* 🏛, *File Explorer* 📂, *View Palette* 🖼, *Appearances, Scenes, and Decals* 🔵 and *Custom Properties* 📄.

SolidWorks Resources

The basic SolidWorks Resources ⌂ menu displays the following default selections: *Getting Started*, *Community*, *Online Resources* and *Tip of the Day*.

Other user interfaces are available during the initial software installation selection: *Machine Design*, *Mold Design* or *Consumer Products Design*.

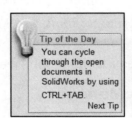

Design Library

The Design Library 🏛 contains reusable parts, assemblies, and other elements, including library features. The Design Library tab contains four default selections. Each default selection contains additional sub categories. The default selections are: *Design Library, Toolbox, 3D ContentCentral* and *SolidWorks Content*.

🔆 To active the SolidWorks Toolbox, click **Tools**, **Add-Ins**... from the Main menu. Check the **SolidWorks Toolbox** and the **SolidWorks Toolbox Browser** boxes from the Add-Ins dialog box. Click **OK**.

To access the Design Library folders in a non-network environment, click **Add File Location** 🏛 and browse to the needed path. Paths will vary depending on your SolidWorks version and Windows setup. In a network environment, contact your IT department for system details.

File Explorer

File Explorer 📂 in the Task Pane duplicates Windows Explorer from your local computer and displays the following directories: *Recent Documents, Samples, Open in SolidWorks and Desktop.*

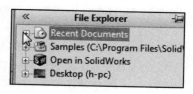

Search

SolidWorks Search box is displayed in the upper right corner of the SolidWorks Graphics window. Enter the text or key words to search.

New search modes have been added to SolidWorks Search. In addition to searching for files and models, you can search *SolidWorks Help*, the *Knowledge Base*, or the *Community Forums*. Internet access is required for the Community Forums and Knowledge Base.

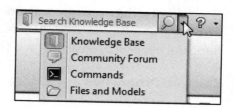

View Palette

The View Palette 🖼 tab located in the Task Pane provides the ability to insert drawing views of an active document, or click the Browse button to locate the desired document.

Drag and drop the view from the View Palette into an active drawing sheet to create a drawing view.

The selected model is View Palette 13-1 in the illustration. The (**A**) **Front** and (**A**) **Top** drawing views are displayed with DimXpert Annotations which were applied at the part level.

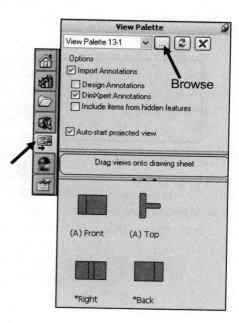

Appearances, Scenes, and Decals

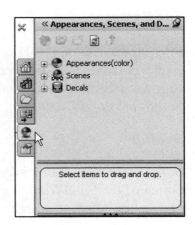

Appearances, Scenes, and Decals 🌐 provide a simplified way to display models in a photo-realistic setting using a library of Appearances, Scenes, and Decals.

An appearance defines the visual properties of a model, including color and texture. Appearances do not affect physical properties, which are defined by materials.

Scenes provide a visual backdrop behind a model. In SolidWorks, they provide reflections on the model. PhotoView 360 is an Add-In. Drag and drop a selected appearance, scene, or decal on a feature, part, or assembly.

Custom Properties

The Custom Properties 📋 tool provides the ability to enter custom and configuration specific properties directly into SolidWorks files. In assemblies, you can assign properties to multiple parts at the same time. If you select a lightweight component in an assembly, you can view the component's custom properties in the Task Pane without resolving the component. If you edit a value, you are prompted to resolve the component so the change can be saved.

Document Recovery

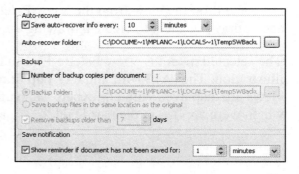

If auto recovery is initiated in the System Options section and the system terminates unexpectedly with an active document, the saved information files are available on the Task Pane Document Recovery tab the next time you start a SolidWorks session.

💡 Run DFMXpress from the Evaluate tab or from Tools, DFMXpress in the Menu bar menu. The DFMXpress icon is displayed in the Task Pane.

Motion Study tab

Motion Studies are graphical simulations of motion for an assembly. Access MotionManager from the Motion Study tab. The Motion Study tab is located in the bottom left corner of the Graphics window.

Incorporate visual properties such as lighting and camera perspective. Click the Motion Study tab to view the MotionManager. Click the Model tab to return to the FeatureManager design tree.

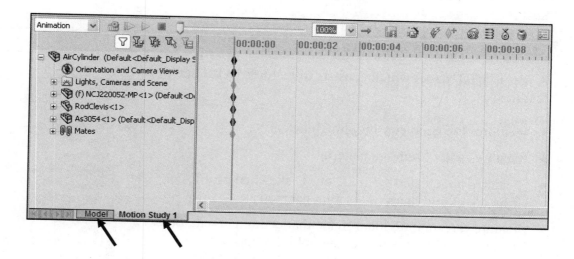

The MotionManager display a timeline-based interface, and provide the following selections from the drop-down menu as illustrated:

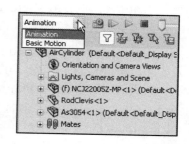

- *Animation:* Apply Animation to animate the motion of an assembly. Add a motor and insert positions of assembly components at various times using set key points. Use the Animation option to create animations for motion that do **not** require accounting for mass or gravity.

- *Basic Motion:* Apply Basic Motion for approximating the effects of motors, springs, collisions and gravity on assemblies. Basic Motion takes mass into account in calculating motion. Basic Motion computation is relatively fast, so you can use this for creating presentation animations using physics-based simulations. Use the Basic Motion option to create simulations of motion that account for mass, collisions or gravity.

If the Motion Study tab is not displayed in the Graphics window, click **View, MotionManager** from the Menu bar.

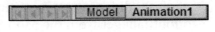

🔅 For older assemblies created before 2008, the Animation1 tab maybe displayed. View the Assembly Chapter for additional information.

🔅 To create a new Motion Study, click **Insert, New Motion Study** from the Menu bar.

🔅 View SolidWorks Help for additional information on Motion Study.

Activity: Create a new 3D Part

A part is a 3D model which consists of features. What are features?

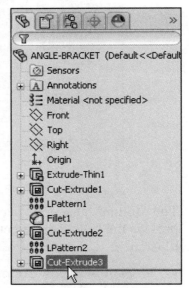

- Features are geometry building blocks.

- Features add or remove material.

- Features are created from 2D or 3D sketched profiles or from edges and faces of existing geometry.

- Features are an individual shape that combined with other features, makes up a part or assembly. Some features, such as bosses and cuts, originate as sketches. Other features, such as shells and fillets, modify a feature's geometry.

- Features are displayed in the FeatureManager as illustrated (Extrude-Thin1, Cut-Extrude1, LPattern1, Fillet1, Cut-Extrude2, Lpatern2, and Cut-Extrude3).

You can suppress a feature. A suppressed feature is displayed in light gray.

🔅 The first sketch of a part is called the Base Sketch. The Base sketch is the foundation for the 3D model. The book focuses on 2D sketches and 3D features.

🔅 During the initial SolidWorks installation, you were requested to select either the ISO or ANSI drafting standard. ISO is typically; a European drafting standard and uses First Angle Projection. The book is written using the ANSI (US) overall drafting standard and Third Angle Projection for drawings.

The first sketch of a part is the Base sketch. The Base sketch is the foundation for the 3D model.

Create a new part.

16) Click **New** ⬜ from the Menu bar. The New SolidWorks Document dialog box is displayed.

Select Advanced Mode.

17) Click the **Advanced** button to display the New SolidWorks Document dialog box in Advance mode.

18) The Templates tab is the default tab. Part is the default template from the New SolidWorks Document dialog box. Click **OK**.

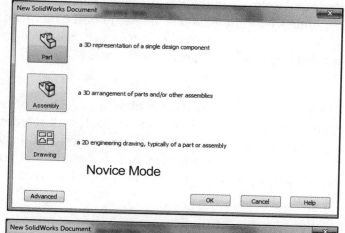

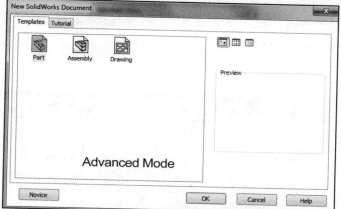

The *Advanced* mode remains selected for all new documents in the current SolidWorks session. When you exit SolidWorks, the *Advanced* mode setting is saved.

The default SolidWorks installation contains two tabs in the New SolidWorks Document dialog box: *Templates* and *Tutorial*. The *Templates* tab corresponds to the default SolidWorks templates. The *Tutorial* tab corresponds to the templates utilized in the SolidWorks Tutorials.

During the initial SolidWorks installation, you are requested to select either the ISO or ANSI drafting standard. ISO is typically a European drafting standard and uses First Angle Projection. The book is written using the ANSI (US) overall drafting standard and Third Angle Projection for all drawing documents.

Part1 is displayed in the FeatureManager and is the name of the document. Part1 is the default part window name. The Menu bar, CommandManager, FeatureManager, Heads-up View toolbar, SolidWorks Resources, SolidWorks Search, Task Pane, and the Origin are displayed in the Graphics window.

The Origin ⤱ is displayed in blue in the center of the Graphics window. The Origin represents the intersection of the three default reference planes: *Front Plane*, *Top Plane* and *Right Plane*. The positive X-axis is horizontal and points to the right of the Origin in the Front view. The positive Y-axis is vertical and point upward in the Front view. The FeatureManager contains a list of features, reference geometry, and settings utilized in the part.

☼ You can now edit the document units directly from the Graphics window.

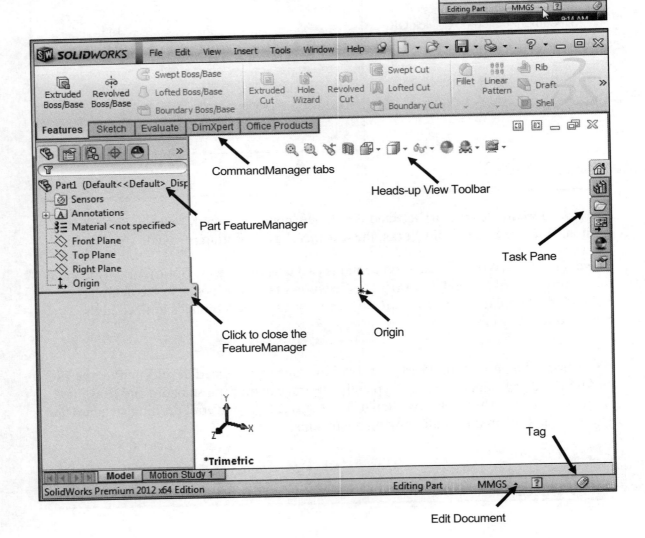

In this book, Reference planes and Grid/Snaps are deactivated in the Graphics window for improved model clarity.

Activity: Menu Bar toolbar, Menu Bar menu, Heads-up View toolbar

Display tools and tool tips.

19) Position the **mouse pointer** over the Heads-up View toolbar and view the tool tips.

20) **Read** the large tool tip.

21) Select the **drop-down arrow** ▼ to view the available view tools.

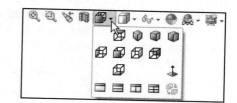

Display the View toolbar and the Menu bar.

22) Right-click in the **gray area** of the Menu bar.

23) Click **View**. The View toolbar is displayed.

24) Click and drag the **View toolbar** off the Graphics window.

25) Click **SolidWorks** as illustrated to expand the Menu bar menu.

26) **Pin** the Menu bar as illustrated. Use both the Menu bar menu and the Menu bar toolbar in this book.

The SolidWorks Help Topics contains step-by-step instructions for various commands. The Help ⍰ icon is displayed in the dialog box or in the PropertyManager for each feature. Display the SolidWorks Help Home Page. Use SolidWorks Help to locate information on What's New, Sketches, Features, Assemblies and more.

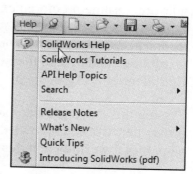

27) Click **Help** from the Menu bar.

28) Click **SolidWorks Help**. The SolidWorks Help Home Web Page is displayed by default. (Use SolidWorks Web Help is selected by default). View your options and features.

29) Click the **Home Page** 🏠 icon to return to the Home Page.

30) **Close** ❌ the SolidWorks Home Page dialog box.

Display and explore the SolidWorks tutorials.

31) Click **Help** from the Menu bar.

32) Click **SolidWorks Tutorials**. The SolidWorks Tutorials are displayed. The SolidWorks Tutorials are presented by category.

33) Click the **Getting Started** category. The Getting Started category provides three 30 minute lessons on parts, assemblies, and drawings. This section also provides information for users who are switching from AutoCAD to SolidWorks. The tutorials also provide links to the CSWP and CSWA Certification programs and a new What's New Tutorials for 2012.

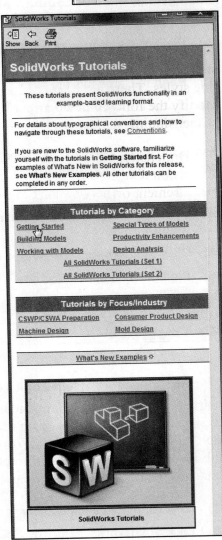

SolidWorks Corporation offers various levels of certification representing increasing levels of expertise in 3D CAD design as it applies to engineering.

The *Certified SolidWorks Associate* CSWA certification indicates a foundation in and apprentice knowledge of 3D CAD design and engineering practices and principles.

The main requirement for obtaining the CSWA certification is to take and pass the three hour, seven question on-line proctored exam at a Certified SolidWorks CSWA Provider, "university, college, technical, vocational or secondary educational institution" and to sign the SolidWorks Confidentiality Agreement.

Passing this exam provides students the chance to prove their working knowledge and expertise and to be part of a worldwide industry certification standard.

34) **Close** the ⊠ Online Tutorial dialog box. Return to the SolidWorks Graphics window.

Part Template

The Part Template is the foundation for a SolidWorks part. Part1 displayed in the FeatureManager utilizes the *Part.prtdot* default template located in the New SolidWorks dialog box.

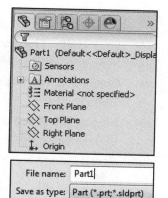

Document properties contain the default settings for the Part Template. The document properties include the dimensioning standard, units, dimension decimal display, grids, note font, and line styles. There are hundreds of document properties. You will modify the following document properties in this Project: Dimensioning standard, unit, and decimal places.

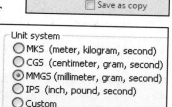

The Dimensioning (drafting) standard determines the display of dimension text, arrows, symbols, and spacing. Units are the measurement of physical quantities. MMGS, (millimeter, gram, second) and IPS, (inch, pound, second) are the two most common unit systems specified for engineering parts and drawings.

Document properties are stored with the document. Apply the document properties to the Part Template. Create two Part Templates: PART-IN-ANSI and PART-MM-ISO. Save the Part Templates in the MY-TEMPLATE folder.

System Options are stored in the registry of your computer. The File Locations option controls the file folder location of SolidWorks documents.

Utilize the File Locations option to reference your Part Templates in the MY-TEMPLATES folder. Add the SOLIDWORKS-MODELS 2012\MY-TEMPLATES folder path name to the Document Templates File Locations list.

Activity: Create the PART-IN-ANSI and PART-MM-ISO Part Template

Create a **PART-IN-ANSI** Part template.

35) Click **Options** , **Document Properties** tab from the Menu bar. The Document Properties - Drafting Standard dialog box is displayed.

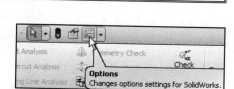

36) Select **ANSI** from the Overall drafting standard drop-down box.

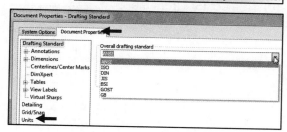

Set the part units.

37) Click **Units**. The Document Properties - Unit dialog box is displayed.

38) Select **IPS, (inch, pound, second)** for Unit system.

39) Select **.123** (three decimal places) for Length basic units.

40) Select **None** for Angular units Decimal places.

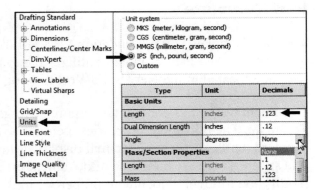

Set the Grid/Snap option.

41) Click **Grid/Snap**. The Document Properties Grid/Snap dialog box is displayed.

42) Un-check the **Display grid** box.

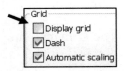

Return to the SolidWorks Graphics window.

43) Click **OK** from the Document Properties Grid/Snap dialog box.

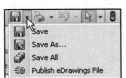

Save the Part Template.

44) Click **Save As** from the Menu bar drop-down menu. The Save As dialog box is displayed.

45) Select **Part Templates (*.prtdot)** from the Save as type box.

46) Select the **SOLIDWORKS-MODELS 2012/MY-TEMPLATES** folder.

47) Enter **PART-IN-ANSI** in the File name box.

48) Click **Save** from the Save As dialog box.

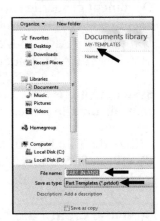

Create the PART-MM-ISO Part Template.

49) Click **Options** , **Document Properties** tab from the Menu bar. The Document Properties - Drafting Standard dialog box is displayed.

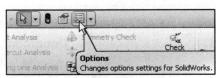

50) Select **ISO** from the Overall drafting standard drop-down box.

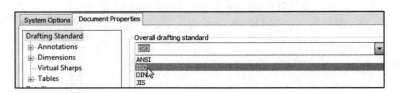

Set the part units.

51) Click **Units**. The Document Properties - Unit dialog box is displayed.

52) Select **MMGS, (millimeter, gram, second)** for Unit system.

53) Select **.12** (two decimal places) for Length basic units.

54) Select **None** for Angular units Decimal places.

55) Click **OK**.

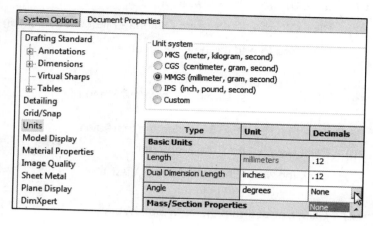

Save the Part Template.

56) Click **Save As** from the Menu bar drop-down menu. The Save As dialog box is displayed.

57) Select **Part Templates (*.prtdot)** from the Save as type box.

58) Select the **SOLIDWORKS-MODELS 2012/MY-TEMPLATES** folder.

59) Enter **PART-MM-ISO** in the File name box.

60) Click **Save**.

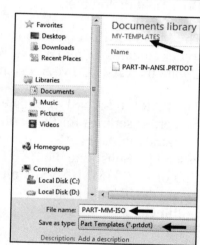

Set System Options to add the two Part Templates.

61) Click **Options** from the Menu bar. The System Options - General dialog box is displayed.

62) Click **File Locations** from the System Options tab.

63) Select **Document Templates** from Show folders for.

64) Click the **Add** button.

65) Select the **SOLIDWORKS-MODELS 2012/MY-TEMPLATES** folder.

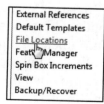

66)   Click **OK** from the Browse for Folder.

67)   Click **OK** from the System Options - File Location dialog box.

68)   Click **Yes** to add the new file location.

Close All documents.
69)   Click **Windows**, **Close All** from the Menu bar.

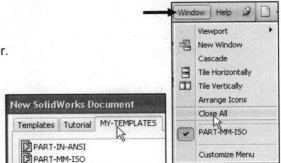

Display the MY-TEMPLATES folder and templates.
70)   Click **New** from the Menu bar.

71)   Click the **MY-TEMPLATES** tab. View the two new Part Templates.

72)   Click **Cancel** from the New SolidWorks Document dialog box.

Each folder listed in the System Options, File Locations, Document Templates, Show Folders For option produces a corresponding tab in the New SolidWorks Document dialog box.

The MY-TEMPLATES tab is only visible when the folder contains a SolidWorks Template document. Create the PART-MM-ANSI template as an exercise.

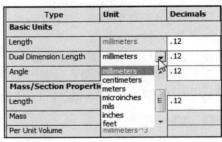

The PART-IN-ANSI Template contains document properties settings for the parts contained in the FLASHLIGHT assembly. Substitute the PART-MM-ISO or PART-MM-ANSI template to create the identical parts in millimeters.

The primary units in this book are IPS, (inch, pound, second). The optional secondary units are MMGS, (millimeter, gram, second) and are indicated in brackets [ ]. Illustrations are provided in both inches and millimeters.

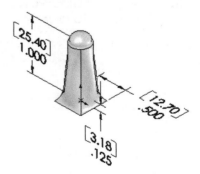

Additional information on System Options, Document Properties, File Locations, and Templates is located in SolidWorks Help Topics. Keywords: Options (detailing, units), Templates, Files (locations), menus and toolbars (features, sketch).

Review of the SolidWorks 2012 User Interface and Part Templates

The SolidWorks 2012 User Interface and CommandManager consist of the following options: Menu bar toolbar, Menu bar menu, Drop-down menus, Context toolbars, Consolidated fly-out menus, System feedback icons, Confirmation Corner and Heads-up View toolbar.

The default CommandManager Part tabs control the display of the *Features, Sketch, Evaluate, DimXpert* and *Office Products* toolbars.

The FeatureManager consist of five default tabs:

- *FeatureManager design tree*

- *PropertyManager*

- *ConfigurationManager*

- *DimXpertManager*

- *DisplayManager*

The Task Pane is displayed when a SolidWorks session starts. The Task Pane can be displayed in the following states: *visible or hidden, expanded or collapsed, pinned or unpinned, docked or floating.* The Task Pane contains the following default tabs: *SolidWorks Resources* , *Design Library* , *File Explorer* , *View Palette* , *Appearances, Scenes, and Decals* and *Custom Properties* .

You created two Part Templates: **PART-MM-ISO** and **PART-IN-ANSI**. The document properties Overall drafting standard, units and decimal places were stored in the Part Templates. The File Locations System Option, Document Templates option controls the reference to the MY-TEMPLATES folder.

Note: In some network locations and school environments, the File Locations option must be set to MY-TEMPLATES for each session of SolidWorks. You can exit SolidWorks at any time during this chapter. Save your document. Select File, Exit from the Menu bar.

BATTERY Part

The BATTERY is a simplified representation of a purchased OEM part. Represent the battery terminals as cylindrical extrusions. The battery dimensions are obtained from the ANSI standard 908D.

A 6-Volt lantern battery weighs approximately 1.38 pounds, (0.62kg). Locate the center of gravity closest to the center of the battery.

Create the BATTERY part. Use features to create parts. Features are building blocks that add or remove material.

Utilize the Instant3D tool to create the Extruded Boss/Base (Boss-Extrude1) feature. The Extrude Boss/Base feature adds material. The Base feature is the first feature of the part.

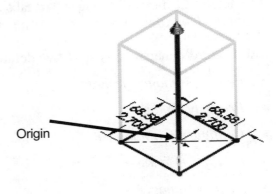

Apply symmetry. Use the Center Rectangle Sketch tool on the Top Plane. The 2D Sketch profile is centered at the Origin.

Extend the profile perpendicular (⊥) to the Top Plane.

Utilize the Fillet feature to round the four vertical edges.

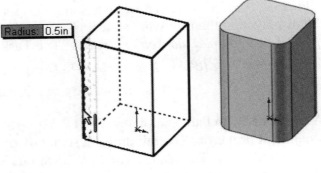

The Extruded Cut feature removes material from the top face. Utilize the top face for the Sketch plane. Utilize the Offset Entity Sketch tool to create the profile.

🔅 Tangent Edges and Origin are displayed for educational purposes.

🔅 When you create a new part or assembly, the three default Planes (Front, Right and Top) are aligned with specific views. The Plane you select for the Base sketch determines the orientation of the part.

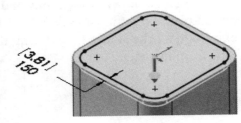

Utilize the Fillet feature 🍥 to round the top narrow face.

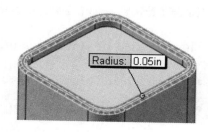

💡 Fillet/Round features creates a rounded internal or external face on the part. You can fillet all edges of a face, selected sets of faces, selected edges, or edge loops.

💡 Add larger fillets before smaller ones. When several fillets converge at a vertex, create the larger fillets first

The Extruded Boss/Base 🗔 feature adds material. Conserve design time. Represent each of the terminals as a cylindrical Extruded Boss (Boss-Extrude2) feature.

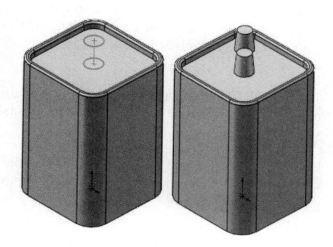

💡 Extrude Features creates a feature by extruding a 3D object from a 2D sketch, essentially adding the third dimension. An extrusion can be a base, a boss (which adds material, often on another extrusion), or a cut (which removes material).

💡 View the additional Power point presentations on the enclosed DVD for supplementary information.

Avi folder
History of Engineering Graphics
Visualization, Arrangement of Views, and Primitives
General Tolerancing and Fits
Open a Drawing Document 2012
Open an Assembly Document 2012
SolidWorks Basic Concepts
Boolean Operation
Alphabet of lines and Precedent of Line Types
Fastners in General
Fundamental ASME Y14.5 Dimensioning Rules
Part and Drawing Dimensioning
Design Intent

BATTERY Part-Extruded Boss/Base Feature

The Extruded Boss/Base feature requires:

- Sketch plane (Top)
- Sketch profile (Rectangle)
  - Geometric relations and dimensions
- End Condition Depth (Blind) in Direction 1

Create a new part named, BATTERY. Insert an Extruded Boss/Base feature. Extruded features require a Sketch plane. The Sketch plane determines the orientation of the Extruded Base feature. The Sketch plane locates the Sketch profile on any plane or face.

The Top Plane is the Sketch plane. The Sketch profile is a rectangle. The rectangle consists of two horizontal lines and two vertical lines.

Geometric relations and dimensions constrain the sketch in 3D space. The Blind End Condition in Direction 1 requires a depth value to extrude the 2D Sketch profile and to complete the 3D feature.

Alternate between the Features tab and the Sketch tab in the CommandManager to display the available Feature and Sketch tools for the Part document.

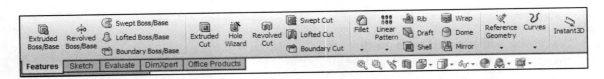

When you create a new part or assembly, the three default Planes (Front, Right and Top) are aligned with specific views. The Plane you select for the Base sketch determines the orientation of the part.

Activity: BATTERY Part-Create the Extruded Boss/Base Feature

Create a new part.

73) Click **New** from the Menu bar.

74) Click the **MY-TEMPLATES** tab.

75) Double-click **PART-IN-ANSI**, [**PART-MM-ISO**].

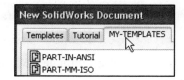

Save the empty part.

76) Click **Save** .

77) Select **PROJECTS** for Save in folder.

78) Enter **BATTERY** for File name.

79) Enter **BATTERY, 6-VOLT** for Description.

80) Click **Save**. The Battery FeatureManager is displayed.

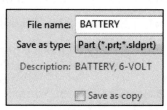

Select the Sketch plane.

81) Right-click **Top Plane** from the FeatureManager. This is your Sketch plane.

Sketch the 2D Sketch profile centered at the Origin.

82) Click **Sketch** from the Context toolbar. The Sketch toolbar is displayed.

83) Click the **Center Rectangle** Sketch tool. The

Center Rectangle icon is displayed.

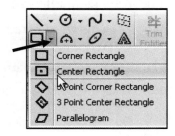

84) Click the **Origin**. This is your first point.

85) Drag and click the **second point** in the upper right quadrant as illustrated. The Origin is located in the center of the sketch profile. The Center Rectangle Sketch tool automatically applies equal relations to the two horizontal and two vertical lines. A midpoint relation is automatically applied to the Origin.

The book is design to expose the new user to different tools and procedures.

Click **View**, **Sketch Relations** from the Menu bar to display the relations in the Graphics window.

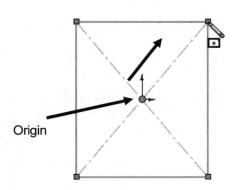

Origin

Dimension the sketch.

86)   Click the **Smart Dimension**  Sketch tool.

87)   Click the **top horizontal line**.

88)   Click a **position** above the horizontal line.

89)   Enter **2.700**in, **[68.58]** for width.

90)   Click the **Green Check mark**  in the Modify dialog box.

91)   Enter **2.700**in, [68.58] for height as illustrated.

92)   Click the **Green Check mark** ✓ in the Modify dialog box. The black Sketch status is fully defined

93)   Click **OK** ✓ from the Dimension PropertyManager.

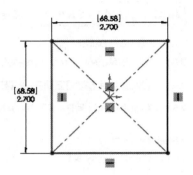

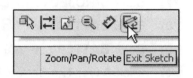

Exit the Sketch.

94)   Click **Exit Sketch**.

Insert an Extruded Boss/Base feature. Apply the Instant3D tool. The Instant3D tool provides the ability to drag geometry and dimension manipulator points to resize or to create features directly in the Graphics window.

Use the On-screen ruler.

95)   Click **Isometric view** from the Heads-up View toolbar.

96)   Click the **front horizontal line** as illustrated. A green arrow is displayed.

97)   Click and drag the **green/red arrow** upward.

98)   Click the on-screen ruler at **4.1**in, [104.14] as illustrated. This is the depth in direction 1. The extrude direction is upwards. Boss-Extrude1 is displayed in the FeatureManager.

Check the Boss-Extrude1 feature depth dimension.

99)   Right-click **Boss-Extrude1** from the FeatureManager.

100) Click **Edit Feature** 🗔 from the Context toolbar. 4.100in is displayed for depth. Blind is the default End Condition. Note: If you did not select the correct depth, input the depth in the Boss-Extrude1 PropertyManager.

101) Click **OK** ✓ from the Boss-Extrude1 PropertyManager.

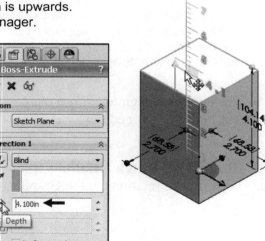

Modify the **Spin Box Increments** in System Options to display different increments in the on-screen ruler.

Fit the part to the Graphics window.
102) Press the **f** key.

Rename the Boss-Extrude1 feature.
103) Rename **Boss-Extrude1** to **Base Extrude**.

Save the BATTERY.
104) Click **Save** 💾.

Modify the BATTERY.
105) Click **Base Extrude** from the FeatureManager. Note: Instant3D is activated by default.

106) Drag the **manipulator point** upward and click the on-screen ruler to create a **5.000in, [127]** depth as illustrated. Blind is the default End Condition.

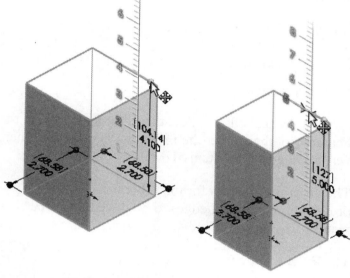

Return to the 4.100 depth.

107) Click the **Undo** ↺ button from the Menu bar. The depth of the model is 4.100in, [104.14]. Blind is the default End Condition. Practice may be needed to select the correct on-screen ruler dimension.

The color of the sketch indicates the sketch status.

- Light Blue - Currently selected

- Blue - Under defined, requires additional geometric relations and dimensions

- Black - Fully defined

- Red - Over defined, requires geometric relations or dimensions to be deleted or redefined to solve the sketch

The Instant3D tool is active by default in the Features toolbar located in the CommandManager.

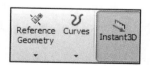

BATTERY Part-Fillet Feature Edge

Fillet features remove sharp edges. Utilize Hidden Lines Visible 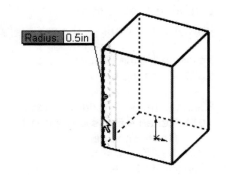 from the Heads-up View toolbar to display hidden edges.

An edge Fillet feature requires:

- A selected edge
- Fillet radius

Select a vertical edge. Select the Fillet feature from the Features toolbar. Enter the Fillet radius. Add the other vertical edges to the Items To Fillet option.

The order of selection for the Fillet feature is not predetermined. Select edges to produce the correct result.

The Fillet feature uses the Fillet PropertyManager. The Fillet PropertyManager provides the ability to select either the *Manual* or *FilletXpert* tab.

Each tab has a separate menu and PropertyManager. The Fillet PropertyManager and FilletXpert PropertyManager displays the appropriate selections based on the type of fillet you create.

The FilletXpert automatically manages, organizes and reorders your fillets in the FeatureManager design tree. The FilletXpert PropertyManager provides the ability to add, change or corner fillets in your model. The PropertyManager remembers its last used state. View the SolidWorks tutorials for additional information on fillets.

The FilletXpert can ONLY create and edit Constant radius fillets.

Activity: BATTERY Part-Fillet Feature Edge

Display the hidden edges.

108) Click **Hidden Lines Visible** 🔲 from the Heads-up View toolbar.

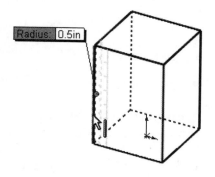

Insert a Fillet feature.

109) Click the **left front vertical edge** as illustrated. Note the mouse pointer edge 🖱 icon.

110) Click the **Fillet** 🔵 feature tool. The Fillet PropertyManager is displayed.

111) Click the **Manual** tab. Edge<1> is displayed in the Items To Fillet box. Constant radius is the default Fillet Type.

112) Click the remaining **3 vertical edges**. The selected entities are displayed in the Items To Fillet box

113) Enter **.500**in, **[12.7]** for Radius. Accept the default settings.

114) Click **OK** ✔ from the Fillet PropertyManager. Fillet1 is displayed in the FeatureManager.

115) Click **Isometric view** 🔲 from the Heads-up View toolbar.

116) Click **Shaded With Edges** 🔲 from the Heads-up View toolbar.

Rename the feature.

117) Rename **Fillet1** to **Side Fillets** in the FeatureManager.

Save the BATTERY.

118) Click **Save** 💾.

BATTERY Part-Extruded Cut Feature

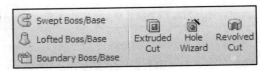

An Extruded Cut feature removes material. An Extruded Cut feature requires:

- Sketch plane (Top face)
- Sketch profile (Offset Entities)
- End Condition depth (Blind) in Direction 1

The Offset Entity Sketch tool uses existing geometry, extracts an edge or face and locates the geometry on the current Sketch plane.

Offset the existing Top face for the 2D sketch. Utilize the default Blind End Condition in Direction 1.

Activity: BATTERY Part-Extruded Cut Feature

Select the Sketch plane.

119) Right-click the **Top face** of the BATTERY in the Graphics window. Base Extruded is highlighted in the FeatureManager.

Create a sketch.

120) Click **Sketch** ✏ from the Context toolbar. The Sketch toolbar is displayed.

Display the face.

121) Click **Top view** ⧉ from the Heads-up View toolbar.

Offset the existing geometry from the boundary of the Sketch plane.

122) Click the **Offset Entities** ⟩ Sketch tool. The Offset Entities PropertyManager is displayed.

123) Enter **.150**in, **[3.81]** for the Offset Distance.

124) Check the **Reverse** box. The new Offset yellow profile displays inside the original profile.

125) Click **OK** ✔ from the Offset Entities PropertyManager.

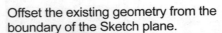

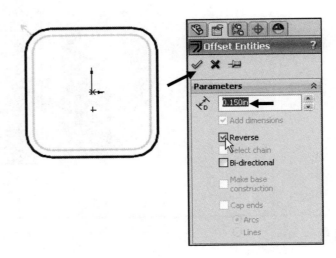

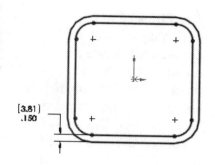

☀ A leading zero is displayed in the spin box. For inch dimensions less than 1, the leading zero is not displayed in the part dimension in the ANSI standard.

Display the profile.

126) Click **Isometric view** 🔲 from the Heads-up View toolbar.

127) Click **Hidden Lines Removed** 🔲 from the Heads-up View toolbar.

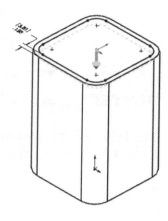

Insert an Extruded Cut feature. As an exercise, use the Instant3D tool to create the Extruded Cut feature. In this section, the Extruded-Cut PropertyManager is used. Note: With the Instant3D tool, you may loose the design intent of the model.

128) Click the **Extruded Cut** 🔲 feature tool. The Cut-Extrude PropertyManager is displayed.

129) Enter **.200**in, **[5.08]** for Depth in Direction 1. Accept the default settings.

130) Click **OK** ✓ from the Cut-Extrude PropertyManager. Cut-Extrude1 is displayed in the FeatureManager.

Rename the feature.
131) Rename **Cut-Extrude1** to **Top Cut** in the FeatureManager.

Save the BATTERY
132) Click **Save** 💾.

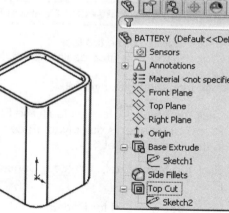

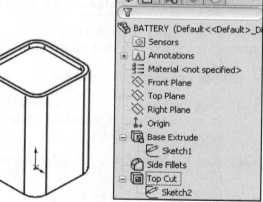

The Cut-Extrude PropertyManager contains numerous options. The Reverse Direction option determines the direction of the Extrude. The Extruded Cut feature is valid only when the direction arrow points into material to be removed.

☀ Think design intent. When do you use the various End Conditions and Geometric sketch relations? What are you trying to do with the design? How does the component fit into an Assembly?

Cut direction not valid, no material to remove

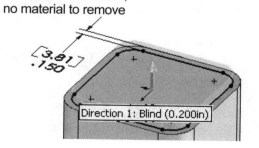

The Flip side to cut option determines if the cut is to the inside or outside of the Sketch profile. The Flip side to cut arrow points outward. The Extruded Cut feature occurs on the outside of the BATTERY.

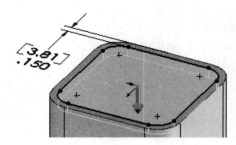

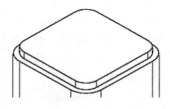

Extruded Cut with Flip side to cut option checked

BATTERY Part-Fillet Feature

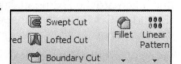

The Fillet feature tool rounds sharp edges by selecting a face. A Fillet requires a:

- A selected face
- Fillet radius

Activity: BATTERY Part-Fillet Feature Face

Insert a Fillet feature on the top face.

133) Click the **top thin face** as illustrated. Note: The face icon feedback symbol.

134) Click the **Fillet** feature tool. The Fillet PropertyManager is displayed. Face<1> is displayed in the Items To Fillet box.

135) Click the **Manual** tab. Create a Constant radius for Fillet Type.

136) Enter **.050**in, **[1.27]** for Radius.

137) Click **OK** from the Fillet PropertyManager. Fillet2 is displayed in the FeatureManager.

Rename the feature.

138) Rename **Fillet2** to **Top Face Fillet**.

139) Press the **f** key.

Save the BATTERY.

140) Click **Save**.

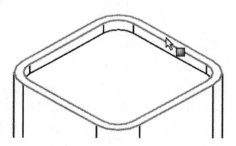

View the mouse pointer for feedback to select Edges or Faces for the fillet.

 Do not select a fillet radius which is larger than the surrounding geometry.

Example: The top edge face width is .150in, [3.81]. The fillet is created on both sides of the face. A common error is to enter a Fillet too large for the existing geometry. A minimum face width of .200in, [5.08] is required for a fillet radius of .100in, [2.54].

The following error occurs when the fillet radius is too large for the existing geometry:

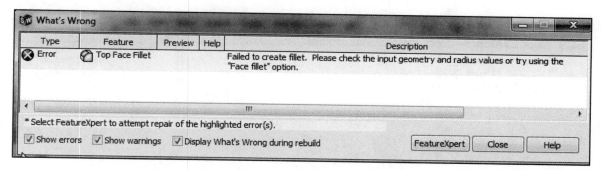

Avoid the fillet rebuild error. Use the FeatureXpert to address a constant radius fillet build error or manually enter a smaller fillet radius size. As an exercise, insert a large Fillet radius and use the FeatureXpert option.

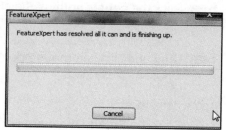

BATTERY Part-Extruded Boss/Base Feature

The Extruded Boss/Base feature requires a truncated cone shape to represent the geometry of the BATTERY terminals. The Draft Angle option creates the tapered shape.

Sketch the first circle on the Top face. Utilize the Ctrl key to copy the first circle.

The dimension between the center points is critical. Dimension the distance between the two center points with an aligned dimension. The dimension text toggles between linear and aligned. An aligned dimension is created when the dimension is positioned between the two circles.

An angular dimension is required between the Right Plane and the centerline. Acute angles are less than 90°. Acute angles are the preferred dimension standard. The overall BATTERY height is a critical dimension. The BATTERY height is 4.500in, [114.30].

Calculate the depth of the extrusion: For inches: 4.500in - (4.100in Base-Extrude height - .200in Offset cut depth) = .600in. The depth of the extrusion is .600in.

For millimeters: 114.3mm - (104.14mm Base-Extrude height - 5.08mm Offset cut depth) = 15.24mm. The depth of the extrusion is 15.24mm.

Activity: BATTERY Part-Extruded Boss/Base Feature

Select the Sketch plane.
141) Right-click the **Top face** of the Top Cut feature in the Graphics window. This is your Sketch plane.

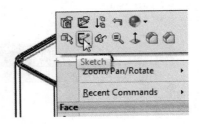

Create the sketch.
142) Click **Sketch** from the Context toolbar. The Sketch toolbar is displayed.

143) Click **Top view** from the Heads-up View toolbar.

Sketch the profile.
144) Click the **Circle** Sketch tool. The Circle PropertyManager is displayed.

145) Click the **center point** of the circle coincident to the Origin.

146) Drag and click the **mouse pointer** to the right of the Origin as illustrated.

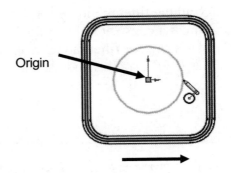

Origin

Add a dimension.
147) Click the **Smart Dimension** Sketch tool.

148) Click the **circumference** of the circle.

149) Click a **position** diagonally to the right.

150) Enter **.500**in, **[12.7]**.

151) Click the **Green Check mark** in the Modify dialog box. The black sketch is fully defined.

Copy the sketched circle.
152) Right-click **Select** to deselect the Smart Dimension Sketch tool.

153) Hold the **Ctrl** key down.

154) Click and drag the **circumference** of the circle to the upper left quadrant as illustrated.

155) Release the **mouse button**.

156) Release the **Ctrl** key. The second circle is selected and is displayed in blue.

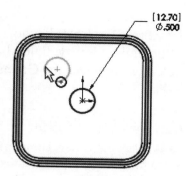

Add an Equal relation.

157) Hold the **Ctrl** key down.

158) Click the **circumference of the first circle**. The Properties PropertyManager is displayed. Both circles are selected and are displayed in green.

159) Release the **Ctrl** key.

160) Right-click **Make Equal** = from the Context toolbar.

161) Click **OK** ✔ from the Properties PropertyManager. The second circle remains selected.

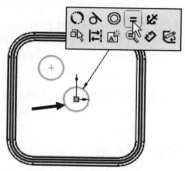

Show the Right Plane for the dimension reference.

162) Click **Right Plane** from the FeatureManager.

163) Click **Show**. The Right Plane is displayed in the Graphics window.

Add an aligned dimension.

164) Click the **Smart Dimension** ✏ Sketch tool.

165) Click the **two center points** of the two circles.

166) Click a **position** off the profile in the upper left corner.

167) Enter **1.000**in, [**25.4**] for the aligned dimension.

168) Click the **Green Check mark** ✔ in the Modify dialog box.

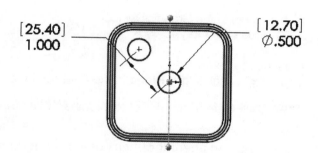

Insert a centerline.

169) Click the **Centerline** ⁝ Sketch tool. The Insert Line PropertyManager is displayed.

170) Sketch a centerline between the **two circle center points** as illustrated.

171) Right-click **Select** to end the line.

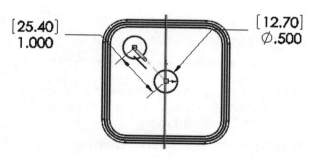

💡 Double-click to end the centerline.

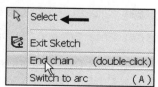

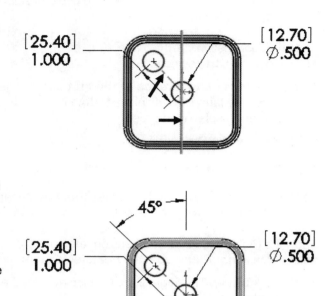

💡 Press the Enter key to accept the value in the Modify dialog box. The Enter key replaces the Green Check mark.

Add an angular dimension.

172) Click the **Smart Dimension** ✎ Sketch tool. Click the **centerline** between the two circles.

173) Click the **Right Plane** (vertical line) in the Graphics window. Note: You can also click Right Plane in the FeatureManager.

174) Click a **position** between the centerline and the Right Plane, off the profile.

175) Enter **45**.

176) Click **OK** ✔ from the Dimension PropertyManager.

Fit the model to the Graphics window.
177) Press the **f** key.

Hide the Right Plane.
178) Right-click **Right Plane** in the FeatureManager.

179) Click **Hide** from the Context toolbar. Click **Save** 💾.

💡 Create an angular dimension between three points or two lines. Sketch a centerline/construction line when an additional point or line is required.

Insert an Extruded Boss/Base feature.

180) Click **Isometric view** 🔲 from the Heads-up View toolbar.

181) Click the **Extruded Boss/Base** 🔳 feature tool. The Boss-Extrude PropertyManager is displayed. Blind is the default End Condition Type.

182) Enter **.600**in, **[15.24]** for Depth in Direction 1.

183) Click the **Draft ON/OFF** button.

184) Enter **5**deg in the Draft Angle box.

185) Click **OK** ✔ from the Boss-Extrude PropertyManager. The Boss-Extrude2 feature is displayed in the FeatureManager.

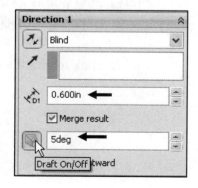

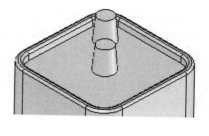

Rename the feature and sketch.

186) Rename **Boss-Extrude2** to **Terminals**.

187) **Expand** Terminals.

188) Rename **Sketch3** to **Sketch-TERMINALS**.

189) Click **Shaded With Edges** 🔲 from the Heads-up View toolbar.

190) Click **Save** 💾.

Each time you create a feature of the same feature type, the feature name is incremented by one. Example: Boss-Extrude1 is the first Extrude feature. Boss-Extrude2 is the second Extrude feature. If you delete a feature, rename a feature or exit a SolidWorks session, the feature numbers will vary from those illustrated in the text.

🔅 Rename your features with descriptive names. Standardize on feature names that are utilized in mating parts. Example: Mounting Holes.

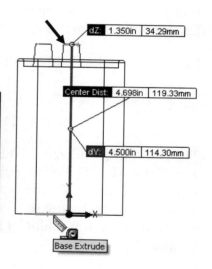

Measure the overall BATTERY height.

191) Click **Front view** 🔲 from the Heads-up View toolbar.

192) Click the **Measure** Measure tool from the Evaluate tab in the CommandManager. The Measure - BATTERY dialog box is displayed.

193) Click the **top edge** of the battery terminal as illustrated.

194) Click the **bottom edge** of the battery. The overall height, Delta Y is 4.500, [114.3]. Apply the Measure tool to insure a proper design.

195) **Close** ✖ the Measure - BATTERY dialog box.

🔅 The Measure tool provides the ability to display custom settings. Click **Units/Precision** from the Measure dialog box. View your options. Click **OK**.

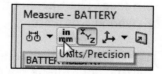

The Selection Filter ⛣ option toggles the Selection Filter toolbar. When Selection Filters are activated, the mouse pointer displays the Filter icon ⛣ . The Clear All Filters ⛣ tool removes the current Selection Filters. The Help ? icon displays the SolidWorks Online Users Guide.

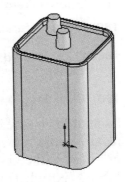

Display the Trimetric view.
196) Click **Trimetric view** ⬛ from the Heads-up View toolbar.

Save the BATTERY.
197) Click **Save** 💾 .

🔍 Additional information on Extruded Boss/Base Extruded Cut and Fillets is located in SolidWorks Help Topics. Keywords: Extruded (Boss/Base, Cut), Fillet (Constant radius fillet), Geometric relations (sketch, equal, midpoint), Sketch (rectangle, circle), Offset Entities and Dimensions (angular).

Refer to the Help, SolidWorks Tutorials, Fillet exercise for additional information.

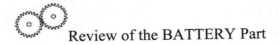 Review of the BATTERY Part

The BATTERY utilized a 2D Sketch profile located on the Top Plane. The 2D Sketch profile utilized the Center Rectangle Sketch tool. The Center Rectangle Sketch tool applied equal geometric relations to the two horizontal and two vertical lines. A midpoint relation was added to the Origin.

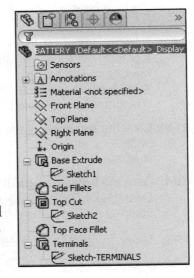

The Extruded Boss/Base feature was created using the Instant3D tool. Blind was the default End Condition. The Fillet feature rounded the sharp edges of the BATTERY. All four edges were selected to combine common geometry into the same Fillet feature. The Fillet feature also rounded the top face. The Sketch Offset Entity created the profile for the Extruded Cut feature.

The Terminals were created with an Extruded Boss/Base feature. You sketched a circular profile and utilized the Ctrl key to copy the sketched geometry. A centerline was required to locate the two holes with an angular dimension. The Draft Angle option tapered the Extruded Boss/Base feature. Features were renamed.

BATTERYPLATE Part

The BATTERYPLATE is a critical plastic part. The BATTERYPLATE:

- Aligns the LENS assembly

- Creates an electrical connection between the BATTERY and LENS

Design the BATTERYPLATE. Utilize features from the BATTERY to develop the BATTERYPLATE. The BATTERYPLATE is manufactured as an injection molded plastic part. Build Draft into the Extruded Boss/Base features.

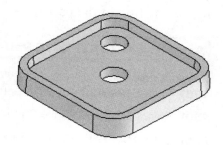

Edit the BATTERY features. Create two holes from the original sketched circles. Apply the Instant3D tool to create an Extruded Cut feature.

Modify the dimensions of the Base feature. Add a 3° draft angle.

🔆 A sand pail contains a draft angle. The draft angle assists the sand to leave the pail when the pail is flipped upside down.

Insert an Extruded Boss/Base feature. Offset the center circular sketch.

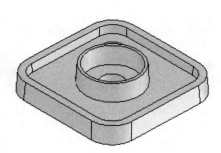

The Extruded Boss/Base feature contains the LENS. Create an inside draft angle. The draft angle assists the LENS into the Holder.

Insert a Face Fillet and a Multi-radius Edge Fillet to remove sharp edges. Plastic parts require smooth edges. Group Fillet features together into a folder.

🔆 Group fillets together into a folder to locate them quickly. Features listed in the FeatureManager must be continuous in order to be placed as a group into a folder.

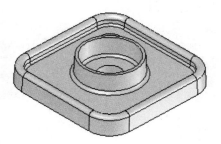

🔆 Tangent Edges and Origin are displayed for educational purposes.

Save As, Delete, Edit Feature and Modify

Create the BATTERYPLATE part from the BATTERY part. Utilize the Save As tool from the Menu bar to copy the BATTERY part to the BATTERYPLATE part.

Reuse existing geometry. Create two holes. Delete the Terminals feature and reuse the circle sketch. Select the sketch in the FeatureManager. Create an Extruded Cut feature from the Sketch–TERMINALS using the Instant3D tool. Blind is the default End Condition. Edit the Base Extrude feature. Modify the overall depth. Rebuild the model.

When you create a new part or assembly, the three default Planes (Front, Right and Top) are aligned with specific views. The Plane you select for the Base sketch determines the orientation of the part.

Activity: BATTERYPLATE Part-Save As, Delete, Modify and Edit Feature

---

Create a new part.

198) Click **Save As** from the Menu bar drop down menu.

199) Select **PROJECTS** for Save In folder.

200) Enter **BATTERYPLATE** for File name.

201) Enter **BATTERY PLATE, FOR 6-VOLT** for Description.

202) Click **Save**. The BATTERYPLATE FeatureManager is displayed. The BATTERY part is closed.

Delete the Terminals feature.

203) Right-click **Terminals** from the FeatureManager.

204) Click **Delete**.

205) Click **Yes** from the Confirm Delete dialog box. Do not delete the two-circle sketch, Sketch-TERMINALS.

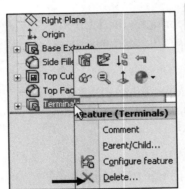

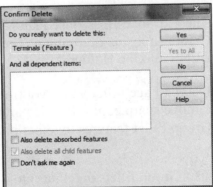

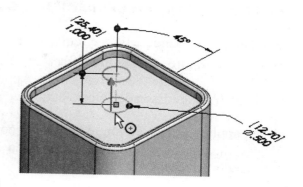

Create an Extruded Cut feature from the Sketch–TERMINALS using Instant3D.

206) Click **Sketch-TERMINALS** from the FeatureManager.

207) Click the **circumference** of the center circle as illustrated. A green arrow is display.

208) Hold the **Alt** key down. Drag the **green arrow** downward below the model to create a hole in Direction 1.

209) Release the mouse button on the **vertex** as illustrated. This insures a Through All End Condition with model dimension changes.

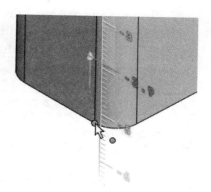

210) Release the **Alt** key. Boss-Extrude1 is displayed in the FeatureManager.

211) Rename the **Boss-Extrude1** feature to **Holes** in the FeatureManager.

Edit the Base Extrude feature.

212) Right-click **Base Extrude** from the FeatureManager.

213) Click **Edit Feature** 🖼 from the Context toolbar. The Base Extrude PropertyManager is displayed.

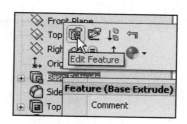

Modify the overall depth.

214) Enter **.400**in, [**10.16**] for Depth in Direction 1.

215) Click the **Draft ON/OFF** button.

216) Enter **3.00**deg in the Angle box.

217) Click **OK** ✓ from the Base Extrude PropertyManager.

Fit the model to the Graphics window.

218) Press the **f** key.

Save the BATTERYPLATE.

219) Click **Save** 💾.

💡 Modify the **Spin Box Increments** in System Options to display different increments for the Instant3D on-screen ruler.

💡 To delete both the feature and the sketch at the same time, select the Also delete absorbed features check box from the Confirm Delete dialog box.

BATTERYPLATE Part-Extruded Boss/Base Feature

The Holder is created with a circular Extruded Boss/Base feature. Utilize the Offset Entities ⁊ Sketch tool to create the second circle. Apply a draft angle of 3° in the Extruded Boss feature.

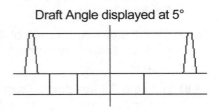
Draft Angle displayed at 5°

When applying the draft angle to the two concentric circles, the outside face tapers inwards and the inside face tapers outwards.

Plastic parts require a draft angle. Rule of thumb; 1° to 5° is the draft angle. The draft angle is created in the direction of pull from the mold. This is defined by geometry, material selection, mold production and cosmetics. Always verify the draft with the mold designer and manufacturer.

Activity BATTERYPLATE Part-Extruded Boss/Base Feature

Select the Sketch plane.
220) Right-click the **top face** of Top Cut. This is your Sketch plane.

Create the sketch.
221) Click **Sketch** ⌁ from the Context toolbar.

222) Click the **top circular edge** of the center hole. Note: Use the keyboard arrow keys or the middle mouse button to rotate the sketch if needed.

223) Click the **Offset Entities** ⁊ Sketch tool. The Offset Entities PropertyManager is displayed.

224) Enter **.300**in, [**7.62**] for Offset Distance. Accept the default settings.

225) Click **OK** ✔ from the Offset Entities PropertyManager.

226) Drag the **dimension** off the model.

Create the second offset circle.
227) Click the **offset circle** in the Graphics window.

228) Click the **Offset Entities** ⁊ Sketch tool. The Offset Entities PropertyManager is displayed.

229) Enter **.100**in, [**2.54**] for Offset Distance.

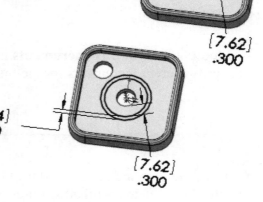

[7.62]
..300

[2.54]
.100

[7.62]
..300

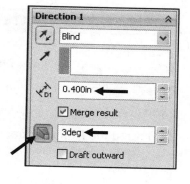

230) Click **OK** ✔ from the Offset Entities PropertyManager. Drag the dimension off the model. Two offset concentric circles define the sketch.

Insert an Extruded Boss/Base feature.

231) Click the **Extruded Boss/Base** 📦 feature tool. The Boss-Extrude PropertyManager is displayed.

232) Enter **.400**in, **[10.16]** for Depth in Direction 1.

233) Click the **Draft ON/OFF** button.

234) Enter **3**deg in the Angle box.

235) Click **OK** ✔ from the Boss-Extrude PropertyManager. The Boss-Extrude2 feature is displayed in the FeatureManager.

Rename the feature.
236) Rename the **Boss-Extrude2** feature to **Holder** in the FeatureManager.

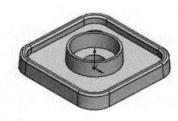

Save the model.
237) Click **Save** 💾.

BATTERYPLATE Part-Fillet Features: Full Round and Multiple Radius Options

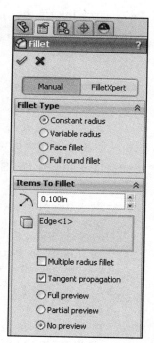

Use the Fillet feature 🪣 tool to smooth rough edges in a model. Plastic parts require fillet features on sharp edges. Create two Fillets. Utilize different techniques. The current Top Face Fillet produced a flat face. Delete the Top Face Fillet. The first Fillet feature is a Full round fillet. Insert a Full round fillet feature on the top face for a smooth rounded transition.

The second Fillet feature is a Multiple radius fillet. Select a different radius value for each edge in the set. Select the inside and outside edge of the Holder. Select all inside tangent edges of the Top Cut. A Multiple radius fillet is utilized next as an exercise. There are machining instances were radius must be reduced or enlarged to accommodate tooling. Note: There are other ways to create Fillets.

🔆 Group Fillet features into a Fillet folder. Placing Fillet features into a folder reduces the time spent for your mold designer or toolmaker to look for each Fillet feature in the FeatureManager.

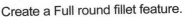

Delete the Top Edge Fillet.

238) Right-click **Top Face Fillet** from the FeatureManager.

239) Click **Delete**.

240) Click **Yes** to confirm delete.

241) Drag the **Rollback** bar below Top Cut in the FeatureManager.

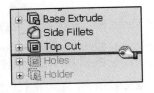

Create a Full round fillet feature.

242) Click **Hidden Lines Visible** ⬚ from the Heads-up View toolbar.

243) Click the **Fillet** ◯ feature tool. The Fillet PropertyManager is displayed.

244) Click the **Manual** tab.

245) Click the **Full round fillet** box for Fillet Type.

246) Click the **inside Top Cut face** for Side Face Set 1 as illustrated.

247) Click **inside** the Center Face Set box.

248) Click the **top face** for Center Face Set as illustrated.

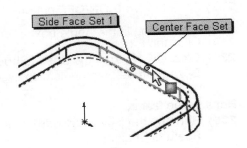

Rotate the part.

249) Press the **Left Arrow** key until you can select the outside Base Extrude face.

250) Click **inside** the Side Face Set 2 box.

251) Click the **outside Base Extrude face** for Side Face Set 2 as illustrated. Accept the default settings.

252) Click **OK** ✓ from the Fillet PropertyManager. Fillet1 is displayed in the FeatureManager.

253) Rename **Fillet1** to **TopFillet**.

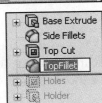

Save the BATTERYPLATE.

254) Click **Isometric view** from the Heads-up View toolbar.

255) Click **Hidden Lines Removed** from the Heads-up View toolbar.

256) Drag the **Rollback bar** to the bottom of the FeatureManager.

257) Click **Save**.

Create a Multiple radius fillet feature.

258) Click the **bottom outside circular edge** of the Holder as illustrated.

259) Click the **Fillet** feature tool. The Fillet PropertyManager is displayed.

260) Click the **Constant radius** box.

261) Enter .050in, [1.27] for Radius.

262) Click the **bottom inside circular edge** of the Top Cut as illustrated.

263) Click the **inside edge** of the Top Cut.

264) Check the **Tangent propagation** box.

265) Check the **Multiple radius fillet** box.

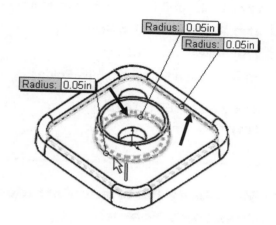

Modify the Fillet values.

266) Click the **Radius** box ⟨Radius: 0.05in⟩ for the Holder outside edge.

267) Enter **0.060**in, [1.52].

268) Click the **Radius** box for the Top Cut inside edge.

269) Enter **0.040**in, [1.02].

270) Click **OK** from the Fillet PropertyManager. Fillet2 is displayed in the FeatureManager.

271) Rename **Fillet2** to **HolderFillet**.

272) Click **Shaded With Edges** from the Heads-up View toolbar. View the results in the Graphics window.

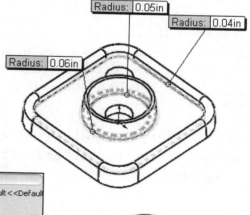

Tangent Edges and Origin are displayed for educational purposes.

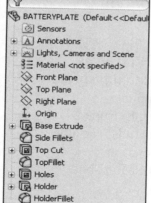

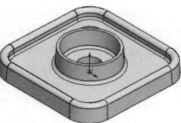

Group the Fillet features into a new folder.

273) Click **TopFillet** from the FeatureManager.

274) Drag the **TopFillet** feature directly above the HolderFillet feature in the FeatureManager.

275) Click **HolderFillet** in the FeatureManager.

276) Hold the **Ctrl** key down.

277) Click **TopFillet** in the FeatureManager.

278) Right-click **Add to New Folder**.

279) Release the **Ctrl** key.

280) Rename **Folder1** to **FilletFolder**.

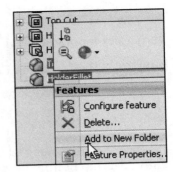

Save the BATTERYPLATE.

281) Click **Save** 🖫.

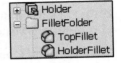

Multi-body Parts and Extruded Boss/Base Feature

A Multi-body part has separate solid bodies within the same part document.

A WRENCH consists of two cylindrical bodies. Each extrusion is a separate body. The oval profile is sketched on the right plane and extruded with the Up to Body option.

The BATTERY consisted of a solid body with one sketched profile. The BATTERY is a single body part.

🔍 Additional information on Save, Extruded Boss/Base, Extruded Cut, Fillets, Copy Sketched Geometry and Multi-body are located in SolidWorks Help Topics. Keywords: Save (save as copy), Extruded (Boss/Base, Cut), Fillet (face blends, variable radius), Chamfer, Geometric relations (sketch), Copy (sketch entities), Multi-body (extrude, modeling techniques).

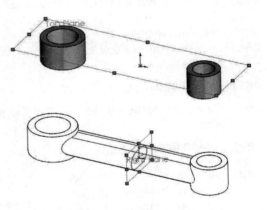

Multi-body part Wrench

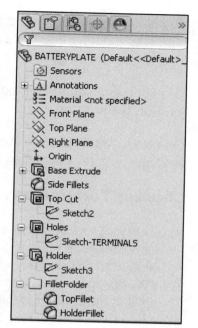

Review of the BATTERYPLATE Part

The Save As option was utilized to copy the BATTERY part to the BATTERYPLATE part. You created a hole in the BATTERYPLATE using the Instant3D tool and modified features using the PropertyManager.

The BATTERYPLATE is a plastic part. The Draft Angle option was added in the Base Extrude feature.

The Holder Extruded Boss feature utilized a circular sketch and the Draft Angle option. The Sketch Offset tool created the circular ring profile. Multi radius Edge Fillets and Face Fillets removed sharp edges. Similar Fillet features were grouped together into a folder. Features were renamed in the FeatureManager. The BATTERY and BATTERYPLATE utilized an Extruded Boss/Base feature.

Design Intent is how your part reacts as parameters are modified. Example: If you have a hole in a part that must always be .125≤ from an edge, you would dimension to the edge rather than to another point on the sketch. As the part size is modified, the hole location remains .125≤ from the edge.

Chapter Summary

SolidWorks is a design software application used to model and create 2D and 3D sketches, 3D parts, 3D assemblies, and 2D drawings. You were introduced to the SolidWorks 2012 User Interface and CommandManager: *Menu bar toolbar, Menu bar menu, Drop-down menus, Context toolbars, Consolidated drop-down menus, System feedback icons, Confirmation Corner, Heads-up View toolbar, Document Properties and more.*

You are designing a FLASHLIGHT assembly that is cost effective, serviceable and flexible for future design revisions. The FLASHLIGHT assembly consists of various parts. The BATTERY and BATTERYPLATE parts were modeled in this chapter.

Folders organized your models and templates. The Part Template is the foundation for all parts in the FLASHLIGHT assembly. You created the *PART-IN-ANSI* and *PART-MM-ISO* Part Template.

This chapter concentrated on the Extruded Boss/Base feature. The Extruded Boss/Base feature required a Sketch plane, Sketch profile and End Condition (Depth). The BATTERY and BATTERYPLATE parts incorporated an Extruded Boss/Base feature:

You also addressed the Extruded Cut, Fillet and Instant3D features. You addressed the following Sketch tools: *Smart Dimension, Line, Centerline, Center Rectangle, Circle, Convert Entities, Offset Entities, and Mirror Entities*.

You addressed additional tools that utilized existing geometry: Add Relations, copy, Save As, Edit feature, and more.

Geometric relations were utilized to build symmetry into the sketches. Practice these concepts with the chapter exercises.

Chapter Terminology

Assembly: An assembly is a document in which parts, features, and other assemblies (sub-assemblies) are put together. A part in an assembly is called a component. Adding a component to an assembly creates a link between the assembly and the component. When SolidWorks opens the assembly, it finds the component file to show it in the assembly. Changes in the component are automatically reflected in the assembly. The filename extension for a SolidWorks assembly file name is .SLDASM. The FLASHLIGHT is an assembly. The BATTERY is a part/component in the FLASHLIGHT assembly.

Chamfer: The chamfer tool creates a beveled feature on selected edges, faces, or a vertex.

CommandManager: The CommandManager is a context-sensitive toolbar that dynamically updates based on the toolbar you want to access. By default, it has toolbars embedded in it based on the document type. When you click a tab below the Command Manager, it updates to show that toolbar. For example, if you click the **Sketches** tab, the Sketch toolbar is displayed.

Convert Entities: A sketch tool that extracts sketch geometry to the current Sketch plane. You can create one or more curves in a sketch by projecting an edge, loop, face, curve, or external sketch contour to the selected Sketch plane.

Cursor Feedback: The system feedback symbol indicates what you are selecting or what the system is expecting you to select. As you move the mouse pointer across your model, system feedback is provided.

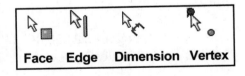

Face   Edge   Dimension   Vertex

Dimension: A value indicating the size of the 2D sketch entity or 3D feature.

Dimensioning Standard: A set of drawing and detailing options developed by national and international organizations. A few key dimensioning standard options are: ANSI, ISO, DIN, JIS, BSI, GOST and GB.

Draft angle: A draft angle is the degree of taper applied to a face. Draft angles are usually applied to molds or castings.

Drawing: A document containing a 2D representation of a 3D part or assembly. The filename extension for a SolidWorks drawing file name is .SLDDRW.

Edit Feature: A tool utilized to modify existing feature parameters. Right-click the feature in the FeatureManager. Click Edit Feature.

Edit Sketch: A tool utilized to modify existing sketch geometry. Right-click the Sketch in the FeatureManager. Click Edit Sketch.

Extruded Boss/Base: A feature that adds material utilizing a 2D sketch profile and a depth perpendicular to the Sketch plane. The Base feature is the first feature in the part.

Extruded Cut: A feature that removes material utilizing a 2D sketch profile and a depth perpendicular to the Sketch plane.

Features: Features are geometry building blocks. Features add or remove material. Features are created from sketched profiles or from edges and faces of existing geometry.

Fillet: A feature that rounds sharp edges or faces by a specified radius.

Geometric relationships: In SolidWorks, Geometric relations between sketch entities and model geometry, in either 2D or 3D sketches, are an important means of building in design intent. Example: Concentric, Tangent, Vertical, etc.

Menus: Menus, (drop-down, pop-out) provides access to the commands that the SolidWorks software offers.

Mirror Entities: A sketch tool that mirrors sketch geometry to the opposite side of a sketched centerline. When you create mirrored entities, the SolidWorks software applies a Symmetric relation between each corresponding pair of sketch points (the ends of mirrored lines, the centers of arcs, and so on). If you change a mirrored entity, its mirror image also changes.

Mouse Buttons: The left, middle, and right mouse buttons have distinct functions in SolidWorks. The left mouse button is utilized to select geometry. The right-mouse button is utilized to invoke commands. The middle button is used to rotate and Zoom in and Zoom out.

Offset Entities: A sketch tool that offsets sketch geometry to the current Sketch plane by a specific amount.

Part: A part is a single 3D object that consists of various features. The filename extension for a SolidWorks part is .SLDPRT.

Plane: Planes are flat and infinite. Planes are represented on the screen with visible edges. The reference plane in chapter 1 is the Top Plane.

Relation: A relation is a geometric constraint between sketch entities or between a sketch entity and a plane, axis, edge or vertex.

Sketch: The name to describe a 2D profile is called a sketch. 2D sketches are created on flat faces and planes within the model. Typical geometry types are lines, arcs, corner rectangles, circles, polygons, and ellipses.

Status of a Sketch: Three states are utilized in this Project: *Fully Defined*: Has complete information, (Black), *Over Defined*: Has duplicate dimensions, (Red), or *Under Defined*: There is inadequate definition of the sketch, (Blue).

Template: A template is the foundation of a SolidWorks document. A Part Template contains the Document Properties such as: Dimensioning Standard, Units, Grid/Snap, Precision, Line Style and Note Font.

Toolbars: The toolbars provide shortcuts enabling you to access the most frequently used commands.

Units: Used in the measurement of physical quantities. Decimal inch dimensioning and Millimeter dimensioning are the two types of common units specified for engineering parts and drawings.

Think design intent. When do you use the various End Conditions and Geometric sketch relations? What are you trying to do with the design? How does the component fit into an Assembly?

Questions

1. Identify and describe the function of the following features:

   - Extruded Boss/Base
   - Fillet
   - Chamfer
   - Extruded Cut

2. Explain the differences between a Template and a Part document.

3. Explain the steps in starting a SolidWorks session.

4. Describe the procedure to develop a new 2D sketch.

5. Explain the procedure required to change part unit dimensions from inches to millimeters.

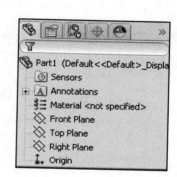

6. Identify the three default Reference planes.

7. What is a Base feature? Provide an example.

8. Describe the differences between an Extruded Boss/Base feature, an Extruded Cut feature and a Fillet feature.

9. The sketch color black indicates a sketch is _____ defined.

10. The sketch color blue indicates a sketch is _____ defined.

11. The sketch color red indicates a sketch is _____ defined.

12. True or False. Folders are utilized to only store part documents.

13. Describe a Symmetric relation.

14. Describe an Angular dimension.

15. Describe is a draft angle. Provide an example.

16. An arc requires _____ points?

17. Identify the properties of a Multi-body part.

18. Identify the name of the following Feature tool icons.

A   B   C   D   E   F

A	B	C	D
E	F		

19. Identify the name of the following Sketch tool icons.

A   B   C   D   E   F   G   H   I   J

A	B	C	D
E	F	G	H
I	J		

Exercises

Exercise 4.1: Identify the Sketch plane for the Boss-Extrude1 feature. View the Origin location. Simplify the number of features.

A: Top Plane

B: Front Plane

C: Right Plane

D: Left Plane

Correct answer _____.

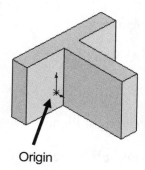

Origin

Exercise 4.2: Identify the Sketch plane for the Boss-Extrude1 feature. View the Origin location. Simplify the number of features.

A: Top Plane

B: Front Plane

C: Right Plane

D: Left Plane

Correct answer _____.

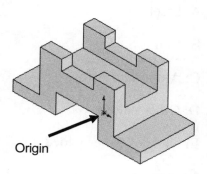

Origin

Exercise 4.3: Identify the Sketch plane for the Boss-Extrude1 feature. View the Origin location. Simplify the number of features.

A: Top Plane

B: Front Plane

C: Right Plane

D: Left Plane

Correct answer _____.

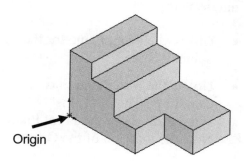

Origin

Exercise 4.4: Identify the Sketch plane for the Boss-Extrude1 feature. View the Origin location. Simplify the number of features.

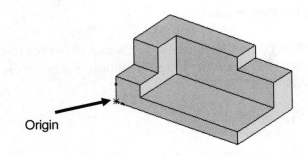

Origin

A: Top Plane

B: Front Plane

C: Right Plane

D: Left Plane

Correct answer _____.

Exercise 4.5: Identify the material category for 6061 Alloy.

A: Steel

B: Iron

C: Aluminum Alloys

D: Other Alloys

E: None of the provided

Correct answer _____.

Exercise 4.6: AXLE

Create an AXLE part as illustrated with dual units. Overall drafting standard - ANSI. IPS is the primary unit system.

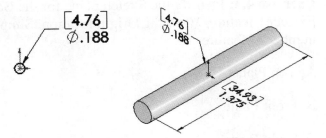

- Utilize the Front Plane for the Sketch plane.

- Utilize the Mid Plane End Condition. The AXLE is symmetric about the Front Plane. Note the location of the Origin.

Exercise 4.7: SHAFT COLLAR

Create a SHAFT COLLAR part as illustrated with dual
system units; MMGS (millimeter, gram, second) and IPS
(inch, pound, second). Overall drafting standard - ANSI.

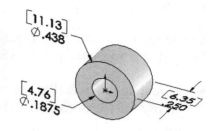

- Utilize the Front Plane for the Sketch plane. Note the
  location of the Origin.

Exercise 4.8: FLAT BAR - 3 HOLE

Create the FLAT BAR - 3 HOLE part as illustrated with
dual system units; MMGS (millimeter, gram, second) and
IPS (inch, pound, second). Overall drafting standard - ANSI.

- Utilize the Front Plane for the
  Sketch plane.

- Utilize the Centerpoint Straight
  Slot Sketch tool.

- Utilize a Linear Pattern feature for
  the three holes. The FLAT BAR -
  3 HOLE part is stamped from
  0.060in, [1.5mm] Stainless Steel.

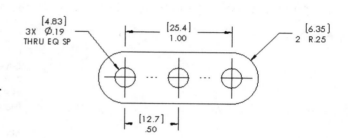

Exercise 4.9: FLAT BAR - 9 HOLE

Create the the FLAT BAR - 9 HOLE part. Overall drafting standard - ANSI.

- The dimensions for hole spacing, height and end arcs are the same as the FLAT
  BAR - 3 HOLE part.

- Utilize the Front Plane for the Sketch plane.

- Utilize the Centerpoint Straight Slot Sketch tool.

- Utilize the Linear
  Pattern feature to
  create the hole
  pattern. The
  FLAT BAR - 9
  HOLE part is
  stamped from
  0.060in, [1.5mm]
  1060 Alloy.

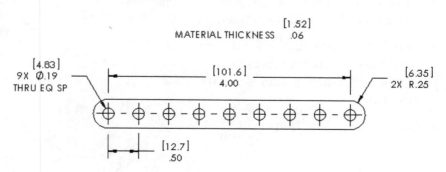

Exercise 4.10

Create the illustrated part. Note the location of the Origin.
Overall drafting standard - ANSI.

- Calculate the overall mass of the illustrated model.
  Apply the Mass Properties tool.

- Think about the steps that
  you would take to build the
  model.

- Review the provided
  information carefully.

- Units are represented in the IPS, (inch,
  pound, second) system.

- A = 3.50in, B = .70in

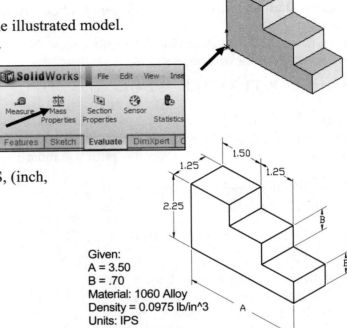

Given:
A = 3.50
B = .70
Material: 1060 Alloy
Density = 0.0975 lb/in^3
Units: IPS
Decimal places = 2

Exercise 4.11

Create the illustrated part. Note the
location of the Origin. Overall
drafting standard - ANSI.

- Calculate the overall mass of the
  illustrated model. Apply the
  Mass Properties tool.

- Think about the steps that you
  would take to build the model.

- Review the provided information
  carefully. Units are represented
  in the IPS, (inch, pound, second)
  system.

- A = 3.00in, B = .75in

Given:
A = 3.00
B = .75
Material: Copper
Density = 0.321 lb/in^3
Units: IPS
Decimal places = 2

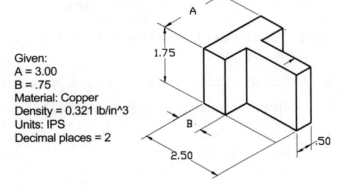

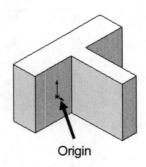

Origin

Exercise 4.12

Create the illustrated part. Note the location of the Origin. Overall drafting standard - ANSI.

- Calculate the volume of the part and locate the Center of mass with the provided information.

- Apply the Mass Properties tool.

- Think about the steps that you would take to build the model.

- Review the provided information carefully.

Given:
A = 3.30
B = 2.00
Material: 2014 Alloy
Density = .101 lb/in^3
Units: IPS
Decimal places = 2

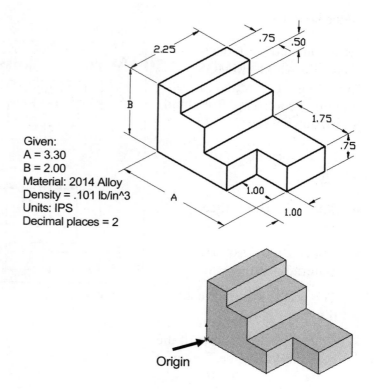

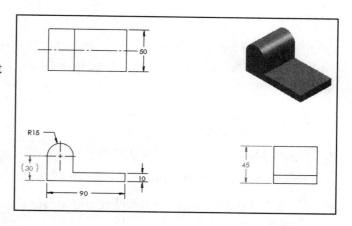

Origin

Exercise 4.13

Create the part from the illustrated ANSI - MMGS Third Angle Projection drawing: Front, Top, Right and Isometric views.

Note: The location of the Origin.

- Apply 1060 Alloy for material.

- Calculate the Volume of the part and locate the Center of mass.

- Think about the steps that you would take to build the model. The part is symmetric about the Front Plane.

Exercise 4.14

Create the part from the illustrated ANSI - MMGS Third Angle Projection drawing: Front, Top, Right and Isometric views.

Note: The location of the Origin.

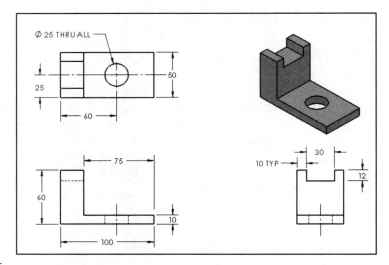

- Apply the Hole Wizard feature.

- Apply 1060 Alloy for material.

- The part is symmetric about the Front Plane.

- Calculate the Volume of the part and locate the Center of mass.

- Think about the steps that you would take to build the model.

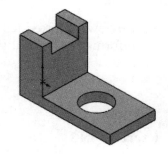

Exercise 4.15

Create the part from the illustrated ANSI - MMGS Third Angle Projection drawing: Front, Top, Right and Isometric views.

Note: The location of the Origin.

- Apply Cast Alloy steel for material.

- The part is symmetric about the Front Plane.

- Calculate the Volume of the part and locate the Center of mass.

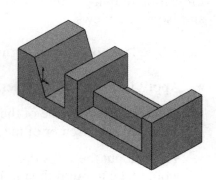

Think about the steps that you would take to build the model. Do you need the Right view for manufacturing? Does it add any information?

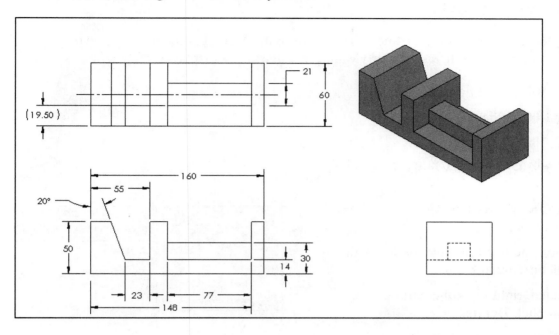

Exercise 4.16: Cosmetic Threads

Apply a Cosmetic thread: 1/4-20x2 UNC. A cosmetic thread represents the inner diameter of a thread on a boss or the outer diameter of a thread.

Copy and open the Cosmetic thread part model from the Chapter 4 Homework folder on the DVD.

Note: Copy the model to your folder on the computer. Do not work from the DVD.

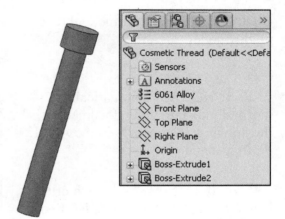

Create a Cosmetic thread.

Click the bottom edge of the part as illustrated.

Click Insert, Annotations, Cosmetic Thread from the Menu bar menu. View the Cosmetic Thread PropertyManager. Edge<1> is displayed.

Select **Blind** for End Condition.

Enter **1.00** for depth.

Enter **.200** for min diameter.

Enter **¼-20-2 UNC 2A** in the Thread Callout box.

Click **OK** ✓ from the Cosmetic Thread FeatureManager.

Expand the FeatureManager. View the Cosmetic Thread feature.

If needed, right-click the **Annotations** folder, click **Details**.

Check the **Cosmetic threads**, and **Shaded cosmetic threads** box.

Click **OK**. View the cosmetic thread on the model.

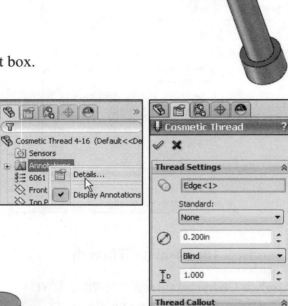

☼ The Thread Callout: ¼-20-2UNC 2A is automatically inserted into a drawing document, if the drawing document is in the ANSI drafting standard.

☼ ¼-20-2 UNC 2A – ¼ inch drill diameter - 20 threads / inch, Unified National Coarse thread series, Class 2 (General Thread), A -External threads

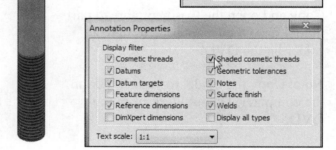

Exercise 4.17: 3D SKETCH / HOLE WIZARD

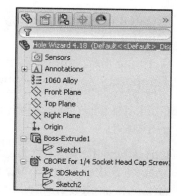

Create a part using the Hole Wizard feature. Apply the 3D sketch placement method as illustrated in the FeatureManager. Insert and dimension a hole on a cylindrical face.

Copy and open the Hole Wizard 4.18 model from the Chapter 4 Homework folder on the DVD.

Click the Hole Wizard 📷 Features tool. The Hole Specification PropertyManager is displayed.

Select the Counterbore Hole Type.

Select ANSI Inch for Standard.

Select Socket Head Cap Screw for fastener Type.

Select 1/4 for Size. Select Normal for Fit.

Select Through All for End Condition.

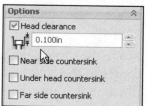

Enter .100 for Head clearance in the Options box.

Click the Positions Tab. The Hole Position PropertyManager is displayed.

Click the 3D Sketch button. SolidWorks

displays a 3D interface with the Point **XYᴌ** tool active.

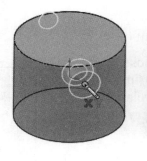

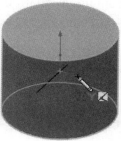

🔅 When the Point tool is active, wherever you click, you will create a point.

Click the cylindrical face of the model as illustrated. The selected face is displayed in orange. This indicates that an OnSurface sketch relations will be created between the sketch point and the cylindrical face. The hole is displayed in the model.

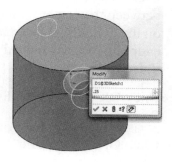

Insert a dimension between the top face and the Sketch point.

Click the Smart Dimension 🔷 Sketch tool.

Click the top flat face of the model and the sketch point.

Enter .25in.

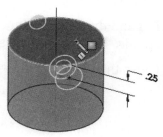

Locate the point angularly around the cylinder. Apply construction geometry.

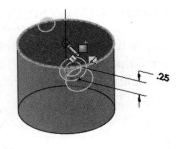

Activate the Temporary Axes. Click View, check the Temporary Axes box from the Menu bar toolbar.

Click the Line ✎ Sketch tool. Note: 3D sketch is still activated.

Ctrl+click the top flat face of the model. This moves the red space handle origin to the selected face. This also constrains any new sketch entities to the top flat face. Note the mouse

pointer ⟋ icon.

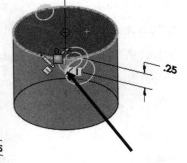

Move the mouse pointer near the center of the activated top flat face as illustrated. View the small black circle. The circle indicates that the end point of the line will pick up a Coincident relation.

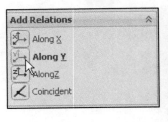

Click the center point of the circle.

Sketch a line so it picks up the **AlongZ sketch relation**. The cursor displays the relation to be applied. **This is a very important step!**

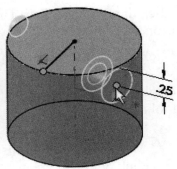

Add Relations	⨉
⤭ Along X	
⤭ **Along Y**	
⤭ AlongZ	
⟨ Coincident	

Create an AlongY sketch relation between the centerpoint of the hole on the cylindrical face and the endpoint of the sketched line as illustrated.

Click OK ✔ from the Properties PropertyManager.

Click OK ✔ from the Hole Position PropertyManager.

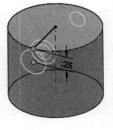

Expand the FeatureManager and view the results. The two sketches are fully defined.

Close the model.

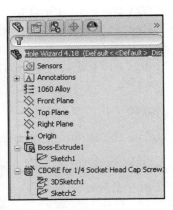

Exercise 4.18: DFMXpress

Apply the DFMXpress Wizard.
DFMXpress is an analysis tool that
validates the manufacturability of
SolidWorks parts. Use DFMXpress
to identify design areas that may
cause problems in fabrication or
increase the costs of production.

- Copy and open the ROD model
  from the Chapter 4 Homework
  folder on the DVD.

- Click the Evaluate tab in the
  CommandManager.

- Click DFMXpress Analysis Wizard.

- Click the RUN button.

- Expand each folder. View the
  results.

- Make any needed changes and
  save the ROD in a DIFFERENT
  file folder.

Notes:

Chapter 5

Revolved Features

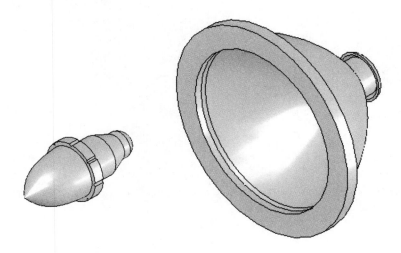

Below are the desired outcomes and usage competencies based on the completion of Chapter 5.

Desired Outcomes:	Usage Competencies:

• Two FLASHLIGHT parts:
   o LENS
   o BULB
	• Specific knowledge and understanding of the following Features: Extruded Boss/Base, Extruded Cut, Revolved Base, Revolved Boss Thin, Revolved Thin Cut, Dome, Shell, Hole Wizard and Circular Pattern.

• Insert the following Geometric relations: Equal, Coincident, Symmetric, Intersection and Perpendicular.	• Ability to insert multiple Geometric relations to a model. • Ability to apply Design Intent in Sketches, Features, Parts and Assemblies.

Notes:

Chapter 5 - Revolved Features

Chapter Overview

This Chapter introduces you to the Revolved
Boss/Base feature. Create two parts for the
FLASHLIGHT assembly in this project:

- LENS

- BULB

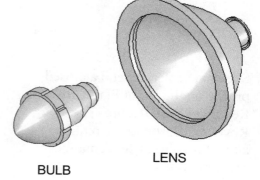

BULB   LENS

A Revolved Boss/Base feature requires a 2D sketch
profile and a centerline. Utilize sketch geometry
and sketch tools to create the following features:

- Revolved Base

- Revolved Boss

- Revolved Boss Thin

- Revolved Cut

Utilize existing faces to create the following features:

- Shell

- Dome

- Hole Wizard

Utilize the Extruded Cut feature, (seed feature) to create the Circular Pattern feature.

After completing the activities in this chapter, you will be able to:

- Utilize the following Sketch tools: Circle, Line, 3 Point Arc, Centerpoint Arc, Spline,
  Mirror, Offset Entities, Trim and Convert Entities.

- Insert the following Geometric relations: Equal, Coincident, Symmetric, Intersection
  and Perpendicular.

- Apply Transparent Optical Properties to the LENS.

- Create and edit the following features: Extruded Boss/Base, Extruded Cut, Revolved
  Base, Revolved Boss, Revolved Boss Thin, Revolved Thin Cut, Shell, Hole Wizard,
  Dome and Circular Pattern.

- Create two parts for the FLASHLIGHT assembly: LENS and BULB.

LENS Part

Create the LENS. The LENS is a purchased part.

The LENS utilizes a Revolved Base feature.

Sketch a centerline and a closed profile on the Right Plane. Insert a Revolved Base feature. The Revolved Base feature requires an axis of revolution and an angle of revolution.

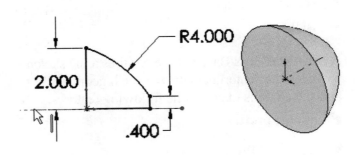

Insert the Shell feature. The Shell feature provides uniform wall thickness. Select the front face as the face to be removed.

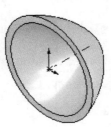

Utilize the Convert Entities sketch tool to extract the back circular edge for the sketched profile. Insert an Extruded Boss feature from the back of the LENS.

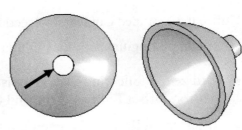

Sketch a single profile. Insert a Revolved Thin feature to connect the LENS to the BATTERYPLATE. The Revolved Thin feature requires a thickness.

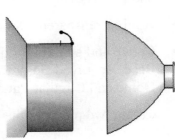

Insert a Counterbore Hole feature using the Hole Wizard feature.

The BULB is located inside the Counterbore Hole.

Insert the front Lens Cover with an Extruded Boss/Base feature. The sketch profile for the Extruded Boss/Base is sketched on the Front Plane. Add a transparent Lens Shield with the Extruded Boss/Base feature.

LENS Part-Revolved Boss/Base Feature

Create the LENS with a Revolved Base feature. The solid Revolved Base feature requires:

- Sketch plane (Right)
- Sketch profile
- Centerline
- Angle of Revolution (360°)

The profile lines reference the Top and Front Planes. Create the curve of the LENS with a 3-point arc.

Activity: LENS Part

Create the new part.

1) Click **New** ☐ from the Menu bar.

2) Click the **MY-TEMPLATES** tab.

3) Double-click **PART-IN-ANSI**, **[PART-MM-ISO]**.

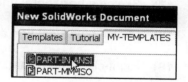

Save the part.

4) Click **Save** 💾.

5) Select the **PROJECTS** folder.

6) Enter **LENS** for File name.

7) Enter **LENS WITH SHIELD** for Description.

8) Click **Save**. The LENS FeatureManager is displayed.

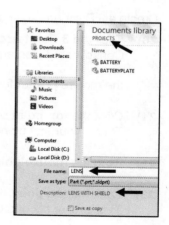

View the Front Plane.

9)   Right click **Front Plane** from the FeatureManager.

10) Click **Show** from the Context toolbar. Note: If you upgraded from 2012, verse a new install, your planes and Origins may be displayed by default. Hide unwanted planes in the FeatureManager.

Create a sketch.

11) Right-click **Right Plane** from the FeatureManager.

12) Click **Sketch** ✎ from the Context toolbar. The Sketch toolbar is displayed.

13) Click the **Centerline** ⦙ Sketch tool. The Insert Line PropertyManager is displayed.

14) Sketch a horizontal **centerline** collinear to the Top plane, through the Origin ⌐ as illustrated.

Sketch the profile. Create three lines.

15) Click the **Line** ＼ Sketch tool. The Insert Line PropertyManager is displayed.

16) Sketch a **vertical line** collinear to the Front Plane coincident with the Origin.

17) Sketch a **horizontal line** coincident with the Top Plane.

18) Sketch a **vertical line** approximately 1/3 the length of the first line.

19) Right-click **End Chain**.

Create a 3 Point Arc. A 3 Point Arc requires three points.

20) Click the **3 Point Arc** ⌢ Sketch tool from the Consolidated Centerpoint Arc toolbar. Note:

the mouse pointer feedback ⌢ icon.

21) Click the **top point** on the left vertical line. This is your first point.

22) Drag the **mouse pointer** to the right.

23) Click the **top point** on the right vertical line. This is your second point.

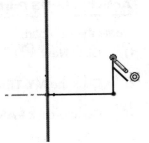

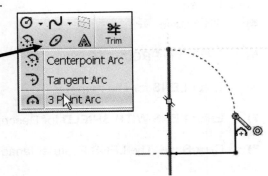

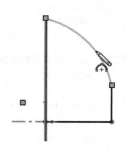

24)   Drag the **mouse pointer** upward.

25)   Click a **position** on the arc.

Deselect the 3 Point Arc Sketch tool and add an Equal relation.

26)   Right-click **Select**.

27)   Click the **left vertical** line.

28)   Hold the **Ctrl** key down.

29)   Click the **horizontal** line. The Properties PropertyManager is displayed. The selected entities are displayed in the Selected Entities box.

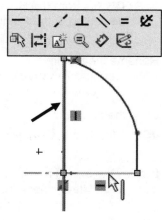

30)   Release the **Ctrl** key.

31)   Right-click **Make Equal** from the Context toolbar.

32)   Click **OK** ✔ from the Properties PropertyManager.

Add dimensions.

33)   Click the **Smart Dimension** ◇ Sketch tool.

34)   Click the **left vertical** line.

35)   Click a **position** to the left of the profile.

36)   Enter **2.000**in, [**50.8**].

37)   Click the **right vertical** line.

38)   Click a **position** to the right of the profile.

39)   Enter **.400**in, [**10.16**].

40)   Click the **arc**.

41)   Click a **position** to the right of the profile.

42)   Enter **4.000**in, [**101.6**]. The black sketch is fully defined.

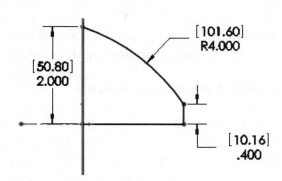

🔅 Utilize **Tools, Sketch Tools, Check Sketch for Feature** option to determine if a sketch is valid for a specific feature and to understand what is wrong with a sketch.

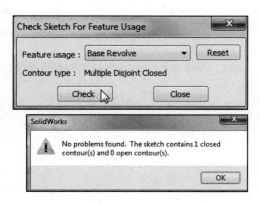

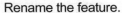
Activity: LENS Part-Revolved Base Feature

Insert the Revolved Base feature.

43) Click the **Revolved Boss/Base** ⊕ feature tool. The Revolve PropertyManager is displayed.

44) If needed, click the **horizontal centerline** for the axis of revolution. Line1 is displayed in the Axis of Revolution box. Note: The direction arrow points clockwise. Click **OK** ✓ from the Revolve PropertyManager. Revolve1 is displayed in the FeatureManager.

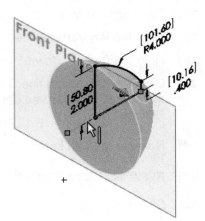

Rename the feature.

45) Rename **Revolve1** to **BaseRevolve**.

Save the model.

46) Click **Save** 💾.

Display the axis of revolution.

47) Click **View**, check **Temporary Axes** from the Menu bar.

Revolve features contain an axis of revolution. The axis of revolution utilizes a sketched centerline, edge or an existing feature/sketch or a Temporary Axis. The solid Revolved feature contains a closed profile. The Revolved thin feature contains an open or closed profile.

LENS Part-Shell Feature

The Revolved Base feature is a solid. Utilize the Shell feature to create a constant wall thickness around the front face. The Shell feature removes face material from a solid. The Shell feature requires a face and thickness. Use the Shell feature to create thin-walled parts.

Activity: LENS Part-Shell Feature

Insert the Shell feature.

48) Click the **front face** of the BaseRevolve feature.

49) Click the **Shell** 🔲 feature tool. The Shell1 PropertyManager is displayed. Face<1> is displayed in the Faces to Remove box.

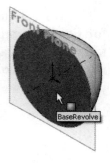

50) Enter **.250**in, **[6.35]** for Thickness.

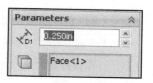

Display the Shell feature.

51) Click **OK** ✔ from the Shell1 PropertyManager. Shell1 is displayed in the FeatureManager.

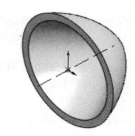

52) Right-click **Front Plane** from the FeatureManager.

53) Click **Hide** from the Context toolbar.

54) Rename **Shell1** to **LensShell**.

55) Click **Save** 💾.

To insert rounded corners inside a shelled part, apply the Fillet feature before the Shell feature. Select the Multi-thickness option to apply different thicknesses.

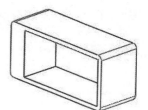

Extruded Boss/Base Feature and Convert Entities Sketch tool

Create the LensNeck. The LensNeck houses the BULB base and is connected to the BATTERYPLATE. Use the Extruded Boss/Base feature. The back face of the Revolved Base feature is the Sketch plane.

Utilize the Convert Entities Sketch tool to extract the back circular face to the Sketch plane. The new curve develops an On Edge relation. Modify the back face, and the extracted curve updates to reflect the change. No sketch dimensions are required.

Activity: Extruded Boss/Base Feature and Convert Entities Sketch tool

Rotate the LENS.

56) **Rotate** the LENS with the middle mouse button to display the back face as illustrated. Note: The Rotate ⟳ icon is displayed.

Sketch the profile.

57) Right-click the **back face** for the Sketch plane. BaseRevolve is highlighted in the FeatureManager. This is your Sketch plane.

58) Click **Sketch** ✏ from the Context toolbar. The Sketch toolbar is displayed.

59) Click the **Convert Entities** ⬚ Sketch tool. The Convert Entities PropertyManager is displayed. Click **OK** ✓ from the Convert Entities PropertyManager.

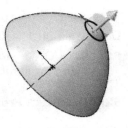

Insert an Extruded Boss/Base feature.

60) Click the **Extruded Boss/Base** 🗔 feature tool. The Boss-Extrude PropertyManager is displayed.

61) Enter **.400**in, [**10.16**] for Depth in Direction 1. Accept the default settings.

62) Click **OK** ✓ from the Boss-Extrude PropertyManager. Boss-Extrude1 is displayed in the FeatureManager.

63) Click **Isometric view** 🔲 from the Heads-up View toolbar.

64) Rename **Boss-Extrude1** to **LensNeck**.

65) Click **Save** 💾.

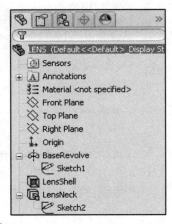

LENS Part-Hole Wizard

The LENS requires a Counterbore hole. Apply the Hole Wizard 🔲 feature. The Hole Wizard feature assists in creating complex and simple holes. The Hole Wizard Hole type categories are: *Counterbore, Countersink, Hole, Tapped, PipeTap*, and *Legacy* (Holes created before SolidWorks 2000).

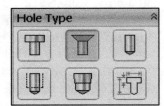

Select the face or plane to locate the hole profile. Specify the user parameters for the custom Counterbore hole. The parameters are: *Description, Standard, Screw Type*, Hole Size, *Fit, Counterbore diameter, Counterbore depth*, and *End Condition*.

Insert a Coincident relation to position the hole center point. Dimensions for the Counterbore hole are provided in both inches and millimeters.

Activity: LENS Part-Hole Wizard Counterbore Hole Feature

Create the Counterbore hole.

66) Click **Front view** 🔲 from the Heads-up View toolbar.

67) Click the **Hole Wizard** 🔲 feature tool. The Hole Specification PropertyManager is displayed. Type is the default tab.

68) Click the **Counterbore** icon as illustrated.

Note: For a metric hole, skip the next few steps.

For inch Counterbore hole:
69) Select **Ansi Inch** for Standard.

70) Select **Hex Bolt** for Type.

71) Select ½ for Size.

72) Check the **Show custom sizing** box.

73) Click inside the **Counterbore Diameter** value box.

74) Enter **.600**in.

75) Click inside the **Counterbore Depth** value box.

76) Enter **.200**in.

77) Select **Through All** for End Condition.

78) Click the **Position** tab.

79) Click the small **inside back face** of the LensShell feature as illustrated. Do not select the Origin. LensShell is highlighted in the FeatureManager. The Point tool ✳ icon is displayed.

80) Click the **origin**. A Coincident relation is displayed.

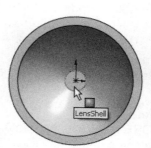

Deselect the Point Sketch tool.
81) Right-click **Select** in the Graphics window. **Go** to step 86.

Note: For an inch hole, skip the next few steps and start on the next page.

For millimeter Counterbore hole:
82) Select **Ansi Metric** for Standard. Enter **Hex Bolt** for Type. Select **M5** for Size. Click **Through All** for End Condition. Check the **Show custom sizing** box. Click inside the **Counterbore Diameter** value box. Enter **15.24**. Click inside the **Counterborebore Depth** value box. Enter **5**. Click the **Position** tab.

83) Click the small **inside back face** of the LensShell feature as illustrated. Do not select the Origin. LensShell is highlighted in the FeatureManager. The Point tool ✳ icon is displayed.

84) Click the **origin**. A Coincident relation is displayed.

85) Right-click **Select** in the Graphics window to deselect the point tool.

86) Click the **Type** tab.

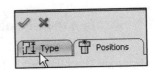

Add the new hole type to your Favorites list.

87) **Expand** the Favorites box.

88) Click the **Add or Update Favorite** button.

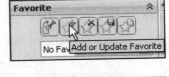

89) Enter **CBORE FOR BULB**.

90) Click **OK** from the Add or Update a Favorite dialog box.

91) Click **Yes**.

92) Click **OK** from the Hole Specification PropertyManager.

Expand the Hole feature.

93) **Expand** the CBORE feature in the FeatureManager. Note: Sketch3 and Sketch4 created the CBORE feature.

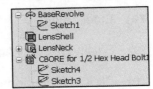

Display the Section view.

94) Click **Right Plane** from the FeatureManager.

95) Click **Section view** from the Heads-up View toolbar.

96) Click **OK** from the Section View PropertyManager.

97) Click **Isometric view** from the Heads-up View toolbar.

Display the Full view.

98) Click **Section view** from the Heads-up View toolbar.

99) Rename **CBORE for ½ Hex Head Bolt1** to **BulbHole**.

100) Click **Save**.

LENS Part-Revolved Boss Thin Feature

Create a Revolved Boss Thin feature. Rotate an open sketched profile around an axis. The sketch profile must be open and cannot cross the axis. A Revolved Boss Thin feature requires:

- Sketch plane (Right Plane)

- Sketch profile (Center point arc)
- Axis of Revolution (Temporary axis)
- Angle of Rotation (360)
- Thickness .100in, [2.54]

Click the Right Plane for the Sketch plane. Sketch a center point arc. The sketched center point arc requires three Geometric relations: *Coincident, Intersection* and *Vertical*.

The three geometric relations insure that the 90° center point of the arc is Coincident with the horizontal silhouette edges of the Revolved feature. A Revolved feature produces silhouette edges in 2D views. A silhouette edge represents the extent of a cylindrical or curved face. Utilize silhouette edges for Geometric relations.

Select the Temporary Axis for Axis of Revolution. Select the Revolved Boss feature. Enter .100in [2.54] for Thickness in the Revolve PropertyManager. Enter 360° for Angle of Revolution. Note: If you cannot select a silhouette edge in Shaded mode, switch to Wireframe mode.

Activity: LENS Part-Revolved Boss Thin Feature

Create a sketch.
101) Right-click **Right Plane** from the FeatureManager.

102) Click **Sketch** ✏ from the Context toolbar.

103) Click **Right view** 🔲 from the Heads-up View toolbar.

104) Zoom in on the LensNeck.

105) Click the **Centerpoint Arc** ⌒ Sketch tool.

106) Click the **top horizontal silhouette edge** of the LensNeck. Do not select the midpoint of the silhouette edge.

107) Click the **top right corner** of the LensNeck.

108) Drag the **mouse pointer** counterclockwise to the left.

109) Click a **position** directly above the first point as illustrated.

Add a dimension.

110) Click the **Smart Dimension** ◇ Sketch tool.

111) Click the **arc**. Click a **position** to the right of the profile.

112) Enter **.100**in, **[2.54]**.

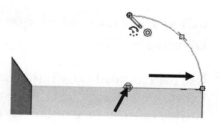

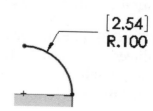

[2.54]
R.100

Insert a Revolved Thin feature.

113) Click the **Revolved Boss/Base** feature tool. The Revolve PropertyManager is displayed.

114) Select **Mid-Plane** for Revolve Type in the Thin Feature box.

115) Enter .050in, [1.27] for Direction1 Thickness.

116) Click the **Temporary Axis** for Axis of Revolution.

117) Click **OK** ✓ from the Revolve PropertyManager.

118) Rename **Revolve-Thin1** to **LensConnector**.

Fit the model to the Graphics window.
119) Press the **f** key.

120) Click **Save** 💾.

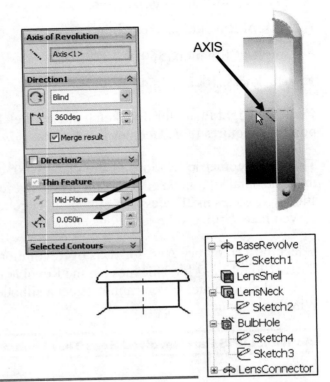

AXIS

💡 A Revolved sketch that remains open results in a Revolve

Thin feature ──┼─┤ . A Revolved sketch that is automatically closed, results in a line drawn from the start point to the end point of the sketch. The sketch is closed and results in a non-

Revolve Thin feature ⟋▪ .

LENS Part-Extruded Boss/Base Feature and Offset Entities

Use the Extruded Boss/Base feature tool to create the front LensCover. Utilize the Offset Entities Sketch tool to offset the outside circular edge of the Revolved feature. The Sketch plane for the Extruded Boss feature is the front circular face.

The Offset Entities Sketch tool requires an Offset Distance and direction. Utilize the Bi-direction option to create a circular sketch in both directions. The extrude direction is away from the Front Plane.

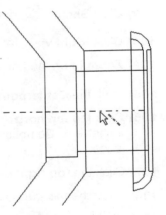

Activity: LENS Part-Extruded Boss Feature and Offset Entities

Create the sketch.

121) Click **Isometric view** from the Heads-up View toolbar.

122) Click **Hidden Lines Removed** from the Heads-up View toolbar.

123) Right-click the **front circular face** for the Sketch plane.

124) Click **Sketch** from the Context toolbar.

125) Click **Front view** from the Heads-up View toolbar.

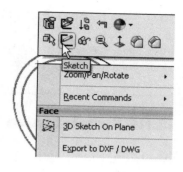

Offset the selected edge.

126) Click the **outside circular edge** of the LENS in the Graphics window.

127) Click the **Offset Entities** Sketch tool. The Offset Entities PropertyManager is displayed.

128) Check the **Bi-directional** box.

129) Enter **.250**in, **[6.35]** for Offset Distance. Accept the default settings.

130) Click **OK** from the Offset Entities PropertyManager.

131) Click **Isometric view** from the Heads-up View toolbar.

132) Click **Shaded With Edges** from the Heads-up View toolbar.

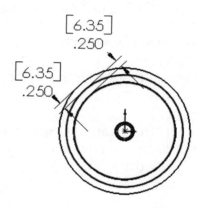

Insert an Extruded Boss feature.

133) Click the **Extruded Boss/Base** feature tool. The Boss-Extrude PropertyManager is displayed.

134) Enter **.250**in, **[6.35]** for Depth in Direction 1. Accept the default settings.

135) Click **OK** from the Boss-Extrude PropertyManager. Boss-Extrude2 is displayed in the PropertyManager.

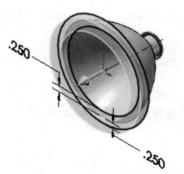

Verify the position of the Boss-Extrude2 feature.

136) Click the **Top view** . View the extruded feature.

137) Rename **Boss-Extrude2** to **LensCover**.

138) Click **Save** .

LENS Part-Extruded Boss Feature and Transparency

Apply the Extruded Boss/Base feature to create the LensShield. Utilize the Convert Entities Sketch tool to extract the inside circular edge of the LensCover and place it on the Front plane.

Apply the Transparent Optical property to the LensShield to control the ability for light to pass through the surface. Transparency is an Optical Property found in the Color PropertyManager. Control the following properties:

- **Diffuse amount, Specular amount, Specular spread, Reflection amount, Transparent amount and Luminous intensity**

Activity: LENS Part-Extruded Boss Feature and Transparency

Create the sketch.

139) Right-click **Front Plane** from the FeatureManager. This is your Sketch plane.

140) Click **Sketch** from the Context toolbar. The Sketch toolbar is displayed.

141) Click **Isometric view** from the Heads-up View toolbar.

142) Click the **front inner circular edge** of the LensCover (Boss-Extrude2) as illustrated.

143) Click **Hidden Line Removed** from the Heads-up View toolbar.

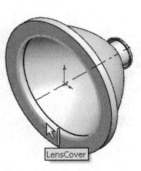

144) Click the **Convert Entities** ⬚ Sketch tool. The circle is projected onto the Front Plane.

145) Click **OK** ✔ from the Convert Entities PropertyManager.

Insert an Extruded Boss feature.

146) Click the **Extruded Boss/Base** 🗔 feature tool. The Boss-Extrude FeatureManager is displayed.

147) Enter .100in, [**2.54**] for Depth in Direction 1.

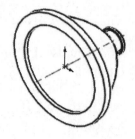

148) Click **OK** ✔ from the Boss-Extrude PropertyManager. Boss-Extrude3 is displayed in the FeatureManager.

149) Rename **Boss-Extrude3** to **LensShield**.

150) Click **Save** 💾.

Apply Transparency of the LensShield

151) Right-click **LensShield** in the FeatureManager.

152) Click **Change Transparency**. View the results.

Save the model.
153) Click **Save** 💾.

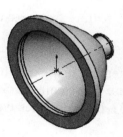

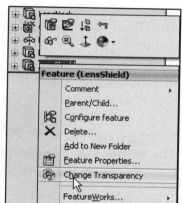

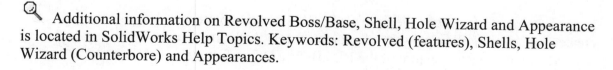

🔍 Additional information on Revolved Boss/Base, Shell, Hole Wizard and Appearance is located in SolidWorks Help Topics. Keywords: Revolved (features), Shells, Hole Wizard (Counterbore) and Appearances.

💡 Design Intent is how your part reacts as parameters are modified. Example: If you have a hole in a part that must always be .125≤ from an edge, you would dimension to the edge rather than to another point on the sketch. As the part size is modified, the hole location remains .125≤ from the edge.

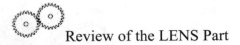 Review of the LENS Part

The LENS feature utilized a Revolved Base feature. A Revolved feature required an axis, profile, and an angle of revolution. The Shell feature created a uniform wall thickness.

You utilized the Convert Entities Sketch tool to create the Extruded Boss feature for the LensNeck. The Counterbore hole was created using the Hole Wizard feature.

The Revolved Thin feature utilized a single 3 Point Arc. Geometric relations were added to the silhouette edge to define the arc. The LensCover and LensShield utilized existing geometry to Offset and Convert the geometry to the sketch. The Color and Optics PropertyManager determined the LensShield transparency.

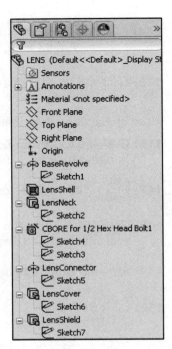

BULB Part

The BULB fits inside the LENS. Use the Revolved feature as the Base feature for the BULB.

Insert the Revolved Base  feature from a sketched profile on the Right plane.

Insert a Revolved Boss  feature using a Spline sketched profile. The profile utilizes a complex curve called a Spline (Non-Uniform Rational B-Spline or NURB). Draw Splines with control points.

Insert a Revolved Cut Thin  feature at the base of the BULB. A Revolved Cut Thin feature removes material by rotating an open sketch profile about an axis.

Insert a Dome 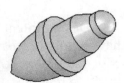 feature at the base of the BULB. A Dome feature creates spherical or elliptical shaped geometry. Use the Dome feature to create the Connector feature of the BULB. The Dome feature requires a face and a height value.

When you create a new part or assembly, the three default Planes (Front, Right and Top) are aligned with specific views. The Plane you select for the Base sketch determines the orientation of the part or assembly.

Apply the Extruded Cut feature 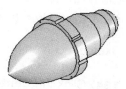 tool. The Extruded Cut feature is the seed feature for the Circular Pattern.

Apply the Circular Pattern feature tool. Insert a Circular Pattern feature from the Extruded Cut.

BULB Part-Revolved Base Feature

Create the new part, BULB. The BULB utilizes a solid Revolved Base feature.

The solid Revolved Base feature requires a:

- Sketch plane (Right Plane)
- Sketch profile (Lines)
- Axis of Revolution (Centerline)
- Angle of Rotation (360°)

Utilize the centerline to create a diameter dimension for the profile. The flange of the BULB is located inside the Counterbore hole of the LENS. Align the bottom of the flange with the Front Plane. The Front Plane mates against the Counterbore face.

Activity: BULB Part-Revolved Base Feature

Create a new part.

154) Click **New** from the Menu bar.

155) Click the **MY-TEMPLATES** tab.

156) Double-click **PART-IN-ANSI**, [**PART-MM-ISO**].

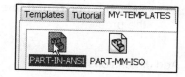

Save the part.

157) Click **Save** .

158) Select **PROJECTS** for Save in folder.

159) Enter **BULB** for File name.

160) Enter **BULB FOR LENS** for Description.

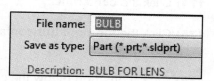

161) Click **Save**. The BULB FeatureManager is displayed.

Select the Sketch plane.

162) Right-click **Right Plane** from the FeatureManager. This is your Sketch plane.

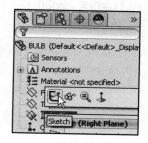

Create the sketch.

163) Click **Sketch** ✏ from the Context toolbar. The Sketch toolbar is displayed.

Sketch a centerline.

164) Click the **Centerline** ⁝ Sketch tool. The Insert Line PropertyManager is displayed.

165) Sketch a horizontal **centerline** coincident through the Origin ⌐.

Create six profile lines.

166) Click the **Line** ＼ Sketch tool. The Insert Line PropertyManager is displayed.

167) Sketch a **vertical line** to the left of the Front Plane.

168) Sketch a **horizontal line** with the endpoint coincident to the Front Plane.

169) Sketch a short **vertical line** towards the centerline, collinear with the Front Plane.

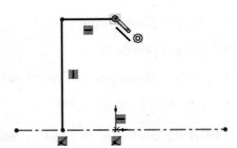

170) Sketch a **horizontal line** to the right.

171) Sketch a **vertical line** with the endpoint collinear with the centerline.

172) Sketch a **horizontal line** to the first point to close the profile.

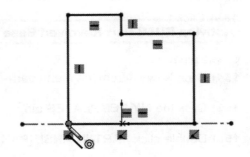

Add dimensions.

173) Click the **Smart Dimension** ✧ Sketch tool.

174) Click the **centerline**.

175) Click the **top right horizontal line** as illustrated.

176) Click a **position below** the centerline and to the right.

177) Enter **.400**in, **[10.016]**.

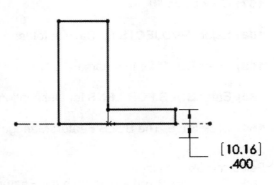

[10.16]
.400

💡 Click **View, Sketch Relations** from the Menu bar to display the relations of the model in the Graphics window.

178) Click the **centerline**.

179) Click the **top left horizontal line**.

180) Click a **position below** the centerline and to the left.

181) Enter **.590**in, [**14.99**].

182) Click the **top left horizontal line**.

183) Click a **position** above the profile.

184) Enter **.100**in, [**2.54**].

185) Click the **top right horizontal line**.

186) Click a **position** above the profile.

187) Enter **.500**in, [**12.7**].

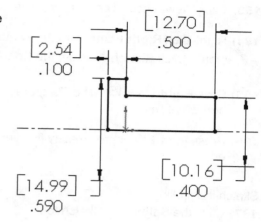

Fit the model to the Graphics window.
188) Press the **f** key.

Insert a Revolved Base feature.
189) Click the **Revolved Boss/Base** ⌖ feature tool. The Revolve PropertyManager is displayed. Accept the default settings.

190) Click **OK** ✔ from the Revolve PropertyManager. Revolve1 is displayed in the FeatureManager.

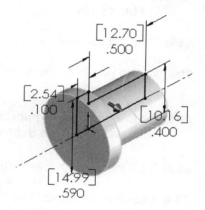

191) Click **Isometric view** ▣ from the Heads-up View toolbar.

192) Click **Save** 💾.

BULB Part-Revolved Boss Feature and Spline Sketch Tool

The BULB requires a second solid Revolved feature. The profile utilizes a complex curve called a Spline (Non-Uniform Rational B-Spline or NURB). Draw Splines with control points. Adjust the shape of the curve by dragging the control points.

🔆 For additional flexibility, deactivate the Snaps option in Document Properties for this model

---

Activity: BULB Part-Revolved Boss Feature and Spline Sketch Tool

---

Create the sketch.
193) Click **View**, check **Temporary Axes** from the Menu bar.

194) Right-click **Right Plane** from the FeatureManager for the Sketch plane. Click **Sketch** from the Context toolbar.

195) Click **Right view** . The Temporary Axis is displayed as a horizontal line.

196) Press the **z** key approximately five times to view the left vertical edge.

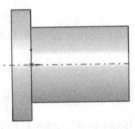

Sketch the profile.
197) Click the **Spline** Sketch tool.

198) Click the **left vertical edge** of the Base feature for the Start point.

199) Drag the **mouse pointer** to the left.

200) Click a **position** above the Temporary Axis for the Control point.

201) Double-click the **Temporary Axis** to create the End point and to end the Spline.

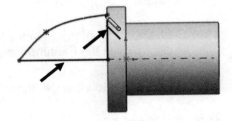

202) Click the **Line** Sketch tool.

203) Sketch a **horizontal line** from the Spline endpoint to the left edge of the Revolved feature.

204) Sketch a **vertical line** to the Spline start point, collinear with the left edge of the Revolved feature. Note: Dimensions are not required to create a feature.

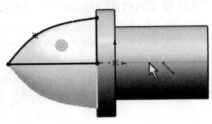

Insert a Revolved Boss feature.
205) De-select the Line Sketch tool. Right-click **Select**.

206) Click the **Temporary Axis** from the Graphics window as illustrated.

207) Click the **Revolved Boss/Base** feature tool. The Revolve PropertyManager is displayed. Accept the default settings.

208) Click **OK** from the Revolve PropertyManager. Revolve2 is displayed in the FeatureManager.

209) Click **Isometric view** 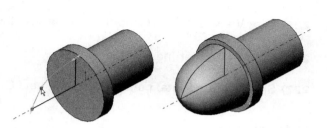 from the Heads-up View toolbar.

210) Click **Save** .

The points of the Spline dictate the shape of the Spline. Edit the control points in the sketch to produce different shapes for the Revolved Boss feature.

BULB Part-Revolved Cut Thin Feature

A Revolved Cut Thin feature removes material by rotating an open sketch profile around an axis. Sketch an open profile on the Right Plane. Add a Coincident relation to the silhouette and vertical edge. Insert dimensions.

Sketch a Centerline to create a diameter dimension for a revolved profile. The Temporary axis does not produce a diameter dimension.

Note: If lines snap to grid intersections, uncheck Tools, Sketch Settings, Enable Snapping for the next activity.

Activity: BULB Part-Revolved Cut Thin Feature

Create the sketch.
211) Right-click **Right Plane** from the FeatureManager.

212) Click **Sketch** from the Context toolbar.

213) Click **Right view** from the Heads-up View toolbar.

214) Click the **Line** Sketch tool.

215) Click the **mid point** of the top silhouette edge.

216) Sketch a **line** downward and to the right as illustrated.

217) Sketch a horizontal **line** to the right vertical edge.

218) De-select the Line Sketch took. Right-click **Select**.

Add a Coincident relation.
219) Click the **end point** of the line.

220) Hold the **Ctrl** key down.

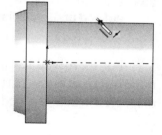

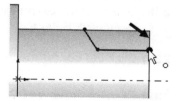

221) Click the right **vertical edge**.

222) Release the **Ctrl** key.

223) Click **Coincident** ⟨.

224) Click **OK** ✔ from the Properties PropertyManager.

Sketch a centerline.
225) Click **View**, uncheck **Temporary Axes** from the Menu bar.

226) Click the **Centerline** ┆ Sketch tool.

227) Sketch a **horizontal centerline** through the Origin.

Add dimensions.
228) Click the **Smart Dimension** ◇ Sketch tool.

229) Click the **horizontal centerline**.

230) Click the **short horizontal line**.

231) Click a **position** below the profile to create a diameter dimension.

232) Enter **.260**in, [**6.6**].

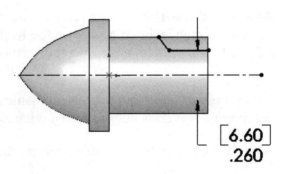

233) Click the **short horizontal line**.

234) Click a **position** above the profile to create a horizontal dimension.

235) Enter **.070**in, [**1.78**]. The Sketch is fully defined and is displayed in black.

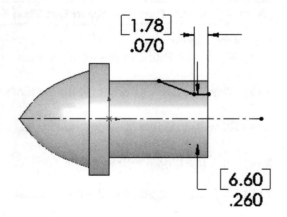

⛯ For Revolved features, the ∅ symbol is not displayed in the part. The ∅ symbol is displayed when inserted into the drawing.

Insert the Revolved Cut Thin feature.
236) De-select the Smart Dimension Sketch tool. Right-click **Select**.

237) Click the **centerline** in the Graphics window.

238) Click the **Revolved Cut** 🔘 feature tool. The Cut-Revolve PropertyManager is displayed.

239) Click **No** to the Warning Message, "Would you like the sketch to be automatically closed?" The Cut-Revolve PropertyManager is displayed.

240) Check the **Thin Feature** box.

241) Enter **.150**in, **[3.81]** for Thickness.

242) Click the **Reverse Direction** box.

243) Click **OK** ✔ from the Cut-Revolve PropertyManager. Cut-Revolve-Thin1 is displayed in the FeatureManager.

244) Click **Save** 🖫.

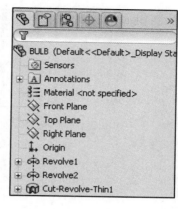

💡 Think design intent. When do you use the various End Conditions and Geometric sketch relations? What are you trying to do with the design? How does the component fit into an Assembly?

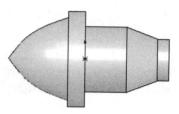

BULB Part-Dome Feature

A Dome 🖰 feature creates spherical or elliptical shaped geometry. Use the Dome feature to create the Connector feature of the BULB. The Dome feature requires a face and a height value.

Activity: BULB Part-Dome Feature

Insert the Dome feature.

245) Click the **back circular face** of Revolve1. Revolve1 is highlighted in the FeatureManager.

246) Click the **Dome** 🖰 feature tool. The Dome PropertyManager is displayed. Face1 is displayed in the Parameters box.

247) Enter **.100**in, **[2.54]** for Distance.

248) Click **OK** ✔ from the Dome PropertyManager. Dome1 is displayed in the FeatureManager.

249) Click **Isometric view** 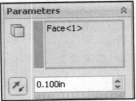 from the Heads-up View toolbar.

250) Click **Save** 💾.

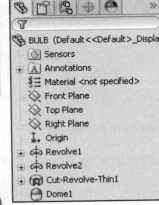

💡 Before creating sketches that use Geometric relations, check the Enable Snapping option in the Document Properties dialog box.

BULB Part-Circular Pattern Feature

A Pattern feature creates one or more instances of a feature or a group of features. The Circular Pattern feature places the instances around an axis of revolution.

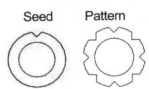

The Circular Pattern ⊕ feature requires a seed feature. The seed feature is the first feature in the pattern. The seed feature in this section is a V-shaped Extruded Cut feature.

Activity: BULB Part-Circular Pattern Feature

Create the Seed Cut feature.
251) Right-click the **front face** of the Base feature, Revolve1 in the Graphics window for the Sketch plane as illustrated. Revolve1 is highlighted in the FeatureManager.

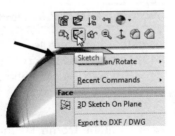

252) Click **Sketch** 🖉 from the Context toolbar. The Sketch toolbar is displayed. Click the **outside circular edge** of the BULB.

253) Click the **Convert Entities** 🗅 Sketch tool.

254) Click **OK** ✔ from the Convert Entities PropertyManager.

255) Click **Front view** 🗗 from the Heads-up View toolbar.

256) **Zoom in** on the top half of the BULB.

Sketch a centerline.

257) Click the **Centerline** ⁝ Sketch tool.

258) Sketch a **vertical centerline** coincident with the top and bottom circular circles and coincident with the Right plane.

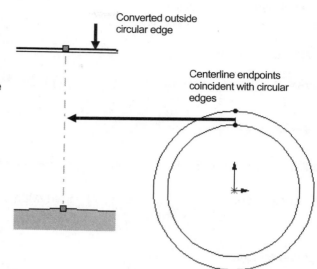

Converted outside circular edge

Centerline endpoints coincident with circular edges

Sketch a V-shaped line.

259) Click **Tools**, **Sketch Tools**, **Dynamic Mirror** from the Menu bar. The Mirror PropertyManager is displayed.

260) Click the **centerline** from the Graphics window.

261) Click the **Line** \ Sketch tool.

262) Click the **midpoint** of the centerline.

263) Click the coincident **outside circle edge to the left** of the centerline.

Deactivate the Dynamic Mirror tool.

264) Click **Tools**, **Sketch Tools**, **Dynamic Mirror** from the Menu bar.

Trim unwanted geometry.

265) Click the **Trim Entities** ✂ Sketch tool. The Trim PropertyManager is displayed.

266) Click **Power trim** from the Options box.

267) Click a **position** in the Graphics window and drag the mouse pointer until it intersects the circle circumference.

268) Click **OK** ✔ from the Trim PropertyManager.

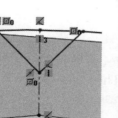

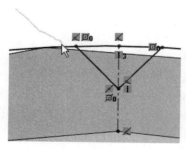

Add a Perpendicular relation.
269) Click the **left** V shape line.

270) Hold the **Ctrl** key down.

271) Click the **right** V shape line. The Properties
PropertyManager is displayed.

272) Release the **Ctrl** key.

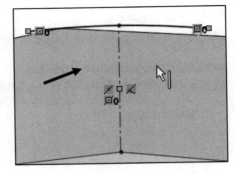

273) Click **Perpendicular** ⊥ from the Add Relations box.

274) Click **OK** ✓ from the Properties PropertyManager. The
sketch is fully defined.

Create an Extruded Cut feature.
275) Click the **Extruded Cut** ▣ feature tool. The Cut-Extrude
PropertyManager is displayed.

276) Click **Through All** for End Condition in Direction 1.
Accept the default settings.

277) Click **OK** ✓ from the Cut-Extrude PropertyManager. The
Cut-Extrude1 feature is displayed in the FeatureManager.

278) Click **Isometric view** 🔲 from the Heads-up View toolbar.

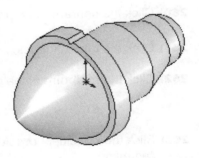

Fit the drawing to the Graphics window.
279) Press the **f** key.

280) Click **Save** 💾.

🔆 Reuse Geometry in the feature. The Cut-Extrude1
feature utilized the centerline, Mirror Entity, and
geometric relations to create a sketch with no dimensions.

The Cut-Extrude1 feature is the seed feature for the
pattern. Create four copies of the seed feature. A copy of a
feature is called an Instance. Modify the four copies Instances
to eight.

🔆 When you create a new part or assembly, the three
default Planes (Front, Right and Top) are aligned with
specific views. The Plane you select for the Base sketch
determines the orientation of the part or assembly.

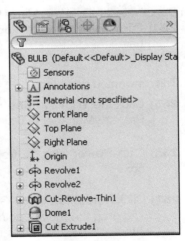

Insert the Circular Pattern feature.
281) Click the **Cut-Extrude1** feature from the FeatureManager.

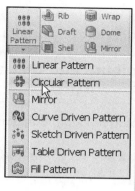

282) Click the **Circular Pattern** feature tool. The Circular Pattern PropertyManager is displayed. Cut-Extrude1 is displayed in the Features to Pattern box.

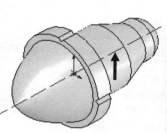

283) Click **View**, check **Temporary Axes** from the Menu bar.

284) Click the **Temporary Axis** from the Graphics window. Axis<1> is displayed in the Pattern Axis box.

285) Enter **4** in the Number of Instances box. Check the **Equal spacing** box.

286) Check the **Geometry pattern** box. Accept the default settings.

287) Click **OK** from the Circular Pattern PropertyManager. CirPattern1 is displayed in the FeatureManager.

Edit the Circular Pattern feature.
288) Right-click **CirPattern1** from the FeatureManager.

289) Click **Edit Feature** from the Context toolbar. The CirPattern1 PropertyManager is displayed.

290) Enter **8** in the Number of Instances box. Click **OK** from the CirPattern1 PropertyManager.

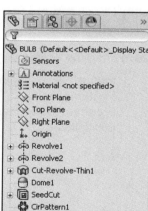

Rename the feature
291) Rename **Cut-Extrude1** to **SeedCut**.

Hide the reference geometry.
292) Click **View**, uncheck **Temporary Axes** from the Menu bar.

Save the model.
293) Click **Save**.

Rename the seed feature of a pattern to locate quickly for future assembly.

Customizing Toolbars and Short Cut Keys

The default toolbars contains numerous icons that represent basic functions. Additional features and functions are available that are not displayed on the default toolbars.

You have utilized the z key for Zoom In/Out, the f key for Zoom to Fit, and Ctrl-C/Ctrl-V to Copy/Paste. Short Cut keys save time.

Assign a key to execute a SolidWorks function. Create a Short Cut key for the Temporary Axis.

Activity: Customizing Toolbars and Short Cut Keys

Customize the toolbar.

294) Click **Tools**, **Customize** from the Menu bar. The Customize dialog box is displayed.

Place the Freeform icon on the Features toolbar.

295) Click the **Commands** tab.

296) Click **Features** from the category text box.

297) Drag the **Freeform** feature icon into the Features toolbar. The Freeform feature option is displayed.

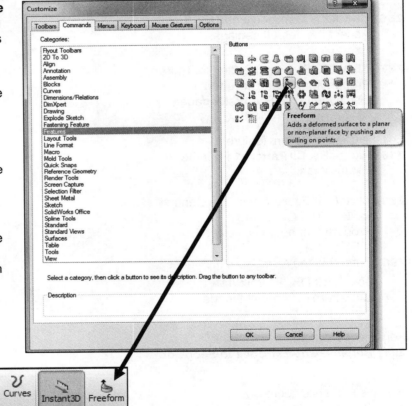

Customize the keyboard for the Temporary Axes.

298) Click the **Keyboard** tab from the Customize dialog box.

299) Select **View** for Categories.

300) Select **Temporary Axes** for Commands.

301) Enter **R** for shortcut key.

302) Click **OK**.

303) Press the **R** key to toggle the display of the Temporary Axes.

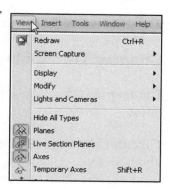

 Test the proposed Short cut key, before you customize your keyboard. Refer to the default Keyboard Short cut table in the Appendix.

Additional information on Revolved Boss/Base, Spline, Circular Pattern, Dome, Line / Arc Sketching is located in SolidWorks Help Topics. Keywords: Revolved (features), Spline, Pattern (Circular) and Dome.

Review of the BULB Part

The Revolved Base feature utilized a sketched profile on the Right plane and a centerline. The Revolved Boss feature utilized a Spline sketched profile. A Spline is a complex curve.

You created the Revolved Cut Thin feature at the base of the BULB to remove material. A centerline was inserted to add a diameter dimension. The Dome feature was inserted on the back face of the BULB. The Circular Pattern feature was created from an Extruded Cut feature. The Extruded Cut feature utilized existing geometry and required no dimensions.

Toolbars and keyboards were customized to save time. Always verify that a Short cut key is not predefined in SolidWorks.

Chapter Summary

You are designing a FLASHLIGHT assembly. You created two parts in this chapter:

- LENS

- BULB

Both parts utilized the Revolved Base feature. The Revolved feature required a Sketch plane, Sketch profile, Axis of revolution, and an Angle of rotation.

You created and edited the following features: Extruded Boss/Base, Extruded Cut, Revolved Base, Revolved Boss, Revolved Boss Thin, Revolved Thin Cut, Shell, Hole Wizard, Dome, and Circular Pattern.

You applied transparent optical properties to the LENS part and established the following Geometric relations: Equal, Coincident, Symmetric, Intersection, and Perpendicular.

The other parts for the FLASHLIGHT assembly are address in the later chapters.

Chapter Terminology

Centerpoint Arc: An arc Sketch tool that requires a centerpoint, start point, and end point.

Circular Pattern: A feature that creates a pattern of features or faces in a circular array about an axis. Use the Circular Pattern feature tool to create multiple instances of one or more features that you can space uniformly around an axis.

CommandManager: The CommandManager is a context-sensitive toolbar that dynamically updates based on the toolbar you want to access. By default, it has toolbars embedded in it based on the document type. When you click a tab below the Command Manager, it updates to show that toolbar. For example, if you click the **Sketches** tab, the Sketch toolbar is displayed.

Convert Entities: A sketch tool that projects one or more curves onto the current sketch plane. Select an edge, loop, face, curve, or external sketch contour, set of edges, or set of sketch curves.

Dome: A feature used to add a spherical or elliptical dome to a selected face.

Hole Wizard: The Hole Wizard feature is used to create specialized holes in a solid. The Hole Wizard creates the following hole types: Counterbore, Countersink, Hole, Straight tap, Tapered tap and Legacy Hole.

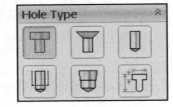

Mirror Entities: A sketch tool that mirrors sketch geometry to the opposite side of a sketched centerline.

Offset Entities: A sketch tool utilized to create sketch curves offset by a specified distance. Select sketch entities, edges, loops, faces, curves, set of edges or a set of curves.

Revolved Boss/Base: A feature used to add material by revolutions. A Revolved feature requires a centerline, a sketch on a Sketch plane and an angle of revolution. The sketch is revolved around the centerline.

Revolved Cut: A feature used to remove material by revolutions. A Revolved Cut requires a centerline, a sketch on a sketch plane and angle of revolution. The sketch is revolved around the centerline.

Shell: A feature used to remove faces of a part by a specified wall thickness.

Silhouette Edge: The imaginary edge of a cylinder or cylindrical face or edge.

Spline: A complex sketch curve. Splines can have two or more points.

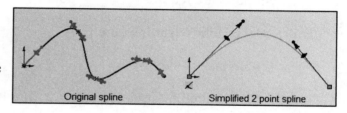

Original spline   Simplified 2 point spline

Thin option: The Thin option for the Revolved Boss and Revolved Cut utilizes an open sketch to add or remove material, respectively.

View the additional Power point presentations on the enclosed DVD for supplementary information.

- Avi folder
- Alphabet of lines and Precedent of Line Types
- Boolean Operation
- Design Intent
- Fastners in General
- Fundamental ASME Y14.5 Dimensioning Rules
- General Tolerancing and Fits
- History of Engineering Graphics
- Measurement and Scale
- Open a Drawing Document 2012
- Open an Assembly Document 2012
- Part and Drawing Dimensioning
- SolidWorks Basic Concepts
- Visualization, Arrangement of Views, and Primitives

Questions

1. Identify and describe the function of the following features:

   - Revolved Base
   - Revolved Boss
   - Revolved Cut
   - Revolved Cut Thin
   - Dome

2. Describe a Symmetric relation.

3. When is the Trim Entity Sketch tool used?

4. Explain the function of the Shell feature.

5. A Center point arc requires _____ points?

6. Describe the Hole Wizard feature.

7. What is a Spline?

8. Identify the required information for a Circular Pattern feature.

9. Name the Pull down menu that lists the Temporary Axis.

10. Describe the procedure to Show/Hide a Plane.

11. Describe the differences between Offset Entities and Convert Entities.

12. Identify the type of line required to utilize Mirror Entities.

13. Identify the geometric relation automatically created between Mirror Entities.

14. True of False. Select the arc center point to dimension an arc to its max condition.

15. True of False. The Transparency Optical Property is located in the Features toolbar.

16. Additional information of the Revolve Boss/Base and Revolve Cut features is located in _____.

17. Identify the following Sketch tool icons.

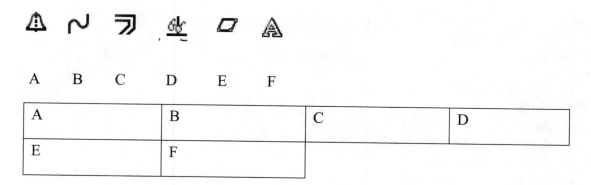

A   B   C   D   E   F

A		B		C		D	
E		F				

Exercises

Exercise 5.1: BUSHING

Create the illustrated BUSINING part on the Front Plane.

- Note the location of the Origin and the provided dimensions. Apply the ANSI overall drafting standard and use the IPS unit system.
- Apply Brass as a material.

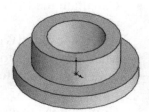

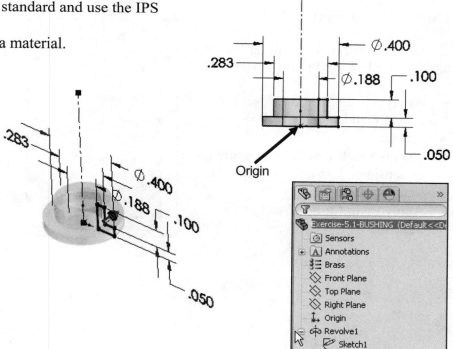

Origin

Exercise 5.2: PIN

Create the illustrated PIN part as illustrated on the Front Plane.

- Note the location of the Origin.
- Apply the ANSI overall drafting standard and use the IPS unit system.
- Apply 6061 Alloy as the material.
- Calculate the total mass of the part. Apply the SolidWorks Mass Properties tool under the Evaluate tab in the CommandManager.

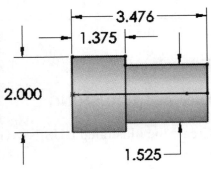

Exercise 5.3: Mounting Nut

Create the illustrated Mounting Nut part as illustrated on the Front Plane.

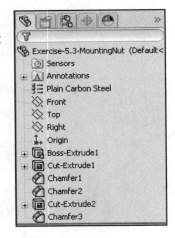

- Apply the ANSI overall drafting standard and use the MMGS unit system.
- Obtain a metric nut and measure the dimensions. Use these dimensions to create the part.
- Apply Plain Carbon Steel as the material.

Exercise 5.4: SCREW

Create the 10-24x3/8 SCREW as illustrated. Apply the ANSI overall drafting standard and use the IPS unit system. Note: A simplified version!

- Sketch a centerline on the Front Plane.
- Sketch a closed profile.

- Utilize a Revolved Feature. For metric size.

- Edit the Revolved Base Sketch. Use the Tangent Arc tool with Trim Entities. Enter an Arc dimension of .304in, [7.72].

- Utilize an Extruded Cut feature with the Mid Plane option to create the Top Cut. Depth = .050in.

- The Top Cut is sketched on the Front Plane.

- Utilize the Convert Entities Sketch tool to extract the left edge of the profile.

- Utilize the Circular Pattern feature and the Temporary Axis to create four Top Cuts.

- Use the Fillet Feature on the top circular edge, .01in to finish the simplified version of the SCREW part.

- Apply Plain Carbon Steel for material.

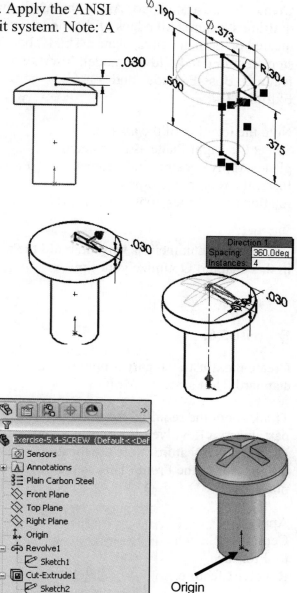

Origin

Exercise 5.5:

Create the illustrated part. Apply the ANSI overall drafting standard. All edges of the model are not located on perpendicular planes. Think about the steps required to build the model. Insert two features: Boss-Extrude1 and Extruded Cut.

Given:
A = 3.00, B = 1.00
Material: 6061 Alloy
Density = .097 lb/in^3
Units: IPS
Decimal places = 2

Note the location of the Origin. Select the Right Plane as the Sketch plane. Apply construction geometry. Insert the required geometric relations and dimensions for Sketch1.

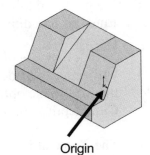

There are numerous ways to build the models in this chapter. Optimize your time.

Origin

Exercise 5.6:

Create the illustrated part. Apply the ANSI overall drafting standard. Unit system - MMGS.

Think about the required steps to build this part. Insert a Revolved Base feature and an Extruded Cut feature. Note the location of the Origin. Select the Front Plane as the Sketch plane.

Apply the Centerline Sketch tool for the Revolve1 feature.

Insert the required geometric relations and dimensions for Sketch1. Sketch1 is the profile for the Revolve1 feature.

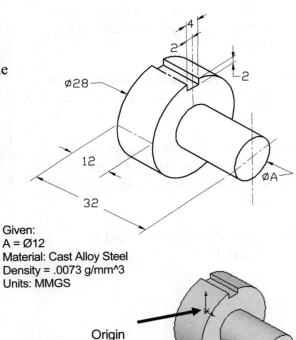

Given:
A = Ø12
Material: Cast Alloy Steel
Density = .0073 g/mm^3
Units: MMGS

Origin

Exercise 5.7:

Create the illustrated part. Calculate the total mass of the part. Apply the SolidWorks
Mass Properties tool under the Evaluate tab in the CommandManager.

Apply the ANSI overall drafting standard. Think about the required steps to build this
part. Insert four features: Boss-Extrude1, Cut-Extrude1, Cut-Extrude2 and Fillet. Note the
location of the Origin. Add 6061 Alloy for material.

Select the Right Plane as the Sketch plane. The
part Origin is located in the lower left corner of
the sketch.

Given:
A = 4.00
B = R.50
Material: 6061 Alloy
Density = .0975 lb/in^3
Units: IPS
Decimal places = 2

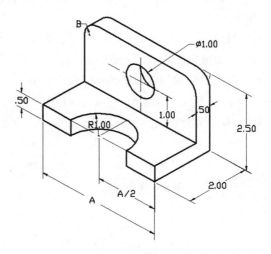

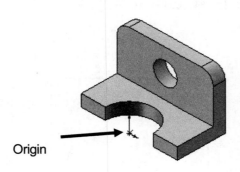

Origin

Exercise 5.8:

Create the Hole - Block (Hole Wizard feature) part as illustrated.
Create the Part document: ANSI drafting standard, IPS unit system
and set precision. Display the Origin and Sketch relations. Create the
Hole - Block part on the Front Plane. Note: All sketches should be
fully defined.

Create a rectangular prism 2 inches wide by 5 inches long by 2
inches high on the top surface of the prism, place four holes, 1
inch apart

- Hole #1: Simple Hole Type: Fractional Drill Size, 7/16
  diameter, End Condition: Blind, 0.75 inch deep.

- Hole #2: Counterbore hole Type: for 3/8 inch diameter Hex
  bolt, End Condition: Through All.

- Hole #3: Countersink hole Type: for 3/8 inch diameter Flat
  head screw, 1.5 inch deep.

- Hole #4: Tapped hole Type, Size ¼-20, End Condition:
  Blind -1.0 inch deep.

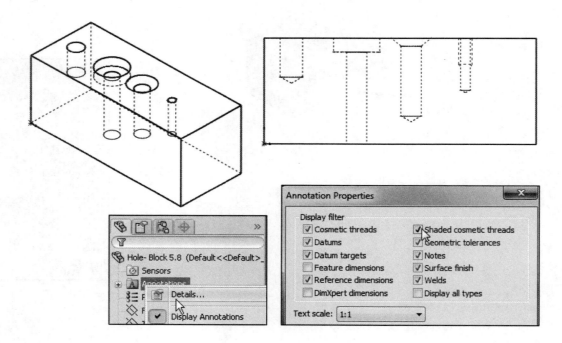

Chapter 6

Swept, Loft and Additional Features

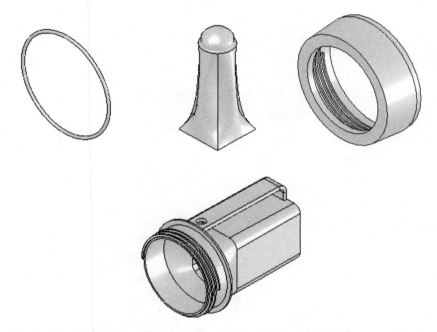

Below are the desired outcomes and usage competencies based on the completion of Chapter 6.

Project Desired Outcomes:	Usage Competencies:

• Create four FLASHLIGHT parts:
  o O-RING
  o SWITCH
  o LENSCAP
  o HOUSING
	• Specific knowledge and understanding of the following Features: Extruded Boss/Base, Extruded Cut, Swept Base, Swept Boss, Loft Base, Loft Boss, Mirror, Draft, Dome, Rib, and Linear Pattern.

• Establish Geometric relations: Pierce, Tangent, Equal, Intersection, Coincident, and Midpoint.	• Ability to apply multiple Geometric relations to a model. • Skill to apply Design Intent to 2D Sketches, 3D Features, Parts and Assemblies.

Notes:

Chapter 6 - Swept, Loft, and Additional Features

Chapter Overview

Create four new parts for the FLASHLIGHT assembly:

- O-RING

- SWITCH

- LENSCAP

- HOUSING

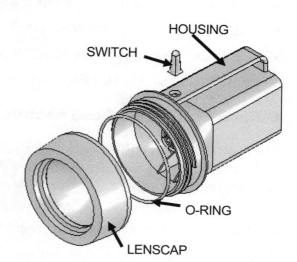

Chapter 6 introduces the Swept and Loft feature tools. The O-RING utilizes a Swept Base feature. The SWITCH utilizes the Loft Base and Dome feature. The LENSCAP and HOUSING utilize the Swept Boss and Loft Boss feature.

A Swept feature requires a minimum of two sketches: *path* and *profile*. Sketch the path (cross-section) and profile on different planes. The profile follows the path to create the following Swept features:

- Swept Base

- Swept Boss

The Loft feature requires a minimum of two profiles sketched on different planes. The profiles are blended together to create the following Loft features:

- Loft Base

- Loft Boss

Utilize existing features to create the Rib, Linear Pattern, and Mirror features. Utilize existing faces to create the Draft and Dome feature. The LENSCAP and HOUSING combines the Extruded Boss/Base, Extruded Cut, Revolved Thin Cut, Shell, and Circular Pattern with the Swept and Loft feature.

After completing the activities in this chapter, you will be able to:

- Utilize the following Sketch tools: Point, Centerline, Convert Entities, Trim Entities and Sketch Fillet.

- Establish the following Geometric relations: Pierce, Tangent, Equal, Intersection, Coincident and Midpoint.

- Create the following features: Swept Boss/Base, Loft Boss/Base, Mirror, Draft, Rib, Dome, and Linear Pattern.

- Review the Extruded Boss/Base, Extruded Cut, Revolve Cut Thin, Shell and Circular Pattern features.

- Suppress and Un-suppress various features.

- Reuse geometry from sketches, features, and other parts to develop new geometry.

- Create four new parts for the FLASHLIGHT assembly:

   o   O-RING, SWITCH, LENSCAP and HOUSING.

O-RING Part-Swept Base Feature

The O-RING part is positioned between the LENSCAP and the LENS. Create the O-RING with a Swept Base feature. The Swept Base feature uses:

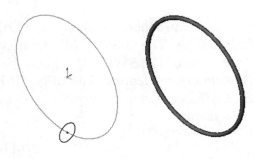

- A circular path sketched on the Front Plane.

- A small cross section profile sketched on the Right Plane.

The Pierce geometric relation positions the center of the cross section profile on the sketched path.

Path & Profile   Swept feature

Utilize the PART-IN-ANSI Template for inch units. Utilize the PART-MM-ISO Template for millimeter units. Millimeter dimensions are provided in brackets [x].

Activity: O-RING Part

Create the new part.

1)   Click **New** ⬜ from the Menu bar.

2)   Click the **MY-TEMPLATES** tab. Double-click **PART-IN-ANSI**, **[PART-MM-ISO]**.

3)   Click **Save**. Select **PROJECTS** for the Save in folder.

4)   Enter **O-RING** for File name. Enter **O-RING FOR LENS** for Description.

5)   Click **Save**. The O-RING FeatureManager is displayed.

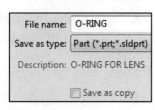

Create the Swept path.

6)   Right-click **Front Plane** from the FeatureManager for the Sketch plane. This is your Sketch plane.

7)   Click **Sketch** ✏ from the Context toolbar. The Sketch toolbar is displayed.

8)   Click the **Circle** ⊘ Sketch tool. The Circle PropertyManager is displayed.

9)   Sketch a circle centered at the **Origin**.

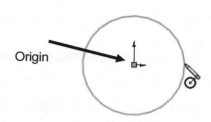

Add a dimension.

10)   Click the **Smart Dimension** Sketch tool.

11)   Click the **circumference** of the circle.

12)   Click a **position** off the profile.

13)   Enter **4.350**in, **[110.49]** as illustrated.

Close the sketch.

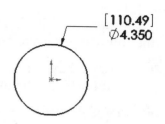

14)   **Rebuild** the model. Sketch1 is displayed in the FeatureManager.

15)   Rename **Sketch1** to **Sketch-path**.

Create the Swept profile.

16)   Click **Isometric view** from the Heads-up View toolbar.

17)   Right-click **Right Plane** from the FeatureManager. This is your Sketch plane.

18)   Click **Sketch** from the Context toolbar. The Sketch toolbar is displayed.

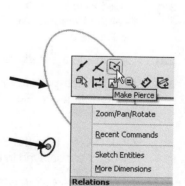

19)   Click the **Circle** Sketch tool.

20)   Create a **small circle** left of the Sketch-path on the Right Plane as illustrated.

Add a Pierce relation.

21)   Right-click **Select** to deselect the Circle Sketch tool.

22)   Click the **small circle centerpoint**.

23)   Hold the **Ctrl** key down.

24)   Click the **large circle circumference**.

25)   Release the **Ctrl** key. The selected entities are displayed in the Selected Entities box.

26)   Right-click **Make Pierce** from the Context toolbar. The centerpoint of the small circle pierces the Sketch-path, (large circle).

27)   Click **OK** from the Properties PropertyManager.

Add a dimension.

28)   Click the **Smart Dimension** Sketch tool.

29)   Click the **circumference** of the small circle.

30)   Click a **position** to the left of the profile.

31)   Enter **.125**in, **[3.18]**.

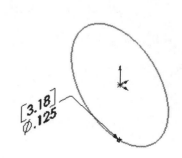

32)   Click **Isometric view** from the Heads-up View toolbar.

Exit the sketch.

33)   Right-click **Exit Sketch**. Sketch2 is displayed in the FeatureManager.

34)   Rename **Sketch2** to **Sketch-profile**.

The FeatureManager displays two sketches: Sketch-path and Sketch-profile. Create the path before the profile. The profile requires a Pierce relation to the path.

☀ Improve visibility. Small profiles are difficult to dimension on large paths. Perform the following steps to create a detailed small profile:

- Create a large cross section profile that contains the required dimensions and relationships. The black profile is fully defined.

- Pierce the profile to the path. Add dimensions to reflect the true size.

- Rename the profile and path to quickly located sketches in the FeatureManager.

Insert the Swept feature.

35)   Click the **Swept Boss/Base** feature tool. The Sweep PropertyManager is displayed.

36)   **Expand** O-RING from the fly-out FeatureManager.

37)   Click **Sketch-profile** from the fly-out FeatureManager. Sketch-profile is displayed in the Profile box.

38)   Click inside the **Path** box.

39)   Click **Sketch-path** from the fly-out FeatureManager. Sketch-path is displayed in the Path box.

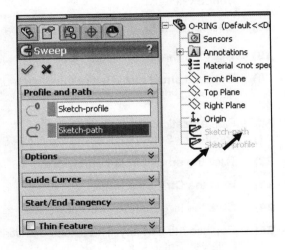

40)   Click **OK** from the Sweep PropertyManager. Sweep1 is displayed in the FeatureManager.

41)   Rename **Sweep1** to **Base-Sweep**.

42)   Click **Isometric view** from the Heads-up View toolbar.

43)   Click **Save** .

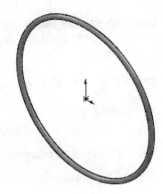

Review of the O-RING Part

The O-RING part utilized a Swept Base feature. The Swept Base feature required a Sketched path and a Sketched profile. The path was a large circle sketched on the Front Plane. The profile was a small circle sketched on the Right Plane. A Pierce Geometric relation was utilized to attach the profile to the path.

The Swept Boss/Base feature required a minimum of two sketches. You created a simple Swept Base feature. Swept features can be simple or complex.

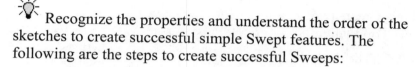 Recognize the properties and understand the order of the sketches to create successful simple Swept features. The following are the steps to create successful Sweeps:

- Create the path as a separate sketch. The path is open or closed. The path is a set curves contained in one sketch. The path can also be one curve or a set of model edges.

- Create each profile as a separate sketch. The profile is closed for a Swept Boss/Base feature.

- Fully define each sketch. The sketch is displayed in black.

- Sketch the profile last before inserting the Swept Boss/Base feature.

- Position the start point of the path on the plane of the profile.

- Path, profile and solid geometry cannot intersect themselves.

Additional information on Swept and Pierce are found in SolidWorks Help Topics. Keywords: Swept, (overview, simple sweeps) and Pierce (relations).

SWITCH Part - Lofted Base Feature

The SWITCH is a purchased part. The SWITCH is a complex assembly. Create the outside casing of the SWITCH as a simplified part.

Create the SWITCH with the Lofted Base  feature.

The orientation of the SWITCH is based on the position in the assembly. The SWITCH is comprised of three cross section profiles. Sketch each profile on a different plane.

The first plane is the Top Plane. Create two reference planes parallel to the Top Plane.

Sketch one profile on each plane. The design intent of the sketch is to reduce the number of dimensions.

Utilize symmetry, construction geometry and Geometric relations to control three sketches with one dimension.

Insert a Lofted feature. Select the profiles to create the Loft feature.

Insert the Dome feature to the top face of the Loft and modify the Loft Base feature. Modify the dimensions to complete the SWITCH.

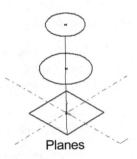

Planes

🔅 You can specify an elliptical dome for cylindrical or conical models or a continuous dome for polygonal models.

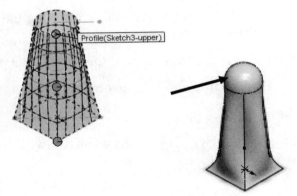

🔅 Tangent Edges and Origin are displayed for educational purposes.

🔅 When you create a new part or assembly, the three default Planes (Front, Right and Top) are aligned with specific views. The Plane you select for the Base sketch determines the orientation of the part or assembly.

Activity: Switch Part-Loft Base Feature

Create a new part.

44) Click **New** ⬜ from the Menu bar.

45) Click the **MY-TEMPLATES** tab.

46) Double-click **PART-IN-ANSI**, [**PART-MM-ISO**].

47) Click **Save**.

48) Select **PROJECTS** for the Save in folder.

49) Enter **SWITCH** for File name.

50) Enter **BUTTON STYLE** for Description.

51) Click **Save**. The SWITCH FeatureManager is displayed.

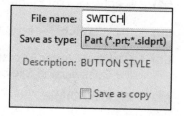

💡 If you upgraded from an over version; your Planes maybe displayed by default. Click View, Planes from the Menu bar to display all reference planes. Hide unwanted planes in the FeatureManager.

Display the Top Plane.

52) Right-click **Top Plane** from the FeatureManager.

53) Click **Show**.

54) Click **Isometric view** 🔲 from the Heads-up View toolbar.

Insert two reference planes.

55) Hold the **Ctrl** key down.

56) Click and drag the **Top Plane** upward. The Plane PropertyManager is displayed.

57) Release the **mouse button**.

58) Release the **Ctrl** key.

59) Enter **.500**in, [**12.7**] for Offset Distance.

60) Enter **2** for # of Planes to Create.

61) Click **OK** ✔ from the Plane PropertyManager. Plane1 and Plane2 is displayed in the FeatureManager.

62) Click **Front view** 🔲 to display the Plane1 and Plane2 offset from the Top plane.

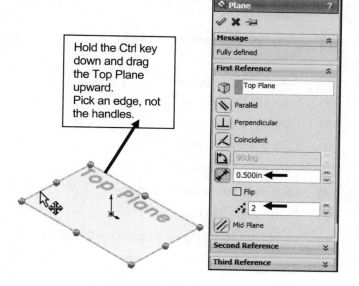

Hold the Ctrl key down and drag the Top Plane upward. Pick an edge, not the handles.

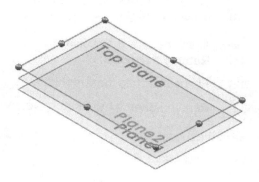

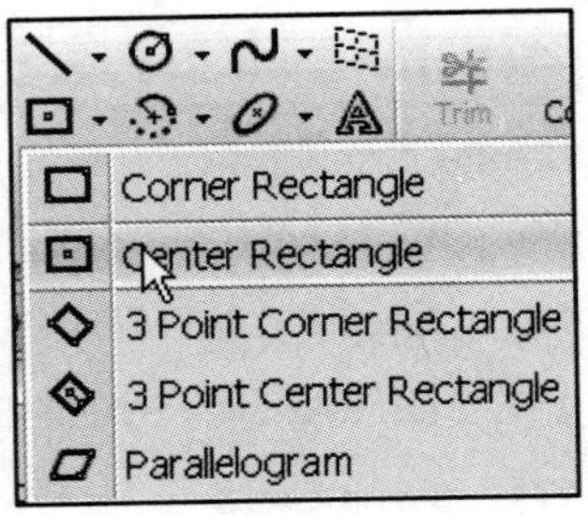

63) Click **Isometric view** from the Heads-up View toolbar.

Insert Sketch1. Sketch1 is a square on the Top Plane centered about the Origin.

64) Right-click **Top Plane** from the FeatureManager.

65) Click **Sketch** from the Context toolbar.

66) Click the **Center Rectangle** tool from the Consolidated Sketch toolbar.

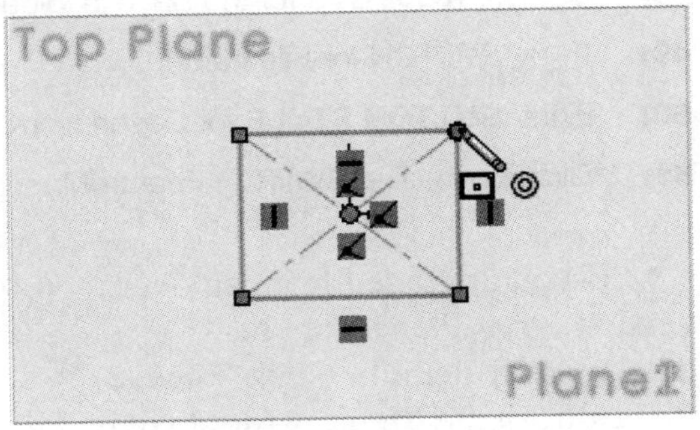

67) Click the **Origin**.

68) Click a **point** in the upper right hand of the window as illustrated. The Center Rectangle Sketch tool automatically inserts a midpoint relation about the Origin and an equal relation between the two horizontal and two vertical lines.

69) Click **OK** from the Rectangle PropertyManager.

Add an Equal relation between the four edges.

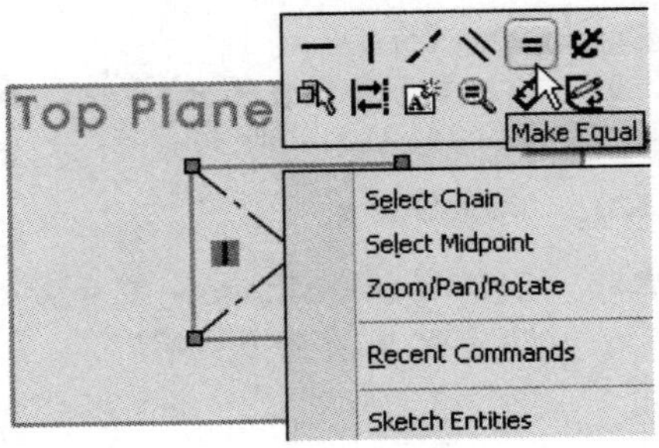

70) Click the **left vertical line**.

71) Hold the **Ctrl** key down.

72) Click the **top horizontal line**.

73) Click the **right vertical line**.

74) Click the **bottom horizontal line**.

75) Release the **Ctrl** key.

76) Right-click **Make Equal** = from the Context toolbar.

Add a dimension.

77) Click the **Smart Dimension** Sketch tool.

78) Click the **top horizontal line**.

79) Click a **position** above the profile.

80) Enter **.500**in, **[12.7]**.

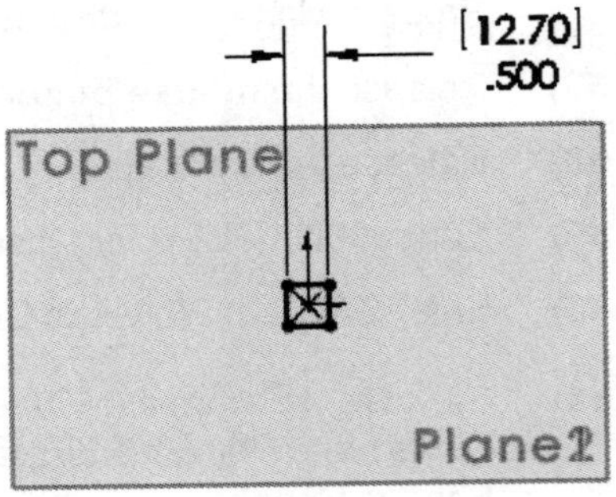

Close, fit and rename the sketch.

81) **Rebuild** the model.

82) **Fit** the sketch to the Graphics area.

83) Rename **Sketch1** to **Sketch1-lower**.

Save the model.

84) Click **Save**.

Insert Sketch2 on Plane1.

85) Click **Top view** from the Heads-up View toolbar.

86) Right-click **Plane1** from the FeatureManager

87) Click **Sketch** from the Context toolbar.

88) Click the **Circle** Sketch tool. The Circle PropertyManager is displayed.

89) Sketch a **Circle** centered at the Origin as illustrated.

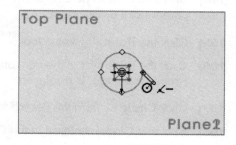

Add a Tangent relation.

90) Right-click **Select** to deselect the Circle Sketch tool.

91) Click the **circumference** of the circle.

92) Hold the **Ctrl** key down.

93) Click the **top horizontal Sketch1-lower line**. The Properties PropertyManager is displayed.

94) Release the **Ctrl** key.

95) Right-click **Make Tangent** from the Context toolbar.

96) Click **OK** from the Properties PropertyManager.

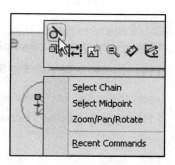

Close and rename the sketch.

97) Right-click **Exit Sketch**.

98) Rename **Sketch2** to **Sketch2-middle**.

99) Click **Isometric view** from the Heads-up View toolbar. View the results.

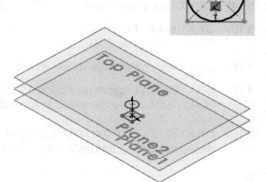

Insert Sketch3 on Plane2.

100) Click **Top view** from the Heads-up View toolbar.

101) Right-click **Plane2** from the FeatureManager.

102) Click **Sketch** from the Context toolbar. The Sketch toolbar is displayed.

103) Click the **Centerline** Sketch tool.

104) Sketch a **centerline** coincident with the Origin and the upper right corner point as illustrated.

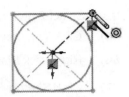

105) Click the **Point** ✳ Sketch tool.

106) Click the **midpoint** of the right diagonal centerline. The Point PropertyManager is displayed.

107) Click **Circle** ⊘ from the Sketch toolbar.

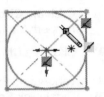

108) Sketch a **Circle** centered at the Origin to the midpoint of the diagonal centerline.

Close and rename the sketch.
109) Right-click **Exit Sketch**.

110) Rename **Sketch3** to **Sketch3-upper**.

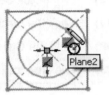

Insert the Lofted Base feature.
111) Click the **Lofted Boss/Base** ⬭ feature tool. The Loft PropertyManager is displayed.

112) Click **Isometric view** 🔲 from the Heads-up View toolbar.

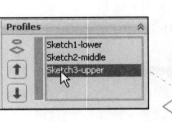

113) Right-click in the **Profiles box**.

114) Click **Clear Selections**.

115) Click the **front corner** of Sketch1-lower as illustrated.

116) Click **Sketch2-middle** as illustrated.

117) Click **Sketch3-upper** as illustrated. The selected sketch entities are displayed in the Profiles box.

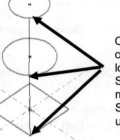

Click the front of Sketch1-lower, Sketch2-middle and Sketch3-upper.

118) Click **OK** ✔ from the Loft PropertyManager. Loft1 is displayed in the FeatureManager.

Rename the feature.
119) Rename **Loft1** to **Base Loft**.

Hide the planes.
120) Click **View**, un-check **Planes** from the Menu bar.

Save the part.
121) Click **Save** 💾.

The system displays a preview curve and loft as you select the profiles. Use the Up button and Down button in the Loft PropertyManager to rearrange the order of the profiles.

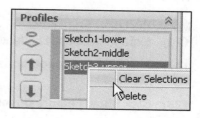

Redefine incorrect selections efficiently. Right-click in the Graphics window, click Clear Selections to remove selected profiles. Select the correct profiles.

SWITCH Part-Dome Feature

Insert the Dome feature on the top face of the Lofted Base feature. The Dome feature forms the top surface of the SWITCH. Note: You can specify an elliptical dome for cylindrical or conical models and a continuous dome for polygonal models. A continuous dome's shape slopes upwards, evenly on all sides.

Activity: SWITCH Part-Dome Feature

Insert the Dome feature.
122) Click the **top face** of the Base Loft feature in the Graphics window.

123) Click the **Dome** feature tool. The Dome PropertyManager is displayed. Face<1> is displayed in the Faces to Dome box.

Enter distance.
124) Enter **.20**in, **[5.08]** for Distance. View the dome feature in the Graphics window.

125) Click **OK** from the Dome PropertyManager.

Experiment with the Dome feature to display different results. Click Insert, Features, Freeform to view the Freeform PropertyManager. As an exercise, replace the Dome feature with a Freeform feature.

In the next section, modify the offset distance between the Top Plane and Plane1.

Modify the Loft Base feature.

126) **Expand** the Base Loft feature.

127) Right-click on **Annotations** in the FeatureManager.

128) Click **Show Feature Dimensions**.

129) Click on the Plane1 offset dimension, **.500**in, [**12.700**].

130) Enter **.125**in, [**3.180**].

131) Click **OK** ✓ from the Dimension PropertyManager.

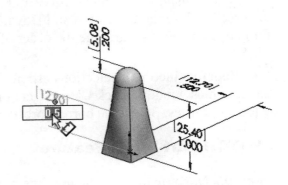

Hide Feature dimensions.

132) Right-click on **Annotations** in the FeatureManager.

133) Un-check **Show Feature Dimensions**.

Display Feature Statistics.

134) Click **Statistics** 🔾 from the Evaluate tab in the CommandManager. View the results.

View the SWITCH feature statistics.

135) Click **Close** from the Feature Statistics dialog box.

136) Click **Save**.

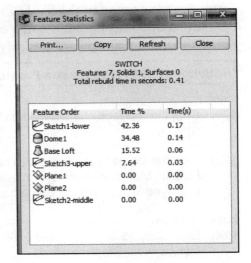

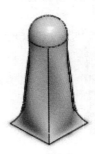

The Dome feature created the top for the SWITCH. The Feature Statistics report displays the rebuild time for the Dome feature and the other SWITCH features. As feature geometry becomes more complex, the rebuild time increases.

 Review of the SWITCH Part

The SWITCH utilized the Lofted Base and Dome feature. The Lofted Base feature required three planes: Sketch1-lower, Sketch2-middle and Sketch3-upper. A profile was sketched on each plane. The three profiles were combined to create the Lofted Base feature.

The Dome feature created the final feature for the SWITCH.

The SWITCH utilized a simple Lofted Base feature. Lofts become more complex with additional Guide Curves. Complex Lofts can contain hundreds of profiles.

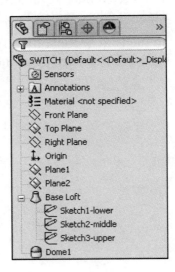

Four Major Categories of Solid Features

The LENSCAP and HOUSING combine the four major categories of solid features:

- Extrude: Requires one profile

- Revolve: Requires one profile and axis of revolution

- Swept: Requires one profile and one path sketched on different planes

- Lofted: Requires two or more profiles sketched on different planes

Identify the simple features of the LENSCAP and HOUSING. Extrude and Revolve are simple features. Only a single sketch profile is required. Swept and Loft are more complex features. Two or more sketches are required.

Example: The O-RING was created as a Swept.

Could the O-RING utilize an Extruded feature?

Answer: No. Extruding a circular profile produces a cylinder.

Can the O-RING utilize a Revolved feature?
Answer: Yes. Revolving a circular profile about a centerline creates the O-RING.

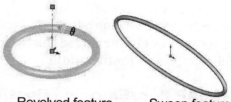

Revolved feature   Sweep feature

A Swept feature is required if the O-RING contained a non-circular path. Example: A Revolved feature does not work with an elliptical path or a more complex curve as in a paper clip. Combine the four major features and additional features to create the LENSCAP and HOUSING.

LENSCAP Part

The LENSCAP is a plastic part used to position the LENS to the HOUSING. The LENSCAP utilizes an Extruded Boss/Base, Extruded Cut, Extruded Thin Cut, Shell, Revolved Cut and Swept features.

The design intent for the LENSCAP requires that the Draft Angle be incorporated into the Extruded Boss/Base and Revolved Cut feature. Create the Revolved Cut feature by referencing the Extrude Base feature geometry. If the Draft angle changes, the Revolved Cut also changes.

Insert an Extruded Boss/Base ⬚ feature with a circular profile on the Front Plane. Use a Draft option in the Boss-Extrude PropertyManager.

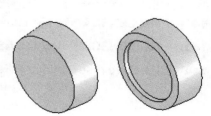

Enter 5° for Draft angle.

Insert an Extruded Cut 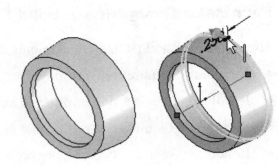 feature. The Extruded Cut feature should be equal to the diameter of the LENS Revolved Base feature.

Insert a Shell  feature. Use the Shell feature for a constant wall thickness.

Insert a Revolved Cut feature on the back face. Sketch a single line on the Silhouette edge of the Extruded Base. Utilize the Thin Feature option in the Cut-Revolve PropertyManager.

Utilize a Swept feature for the thread. Insert a new reference plane for the start of the thread. Insert a Helical Curve for the path. Sketch a trapezoid for the profile.

LENSCAP Part-Extruded Boss/Base, Extruded Cut and Shell Features

Create the LENSCAP. The first feature is a Boss-Extrude feature. Select the Front Plane for the Sketch plane. Sketch a circle centered at the Origin for the profile. Utilize a Draft angle of 5°.

Create an Extruded Cut feature on the front face of the Base feature. The diameter of the Extruded Cut equals the diameter of the Revolved Base feature of the LENS. The Shell feature removes the front and back face from the solid LENSCAP.

Activity: LENSCAP Part-Extruded Base, Extruded Cut and Shell Features

Create a new part.

137) Click **New** ⬜ from the Menu bar.

138) Click the **MY-TEMPLATES** tab.

139) Double-click **PART-IN-ANSI**, [**PART-MM-ISO**].

140) Click **Save**.

141) Select **PROJECTS** for the Save in folder.

142) Enter **LENSCAP** for File name.

143) Enter **LENSCAP for 6V FLASHLIGHT** for Description.

144) Click **Save**. The LENSCAP FeatureManager is displayed.

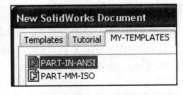

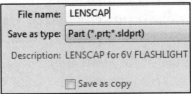

Create the sketch for the Extruded Base feature.

145) Right-click **Front Plane** from the FeatureManager.

146) Click **Sketch** from the Context toolbar.

147) Click the **Circle** Sketch tool. The Circle PropertyManager is displayed.

148) Sketch a **circle** centered at the Origin .

Origin

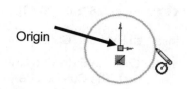

Add a dimension.

149) Click the **Smart Dimension** Sketch tool.

150) Click the **circumference** of the circle.

151) Click a **position** off the profile. Enter **4.900**in, [**124.46**].

[124.46]
⌀4.900

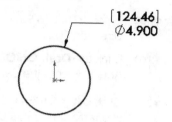

Insert an Extruded Boss/Base feature.

152) Click the **Extruded Boss/Base** feature tool. The Boss-Extrude PropertyManager is displayed. Blind is the default End Condition in Direction 1.

153) Click the **Reverse Direction** box.

154) Enter **1.725**in, [**43.82**] for Depth in Direction 1. Click the **Draft On/Off** button.

155) Enter **5**deg for Angle.

156) Click the **Draft outward** box.

[124.46]
⌀4.900

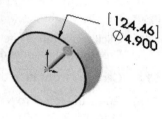

157) Click **OK** from the Boss-Extrude PropertyManager. Boss-Extrude1 is displayed in the FeatureManager.

158) Rename **Boss-Extrude1** to **Base Extrude**.

159) Click **Save**.

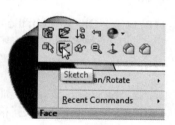

Create the sketch for the Extruded Cut feature.
160) Right-click the **front face** for the Sketch plane.

161) Click **Sketch** from the Context toolbar. The Sketch toolbar is displayed.

162) Click the **Circle** Sketch tool. The Circle PropertyManager is displayed.

163) Sketch a **circle** centered at the Origin .

Add a dimension.

164) Click the **Smart Dimension** Sketch tool.

165) Click the **circumference** of the circle.

166) Click a **position** off the profile.

[98.43]
⌀3.875

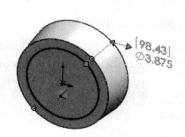

167) Enter **3.875**in, **[98.43]**.

Insert an Extruded Cut feature.

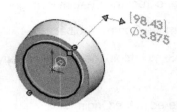

168) Click the **Extruded Cut** 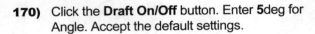 feature tool. The Cut-Extrude PropertyManager is displayed. Blind is the default End Condition.

169) Enter **.275**in, **[6.99]** for Depth in Direction 1.

170) Click the **Draft On/Off** button. Enter **5**deg for Angle. Accept the default settings.

171) Click **OK** ✔ from the Cut-Extrude PropertyManager. Cut-Extrude1 is displayed in the FeatureManager.

172) Rename **Cut-Extrude1** to **Front-Cut**.

Insert the Shell feature.

173) Click the **Shell**  feature tool. The Shell1 PropertyManager is displayed.

174) Click the **front face** of the Front-Cut as illustrated.

175) Press the **left arrow** approximately 8 times to view the back face.

176) Click the **back face** of the Base Extrude.

177) Enter **.150**in, **[3.81]** for Thickness.

178) Click **OK** ✔ from the Shell1 PropertyManager. Shell1 is displayed in the FeatureManager.

179) Click **Isometric view** from the Heads-up View toolbar.

Display the inside of the Shell.

180) Click **Right view** from the Heads-up View toolbar.

181) Click **Hidden Lines Visible** from the Heads-up View toolbar.

182) Click **Save**.

💡 Use the inside gap created by the Shell feature to seat the O-RING in the assembly.

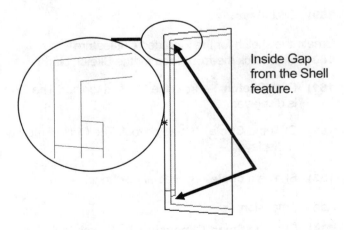

Inside Gap from the Shell feature.

LENSCAP Part-Revolved Thin Cut Feature

The Revolved Thin Cut  feature removes material by rotating a sketched profile around a centerline.

The Right Plane is the Sketch plane. The design intent requires that the Revolved Cut maintains the same Draft angle as the Extruded Base feature.

Utilize the Convert Entities Sketch tool to create the profile. Small thin cuts are utilized in plastic parts. Utilize the Revolved Thin Cut feature for cylindrical geometry in the next activity.

Utilize a Swept Cut for non-cylindrical geometry. The semi-circular Swept Cut profile is explored in the chapter exercises.

Sweep Cut Example

Activity: LENSCAP Part-Revolved Thin Cut Feature

Create the sketch.

183) Right-click **Right Plane** from the FeatureManager. This is your Sketch plane.

184) Click **Sketch** ✎ from the Context toolbar. The Sketch toolbar is displayed.

Sketch a centerline.

185) Click the **Centerline** ┊ Sketch tool. The Insert Line PropertyManager is displayed.

186) Sketch a **horizontal centerline** through the Origin as illustrated.

Create the profile.

187) Right-click **Select** to deselect the Centerline Sketch tool.

188) Click the **top silhouette outside edge** of Base Extrude as illustrated.

189) Click the **Convert Entities** 🗗 Sketch tool.

190) Click **OK** ✔ from the Convert Entities PropertyManager.

191) Click and drag the **left endpoint 2/3** towards the right endpoint.

192) Release the **mouse button**.

Add a dimension.

193) Click the **Smart Dimension** ⬨ Sketch tool.

194) Click the **line**. The aligned dimension arrows are parallel to the profile line.

195) Drag the **text upward** and to the right.

196) Enter **.250**in, **[6.35]**.

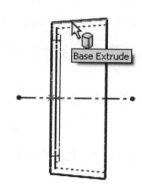

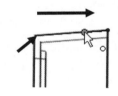

[6.35]
.250

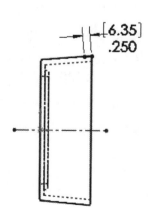

Insert a Revolved Cut feature.

197) Click **Revolved Cut** from the Features toolbar. Do not close the Sketch. The warning message states; "The sketch is currently open."

198) Click **No**. The Cut-Revolve PropertyManager is displayed.

199) Click the **Reverse Direction** box in the Thin Feature box. The arrow points counterclockwise.

200) Enter **.050**in, **[1.27]** for Direction 1 Thickness.

201) Click **OK** from the Cut-Revolve PropertyManager. Cut-Revolve-Thin1 is

displayed in the FeatureManager.

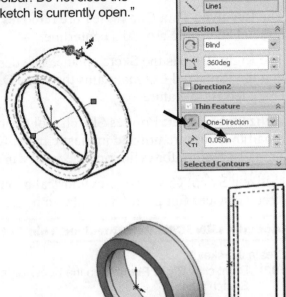

Display the Revolved Thin Cut feature.
202) **Rotate** the part to view the back face.

203) Click **Isometric view**.

204) Click **Shaded With Edges**.

205) Rename **Cut-Revolve-Thin1** to **BackCut**.

206) Click **Save**.

LENSCAP Part-Thread, Swept Feature, and Helix/Spiral Curve

Utilize the Swept feature to create the required threads. The thread requires a spiral path. This path is called the ThreadPath. The thread requires a Sketched profile. This cross section profile is called the ThreadSection.

The plastic thread on the LENSCAP requires a smooth lead in. The thread is not flush with the back face. Use an Offset plane to start the thread. There are numerous steps required to create a thread:

- Create a new plane for the start of the thread.

- Create the Thread path. Utilize Convert Entities and Insert, Curve, Helix/Spiral.

- Create a large thread cross section profile for improve visibility.

- Insert the Swept feature.

- Reduce the size of the thread cross section.

Obtain the Helix and Spiral tool from the Features tab, Curves Consolidated toolbar.

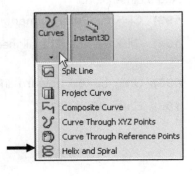

Activity: LENSCAP Part-Thread, Swept Feature, and Helix/Spiral Curve

Create the offset plane.

207) **Rotate** ⟳ and **Zoom to Area** 🔍 on the back face of the LENSCAP.

208) Click the **narrow back face** of the Base Extrude feature. Note the mouse feedback 🔲 icon.

209) Click **Insert**, **Reference Geometry**, **Plane** from the Menu bar. The Plane PropertyManager is displayed.

210) Enter .450in, [11.43] for Distance.

211) Click the **Flip** box.

212) Click **OK** ✔ from the Plane PropertyManager. Plane1 is displayed in the FeatureManager.

213) Rename **Plane1** to **ThreadPlane**.

Display the Isometric view with Hidden Lines Removed.

214) Click **Isometric view** 🟦.

215) Click **Hidden Lines Removed** 🟦.

216) Click **Save**.

Utilize the Convent Entities Sketch tool to extract the back circular edge of the LENSCAP to the ThreadPlane.

Create the Thread path.

217) Right-click **ThreadPlane** from the FeatureManager.

218) Click **Sketch** ✏ from the Context toolbar.

219) Click the **back inside circular edge** of the Shell as illustrated.

220) Click the **Convert Entities** 🔲 Sketch tool.

221) Click **OK** ✔ from the Convert Entities PropertyManager.

222) Click **Top view** 🔲. The circular edge is displayed on the ThreadPlane.

223) Click **Hidden Lines Visible** 🔲 from the Heads-up View toolbar. View the results.

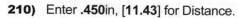

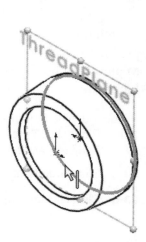

💡 Access the Plane tool from the Consolidated Reference Geometry toolbar.

Insert the Helix/Spiral curve path.

224) Click **Insert, Curve, Helix/Spiral** from the Menu bar. The Helix/Spiral PropertyManager is displayed.

225) Enter **.250**in, **[6.35]** for Pitch.

226) Check the **Reverse direction** box.

227) Enter **2.5** for Revolutions.

228) Enter **0**deg for Starting angle. The Helix start point and end point are Coincident with the Top Plane.

229) Click the **Clockwise** box.

230) Click the **Taper Helix** box.

231) Enter **5**deg for Angle.

232) Uncheck the **Taper outward** box.

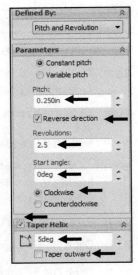

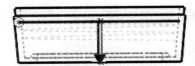

233) Click **OK** ✔ from the Helix/Spiral PropertyManager.

234) Rename **Helix/Spiral1** to **ThreadPath**.

235) Click **Save** 🖫.

The Helix tapers with the inside wall of the LENSCAP. Position the Helix within the wall thickness to prevent errors in the Swept.

Sketch the profile on the Top plane. Position the profile to the Top right of the LENSCAP in order to pierce to the ThreadPath in the correct location.

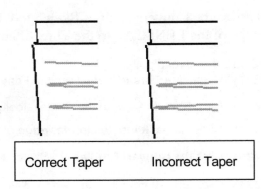

Correct Taper   Incorrect Taper

If required, hide the ThreadPlane.

236) Right-click **ThreadPlane** from the FeatureManager.

237) Click **Hide** from the Context toolbar.

238) Click **Hidden Lines Removed** ▱ from the Heads-up View toolbar.

Sketch to the Top right ➝

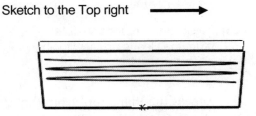

Select the Plane for the Thread.

239) Right-click **Top Plane** from the FeatureManager.

Sketch the profile.

240) Click **Sketch** from the Context toolbar.

241) Click **Top view** from the Heads-up View toolbar.

242) Click the **Centerline** Sketch tool.

243) Create a short **vertical centerline** off to the upper top area of the ThreadPath feature.

244) Create a second **centerline** horizontal from the Midpoint to the left of the vertical line.

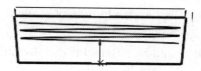

245) Create a third centerline coincident with the left horizontal endpoint. Drag the **centerline upward** until it is approximately the same size as the right vertical line.

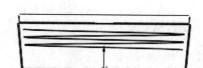

246) Create a fourth **centerline** coincident with the left horizontal endpoint. Drag the **centerline** downward until it is approximately the same size as the left vertical line as illustrated.

Add an Equal relation.

247) Right-click **Select** to deselect the Centerline Sketch tool.

248) Click the **right vertical centerline**.

249) Hold the **Ctrl** key down.

250) Click the **two left vertical centerlines**.

251) Release the **Ctrl** key. The selected sketch entities are displayed in the Selected Entities box.

252) Click **Equal** =

253) Click **OK** from the Properties PropertyManager.

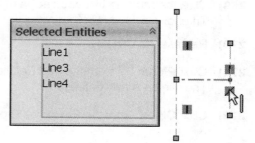

Utilize centerlines and construction geometry with geometric relations to maintain relationships with minimal dimensions.

Check **View, Sketch Relations** from the Menu bar to show/hide sketch relation symbols.

Add a dimension.

254) Click the **Smart Dimension** ✐ Sketch tool.

255) Click the two **left vertical endpoints**.

256) Click a **position** to the left.

257) Enter **.500**in, **[12.7]**.

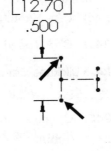

Sketch the profile. The profile is a trapezoid.

258) Click the **Line** ╲ Sketch tool.

259) Click the **endpoints** of the vertical centerlines to create the trapezoid as illustrated.

260) Right-click **Select** to deselect the Line Sketch tool.

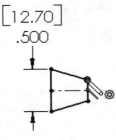

Add an Equal relation.

261) Click the **left vertical line**.

262) Hold the **Ctrl** key down.

263) Click the **top** and **bottom lines** of the trapezoid.

264) Release the **Ctrl** key.

265) Click **Equal** ═ .

266) Click **OK** ✔ from the Properties PropertyManager.

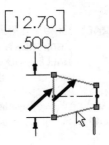

Click and drag the sketch to a position above the top right corner of the LENSCAP.

Add a Pierce relation.

267) Click the **left midpoint** of the trapezoid.

268) Hold the **Ctrl** key down.

269) Click the **starting left back edge** of the ThreadPath.

270) Release the **Ctrl** key.

271) Click **Pierce** ⬝ from the Add Relations box. The sketch is fully defined.

272) Click **OK** ✔ from the Properties PropertyManager.

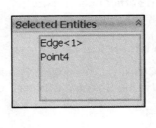

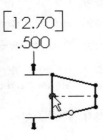

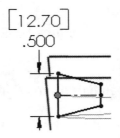

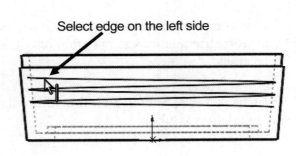

Select edge on the left side

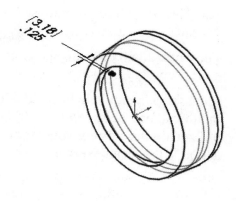

Display the sketch in an Isometric view.

273) Click **Isometric view** from the Heads-up View toolbar.

Modify the dimension.

274) Double-click the **.500** dimension text.

275) Enter **.125**in, **[3.18]**.

Close the sketch.

276) **Rebuild** the model.

277) Rename **Sketch5** to **ThreadSection**.

278) Click **Save**.

Insert the Swept feature.

279) Click the **Swept Boss/Base** feature tool. The Sweep PropertyManager is displayed. If required, click **ThreadSection** for the Profile from the fly-out FeatureManager.

280) Click inside the **Path** box.

281) Click **ThreadPath** from the fly-out FeatureManager. ThreadPath is displayed in the Path box.

282) Click **OK** from the Sweep PropertyManager. Sweep1 is displayed in the FeatureManager.

283) Rename **Sweep1** to **Thread**.

284) Click **Shaded With Edges** from the Heads-up View toolbar.

285) Click **Save**.

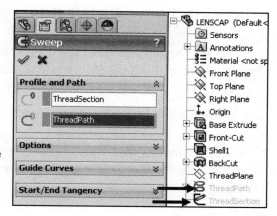

Swept geometry cannot intersect itself. If the ThreadSection geometry intersects itself, the cross section is too large. Reduce the cross section size and recreate the Swept feature.

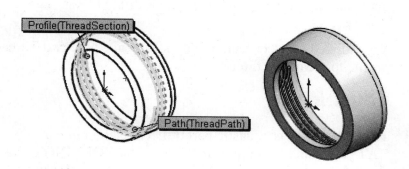

The Thread feature is composed of the following: ThreadSection and ThreadPath.

The ThreadPath contains the circular sketch and the helical curve.

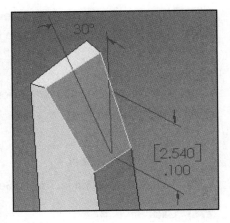

Most threads require a beveled edge or smooth edge for the thread part start point. A 30° Chamfer feature can be utilized on the starting edge of the trapezoid face. This action is left as an exercise.

Create continuous Swept features in a single step. Pierce the cross section profile at the start of the swept path for a continuous Swept feature.

Unsuppress the Pattern feature to resolve both the Pattern feature and the seed feature at the same time.

The LENSCAP is complete. Review the LENSCAP before moving onto the last part of the FLASHLIGHT.

Additional information on Extruded Base/Boss, Extruded Cut, Swept, Helix/Spiral, Circular Pattern, and Reference Planes are found in SolidWorks Help Topics.

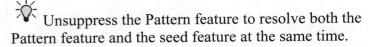   Review of the LENSCAP Part

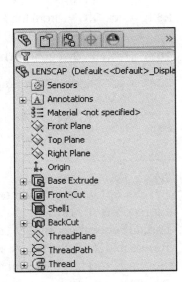

The LENSCAP utilized the Extruded Base feature with the Draft Angle option. The Extruded Cut feature created an opening for the LENS. You utilized the Shell feature with constant wall thickness to remove the front and back faces.

The Revolved Thin Cut feature created the back cut with a single line. The line utilized the Convert Entities tool to maintain the same draft angle as the Extruded Boss/Base feature.

You utilized a Swept feature with a Helical Curve and Thread profile to create the thread.

HOUSING Part

The HOUSING is a plastic part utilized to contain the BATTERY and to support the LENS. The HOUSING utilizes an Extruded Boss/Base, Lofted Boss, Extruded Cut, Draft, Swept, Rib, Mirror and Linear Pattern features.

Insert an Extruded Boss/Base (Boss-Extrude1 ![icon] feature centered at the Origin.

Insert a Lofted Boss feature ![icon]. The first profile is the converted circular edge of the Extruded Base. The second profile is a sketched on the BatteryLoftPlane.

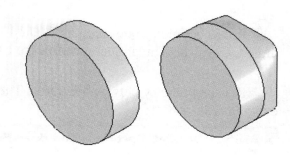

Insert the second Extruded Boss/Base (Boss-Extrude2) ![icon] feature. The sketch is a converted edge from the Loft Boss. The depth is determined from the height of the BATTERY.

Insert a Shell ![icon] feature to create a thin walled part.

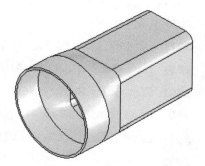

Insert the third Extruded Boss (Boss-Extrude3) ![icon] feature. Create a solid circular ring on the back circular face of the Boss-Extrude1 feature. Insert the Draft feature to add a draft angle to the circular face of the HOUSING. The design intent for the Boss-Extrude1 feature requires you to maintain the same LENSCAP draft angle.

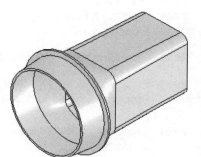

Insert a Swept ![icon] feature for the Thread. Insert a Swept feature for the Handle. Reuse the Thread profile from the LENSCAP part. Insert an Extruded Cut to create the hole for the SWITCH.

Insert the Rib ![icon] feature on the back face of the HOUSING. Insert a Linear Pattern ![icon] feature to create a row of Ribs. Insert a Rib ![icon] feature along the bottom of the HOUSING. Utilize the Mirror ![icon] feature to create the second Rib.

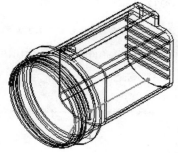

 Reuse geometry between parts. The LENSCAP thread is the same as the HOUSING thread. Copy the ThreadSection from the LENSCAP to the HOUSING.

 Reuse geometry between features. The Linear Pattern and Mirror Pattern utilized existing features.

Activity: HOUSING Part-Extruded Base Feature

Create the new part.

286) Click **New** ⬜ from the Menu bar.

287) Click the **MY-TEMPLATES** tab.

288) Double-click **PART-IN-ANSI**, **[PART-MM-ISO]**.

289) Click **Save**. Select **PROJECTS** for the Save in folder.

290) Enter **HOUSING** for File name.

291) Enter **HOUSING FOR 6VOLT FLASHLIGHT** for Description.

292) Click **Save**. The HOUSING FeatureManager is displayed.

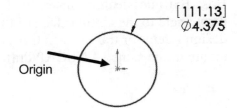

Create the sketch.

293) Right-click **Front Plane** from the FeatureManager. This is your Sketch plane.

294) Click **Sketch** ✏ from the Context toolbar.

295) Click the **Circle** ⊘ Sketch tool. The Circle PropertyManager is displayed.

296) Sketch a circle centered at the **Origin** ↳ as illustrated.

[111.13]
⌀4.375

Origin

Add a dimension.

297) Click the **Smart Dimension** ⬙ Sketch tool.

298) Click the **circumference**. Enter **4.375**in, **[111.13]**.

Insert an Extruded Boss/Base feature.

299) Click the **Extruded Boss/Base** 🗈 feature tool. The Boss-Extrude PropertyManager is displayed.

300) Enter **1.300**, **[33.02]** for Depth in Direction 1. Accept the default settings.

301) Click **OK** ✔ from the Boss-Extrude PropertyManager.

302) Click **Isometric view** 🔲. Note the location of the Origin.

303) Rename **Boss-Extrude1** to **Base Extrude**.

304) Click **Save**.

HOUSING Part-Lofted Boss Feature

The Lofted Boss feature is composed of two profiles. The first sketch is named Sketch-Circle. The second sketch is named Sketch-Square.

Create the first profile from the back face of the Extruded feature. Utilize the Convert Entities sketch tool to extract the circular geometry to the back face.

Create the second profile on an Offset Plane. The FLASHLIGHT components must remain aligned to a common centerline. Insert dimensions that reference the Origin and build symmetry into the sketch. Utilize the Mirror Entities Sketch tool.

Activity: HOUSING Part-Lofted Boss Feature

Create the first profile.

305) Right-click the **back face** of the Base Extrude feature. This is your Sketch plane.

306) Click **Sketch** from the Context toolbar. The Sketch toolbar is displayed.

307) Click the **Convert Entities** Sketch tool to extract the face to the Sketch plane.

308) Click **OK** from the Convert Entities PropertyManager.

Close and rename the sketch.

309) Right-click **Exit Sketch**.

310) Rename **Sketch2** to **SketchCircle**.

Create an offset plane.

311) Click the **back face** of the Base Extrude feature.

312) Click **Plane** from the Consolidated Reference Geometry Features toolbar. The Plane PropertyManager is displayed.

313) Enter **1.300**in, [**33.02**] for Distance.

314) Click **Top view** from the Heads-up View toolbar to verify the Plane position.

315) Click **OK** from the Plane PropertyManager. Plane1 is displayed in the FeatureManager.

316) Rename **Plane1** to **BatteryLoftPlane**.

317) **Rebuild** the model.

318) Click **Save**.

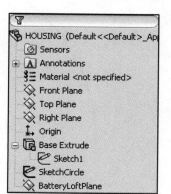

Create the second profile.

319) Right-click **BatteryLoftPlane** in the FeatureManager.

320) Click **Sketch** 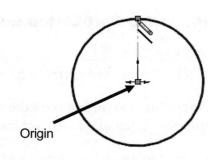 from the Context toolbar. The Sketch toolbar is displayed.

321) Click **Back view** from the Heads-up View toolbar.

322) Click the **circumference** of the circle.

323) Click the **Convert Entities** Sketch tool.

324) Click **OK** from the Convert Entities PropertyManager.

325) Click the **Centerline** Sketch tool.

326) Sketch a **vertical centerline** coincident to the Origin and to the top edge of the circle as illustrated.

327) Click the **Line** Sketch tool.

328) Sketch a **horizontal line** to the right side of the centerline as illustrated.

329) Sketch a **vertical line** down to the circumference.

330) Click the **Sketch Fillet** Sketch tool. The Sketch Fillet PropertyManager is displayed.

331) Click the **horizontal line** in the Graphics window.

332) Click the **vertical line** in the Graphics window.

333) Enter **.1**in [**2.54**] for Radius.

334) Click **OK** from the Sketch Fillet PropertyManager. View the Sketch Fillet in the Graphics window.

Origin

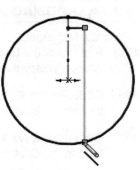

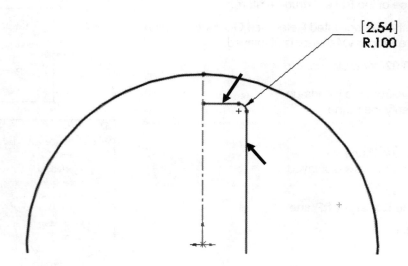

[2.54]
R.100

Mirror the profile.

335) Click the **Mirror Entities** ⚠ Sketch tool. The Mirror PropertyManager is displayed.

336) Click the **horizontal line**, **fillet**, and **vertical line**. The selected entities are displayed in the Entities to mirror box.

337) Click inside the **Mirror about** box.

338) Click the **centerline** from the Graphics window.

339) Click **OK** ✅ from the Mirror PropertyManager.

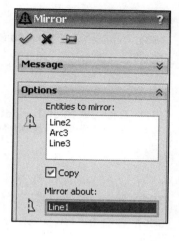

Trim unwanted geometry.

340) Click the **Trim Entities** ✄ Sketch tool. The Trim PropertyManager is displayed.

341) Click **PowerTrim** ⊞ from the Options box.

342) Click a **position** to the far right of the circle.

343) Drag the **mouse pointer** to intersect the circle.

344) Perform the same **actions** on the left side of the circle.

345) Click **OK** ✅ from the Trim PropertyManager.

Add dimensions.

346) Click the **Smart Dimension** ✏ Sketch tool.

Create the horizontal dimension.
347) Click the **left vertical** line.

348) Click the **right vertical** line.

349) Click a **position** above the profile.

350) Enter **3.100**in, [**78.74**]. View the results.

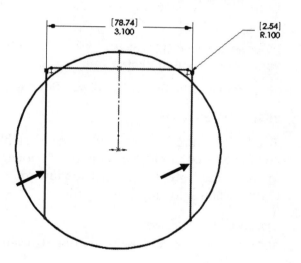

Create the vertical dimension.

351) Click the **Origin** ⌊.

352) Click the **top horizontal** line.

353) Click a **position** to the right of the profile.

354) Enter **1.600**in, [**40.64**].

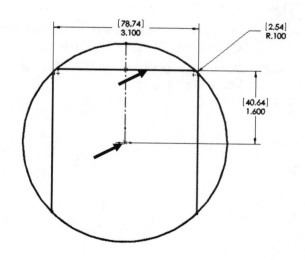

Modify the fillet dimension.
355) Double-click the **.100** fillet dimension.

356) Enter **.500**in, [**12.7**].

Remove all sharp edges.

357) Click the **Sketch Fillet**  Sketch tool. The Sketch Fillet PropertyManager is displayed.

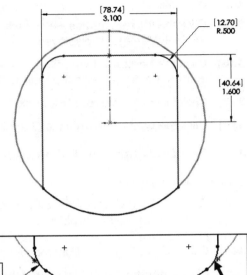

358) Enter **.500**in, [**12.7**] for Radius.

359) Click the **lower left corner point**.

360) Click the **lower right corner point**.

361) Click **OK** ✅ from the Sketch Fillet PropertyManager.

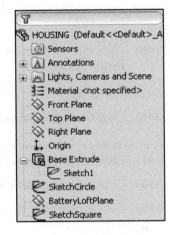

Close and rename the sketch.
362) Right-click **Exit Sketch**.

363) Rename the **Sketch2** to **SketchSquare**.

364) Click **Save** 💾.

The Loft feature is composed of the SketchSquare and the SketchCircle. Select two individual profiles to create the Loft. The Isometric view provides clarity when selecting Loft profiles.

Display an Isometric view.
365) Click **Isometric view** 🧊 from the Heads-up View toolbar.

Insert a Lofted Boss feature.

366) Click the **Lofted Boss/Base** ♌ feature tool. The Loft PropertyManager is displayed.

367) Click the **upper right side** of the SketchCircle as illustrated.

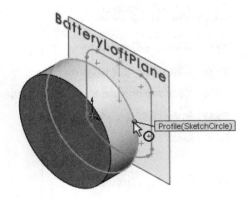

368) Click the **upper right side** of the SketchCircle as illustrated. The selected entities are displayed in the Profiles box.

369) Click **OK** ✔ from the Loft PropertyManager.

370) Rename **Loft1** to **Boss-Loft1**.

371) Click **Save**.

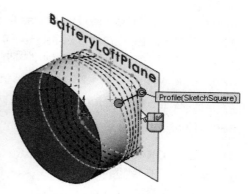

💡 Organize the FeatureManager to locate Loft profiles and planes. Insert the Loft reference planes directly before the Loft feature. Rename the planes, profiles and guide curves with clear descriptive names.

HOUSING Part-Second Extruded Boss/Base Feature

Create the second Extruded Boss/Base feature from the square face of the Loft. How do you estimate the depth of the Extruded Boss/Base feature? Answer: The Extruded Base feature of the BATTERY is 4.100in, [104.14mm].

Ribs are required to support the BATTERY. Design for Rib construction. Ribs add strength to the HOUSING and support the BATTERY. Use a 4.400in, [111.76mm] depth as the first estimate. Adjust the estimated depth dimension later if required in the FLASHLIGHT assembly.

The Extruded Boss/Base feature is symmetric about the Right Plane. Utilize Convert Entities to extract the back face of the Loft Base feature. No sketch dimensions are required.

Activity: HOUSING Part-First Extruded Boss Feature

Select the Sketch plane.
372) **Rotate** the model to view the back.

373) Right-click the **back face** of Boss-Loft1. This is your Sketch plane.

Create the sketch.
374) Click **Sketch** ✏ from the Context toolbar. The Sketch toolbar is displayed.

375) Click the **Convert Entities** 🗋 Sketch tool.

376) Click **OK** ✔ from the Convert Entities PropertyManager.

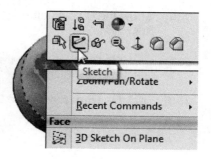

Insert the second Extruded Boss/Base feature.

377) Click the **Extruded Boss/Base** 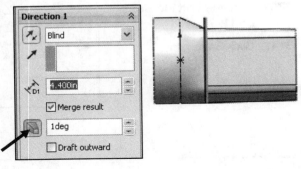 feature tool. The Boss-Extrude PropertyManager is displayed.

378) Enter **4.400**in, [**111.76**] for Depth in Direction 1.

379) Click the **Draft On/Off** box.

380) Enter **1**deg for Draft Angle.

381) Click **OK** ✔ from the Boss-Extrude PropertyManager.

382) Click **Right view** from the Heads-up View toolbar.

383) Rename **Boss-Extrude2** to **Boss-Battery**.

384) Click **Save**.

HOUSING Part-Shell Feature

The Shell feature removes material. Use the Shell feature to remove the front face of the HOUSING. In the injection-molded process, the body wall thickness remains constant.

A dialog box is displayed if the thickness value is greater than the Minimum radius of Curvature.

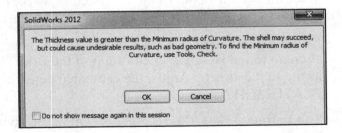

Activity: HOUSING Part-Shell Feature

Insert the Shell feature.

385) Click **Isometric view** from the Heads-up View toolbar. If needed

386) **S**how the BatteryLoftPlane from the FeatureManager as illustrated.

387) Click the **Shell** feature tool. The Shell1 PropertyManager is displayed.

388) Click the **front face** of the Base Extrude feature as illustrated.

389) Enter **.100**in, [**2.54**] for Thickness.

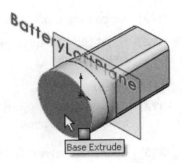

390) Click **OK** ✓ from the Shell1 PropertyManager. Shell1 is displayed in the FeatureManager.

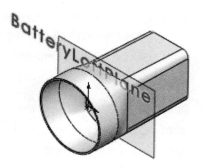

💡 The Shell feature position in the FeatureManager determines the geometry of additional features. Features created before the Shell contained the wall thickness specified in the Thickness option. Position features of different thickness such as the Rib feature and Thread Swept feature after the Shell. Features inserted after the Shell remain solid.

HOUSING Part-Third Extruded Boss/Base Feature

The third Extruded Boss feature creates a solid circular ring on the back circular face of the Extruded Base feature. The solid ring is a cosmetic stop for the LENSCAP and provides rigidity at the transition of the HOUSING. Design for change. The Extruded Boss/Base feature updates if the Shell thickness changes.

Utilize the Front plane for the sketch. Select the inside circular edge of the Shell. Utilize Convert Entities to obtain the inside circle. Utilize the Circle Sketch tool to create the outside circle. Extrude the feature towards the front face.

Activity: HOUSING Part-Third Extruded Boss Feature

Select the Sketch plane.
391) Right-click **Front Plane** from the FeatureManager. This is your Sketch plane.

Create the sketch.
392) Click **Sketch** ✏ from the Context toolbar. The Sketch toolbar is displayed.

393) Zoom-in and click the **front inside circular edge** of Shell1 as illustrated. Note the icon feedback for an edge.

394) Click the **Convert Entities** 🔲 Sketch tool.

395) Click **OK** ✓ from the Convert Entities PropertyManager.

Create the outside circle.
396) Click **Front view** 🔲 from the Heads-up View toolbar.

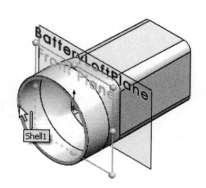

397) Click the **Circle** ⊙ Sketch tool. The Circle PropertyManager is displayed.

398) Sketch a **circle** centered at the Origin.

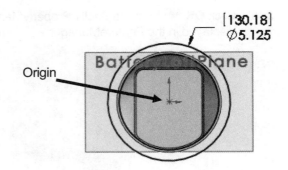

Add a dimension.

399) Click the **Smart Dimension** ✧ Sketch tool.

400) Click the **circumference** of the circle.

401) Enter **5.125**in, **[130.18]**.

Insert an Extruded Boss/Base feature.

402) Click the **Extruded Boss/Base** 🔲 feature tool. The Boss-Extrude PropertyManager is displayed.

403) Enter **.100**in, **[2.54]** for Depth in Direction 1. The extrude arrow points to the front.

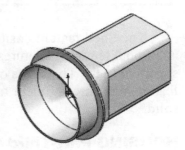

404) Click **OK** ✔ from the Boss-Extrude PropertyManager.

405) Click **Isometric view** 🔲 from the Heads-up View toolbar.

406) Rename **Boss-Extrude3** to **Boss-Stop**.

407) **Hide** the BatteryLoftPlane as illustrated.

408) Click **Save** 💾.

HOUSING Part-Draft Feature

The Draft feature tapers selected model faces by a specified angle by utilizing a Neutral Plane or Parting Line. The Neutral Plane option utilizes a plane or face to determine the pull direction when creating a mold.

The Parting Line option drafts surfaces around a parting line of a mold. Utilize the Parting Line option for non-planar surfaces. Apply the Draft feature to solid and surface models.

A 5° draft is required to insure proper thread mating between the LENSCAP and the HOUSING. The LENSCAP Extruded Base feature has a 5° draft angle.

The outside face of the Extruded Base feature HOUSING requires a 5° draft angle. The inside HOUSING wall does not require a draft angle. The Extruded Base feature has a 5° draft angle. Use the Draft feature to create the draft angle. The front circular face is the Neutral Plane. The outside cylindrical surface is the face to draft.

You created the Extruded Boss/Base and Extruded Cut features with the Draft Angle option. The Draft feature differs from the Extruded feature, Draft Angle option. The Draft feature allows you to select multiple faces to taper.

In order for a model to eject from a mold, all faces must draft away from the parting line which divides the core from the cavity. Cavity side faces display a positive draft and core side faces display a negative draft. Design specifications include a minimum draft angle, usually less than 5°.

For the model to eject successfully, all faces must display a draft angle greater than the minimum specified by the Draft Angle. The Draft feature, Draft Analysis Tools and DraftXpert utilize the draft angle to determine what faces require additional draft base on the direction of pull.

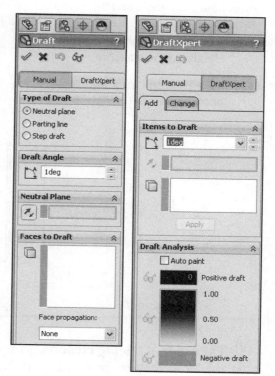

You can apply a draft angle as a part of an Extruded Base/Boss or Extruded Cut feature.

Each tab has a separate menu and option selections. The Draft PropertyManager displays the appropriate selections based on the type of draft you create.

The DraftXpert PropertyManager provides the ability to manage the creation and modification of all Neutral Plane drafts. Select the draft angle and the references to the draft. The DraftXpert manages the rest.

Activity: HOUSING Part-Draft Feature

Insert the Draft feature.

409) Click the **Draft** ⬡ feature tool. The Draft PropertyManager is displayed.

410) Click the **Manual** tab.

411) Zoom-in and click the thin **front circular face** of Base Extrude. The front circular face is displayed in the Neutral Plane box. Note: The face feedback ☐ icon. Face<1> is displayed.

412) Click inside the **Faces to draft** box.

413) Click the **outside cylindrical face** as illustrated.

Note: The face feedback ☐ icon.

414) Enter **5°** for Draft Angle.

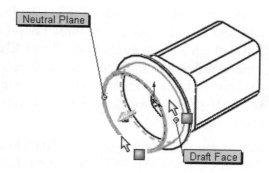

5°

415) Click **OK** ✓ from the Draft
PropertyManager. Draft1 is
displayed in the FeatureManager.

Display the draft angle and the straight
interior.

416) Click **Right view** ⊞ from the
Heads-up View toolbar.

417) Click **Hidden Lines Visible** from the Heads-up View toolbar.

418) Click **Save** 💾.

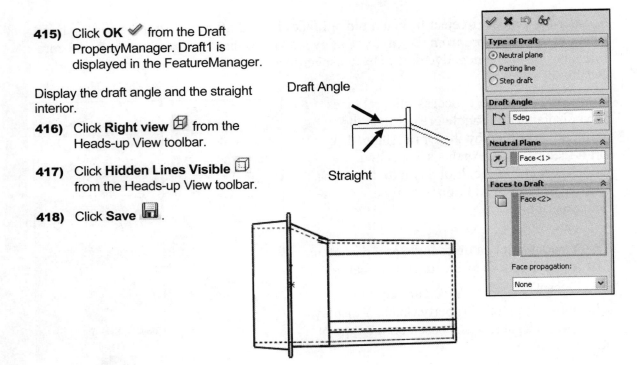

💡 The order of feature creation is important. Apply threads
after the Draft feature for plastic parts to maintain a constant
thread thickness.

HOUSING Part-Thread with Swept Feature

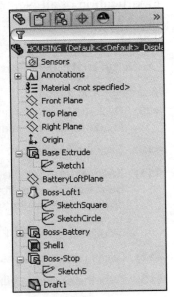

The HOUSING requires a thread. Create the threads for the
HOUSING on the outside of the Draft feature. Create the thread
with the Swept feature. The thread requires two sketches:
ThreadPath and ThreadSection. The LENSCAP and HOUSING
Thread utilize the same technique. Create a ThreadPlane. Utilize
Convert Entities to create a circular sketch referencing the
HOUSING Extruded Base feature. Insert a Helix/Spiral curve to
create the path.

Reuse geometry between parts. The ThreadSection is copied
from the LENSCAP and is inserted into the HOUSING Top
Plane.

Activity: HOUSING Part-Thread with Swept Feature

Insert the ThreadPlane.

419) Click **Isometric view** 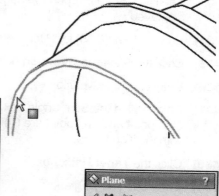 from the Heads-up View toolbar.

420) Click **Hidden Lines Removed** from the Heads-up View toolbar.

421) Click the **thin front circular face**, Base Extrude.

422) Click **Plane** from the Consolidated Reference Geometry toolbar. The Plane PropertyManager is displayed.

423) Check the **Flip** box.

424) Enter **.125**in, [3.18] for Distance. Accept the default settings.

425) Click **OK** from the Plane PropertyManager. Plane2 is displayed in the FeatureManager.

426) Click **Save**.

427) Rename **Plane2** to **ThreadPlane**.

Insert the ThreadPath.

428) Right-click **ThreadPlane** from the FeatureManager. This is your Sketch plane.

429) Click **Sketch** from the Context toolbar. The Sketch toolbar is displayed.

430) Click the **front outside circular edge** of the Base Extrude feature as illustrated.

431) Click the **Convert Entities** Sketch tool. The circular edge is displayed on the ThreadPlane.

432) Click **OK** from the Convert Entities PropertyManager.

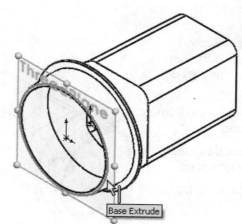

Insert the Helix/Spiral curve.

433) Click the **Helix and Spiral** tool from the Consolidated Curves toolbar as illustrated. The Helix/Spiral PropertyManager is displayed.

434) Enter **.250**in, **[6.35]** for Pitch.

435) Click the **Reverse Direction** box.

436) Enter **2.5** for Revolution.

437) Enter **180** in the Start angle box. The Helix start point and end point are Coincident with the Top Plane.

438) Click the **Taper Helix** box.

439) Enter **5**deg for Angle.

440) Check the **Taper outward** box.

441) Click **OK** ✔ from the Helix/Spiral PropertyManager. Hexlix/Spiral1 is displayed in the FeatureManager.

442) Rename **Helix/Spiral1** to **ThreadPath**.

443) Click **Isometric view** ⬦ from the Heads-up View toolbar.

444) Click **Save** 💾.

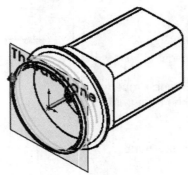

Copy the LENSCAP ThreadSection.

445) **Open** the LENSCAP part. The LENSCAP FeatureManager is displayed.

446) **Expand** the Thread feature from the FeatureManager.

447) Click the **ThreadSection** sketch. ThreadSection is highlighted.

448) Click **Edit**, **Copy** from the Menu bar.

449) **Close** the LENSCAP.

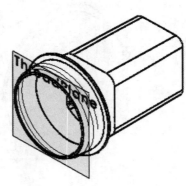

Open the HOUSING.
450) **Return** to the Housing.

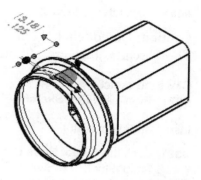

Paste the LENSCAP ThreadSection.
451) Click **Top Plane** from the HOUSING FeatureManager.

452) Click **Edit**, **Paste** from the Menu bar. The ThreadSection is displayed on the Top Plane. The new Sketch7 name is added to the bottom of the FeatureManager.

453) **Hide** ThreadPlane.

454) Rename **Sketch6** to **ThreadSection**.

455) Click **Save** 💾.

Add a Pierce relation.
456) Right-click **ThreadSection** from the FeatureManager.

457) Click **Edit Sketch**.

458) Click **ThreadSection** from the HOUSING FeatureManager.

459) **Zoom in** on the Midpoint of the ThreadSection.

460) Click the **Midpoint** of the ThreadSection.

461) Click **Isometric view** 🧊 from the Heads-up View toolbar.

462) Hold the **Ctrl** key down.

463) Click the **right back edge of the ThreadPath**. Note: Do not click the end point. The Properties PropertyManager is displayed. The selected entities are displayed in the Selected Entities box.

464) Release the **Ctrl** key.

465) Click **Pierce** from the Add Relations box.

🔆 Tangent Edges and Origin are displayed for educational purposes.

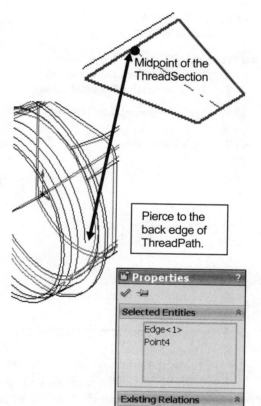

466) Click **OK** ✔ from the Properties PropertyManager.

Caution: Do not click the front edge of the Thread path. The Thread is then created out of the HOUSING.

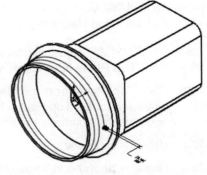

Close the sketch.
467) Right-click **Exit Sketch**.

Insert the Swept feature.

468) Click the **Swept Boss/Base** ⌒ feature tool. The Swept PropertyManager is displayed.

469) **Expand** HOUSING from the fly-out FeatureManager.

470) Click inside the **Profile** box.

471) Click **ThreadSection** from the fly-out FeatureManager.

472) Click **ThreadPath** from the fly-out FeatureManager.

473) Click **OK** ✔ from the Sweep PropertyManager. Sweep1 is displayed in the FeatureManager.

474) Rename **Sweep1** to **Thread**.

475) Click **Save** 🖫.

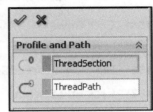
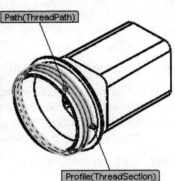

💡 Creating a ThreadPlane provides flexibility to the design. The ThreadPlane allows for a smoother lead. Utilize the ThreadPlane offset dimension to adjust the start of the thread.

HOUSING Part-Handle with Swept Feature

Create the handle with the Swept feature. The Swept feature consists of a sketched path and cross section profile. Sketch the path on the Right Plane. The sketch uses edges from existing features. Sketch the profile on the back circular face of the Boss-Stop feature.

Activity: HOUSING Part-Handle with Swept Feature

Create the Swept path sketch.

476) Right-click **Right Plane** from the FeatureManager.

477) Select **Sketch** ✑ from the Context toolbar. The Sketch toolbar is displayed.

478) Click **Right view** ⊞ from the Heads-up View toolbar.

479) Click **Hidden Lines Removed** ⬚ from the Heads-up View toolbar.

480) Click the **Line** ⟍ Sketch tool.

481) Sketch a **vertical line** from the right top corner of the Housing upward.

482) Sketch a **horizontal line** below the top of the Boss Stop as illustrated.

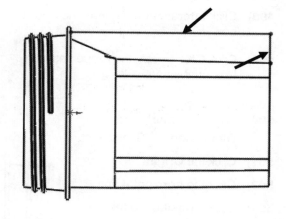

Insert a 2D Fillet.

483) Click the **Sketch Fillet** ⌐ Sketch tool.

484) Click the **right top corner** of the sketch lines as illustrated.

485) Enter **.500**in, **[12.7]** for Radius.

486) Click **OK** ✔ from the Sketch Fillet PropertyManager.

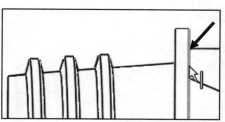

Add a Coincident relation.

487) Click the **left end point** of the horizontal line. Note: the mouse feedback ● icon

488) Hold the **Ctrl** key down.

489) Click the **right vertical edge** of the Boss Stop.

490) Release the **Ctrl** key.

491) Click **Coincident** ⟋ from the Add Relations box.

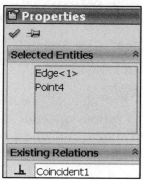

492) Click **OK** ✔ from the Properties PropertyManager.

Add an Intersection relation
493) Click the **bottom end point** of the vertical line.

494) Hold the **Ctrl** key down.

495) Click the **right vertical edge** of the Housing.

496) Click the **horizontal edge** of the Housing.

497) Release the **Ctrl** key.

498) Click **Intersection** ✕ from the Add Relations box.

499) Click **OK** ✔ from the Properties PropertyManager.

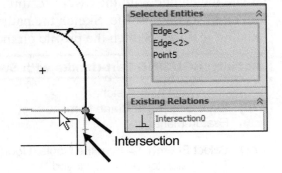

Add a dimension.
500) Click the **Smart Dimension** ✎ Sketch tool.

501) Click the **Origin**.

502) Click the **horizontal line**.

503) Click a **position** to the right.

504) Enter **2.500**in, [**63.5**].

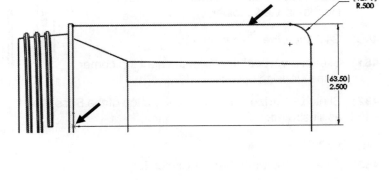

Close and rename the sketch.
505) Right-click **Exit Sketch**.

506) Rename **Sketch7** to **HandlePath**.

507) Click **Save**.

Create the Swept Profile.
508) Click **Back view** ⊞ from the Heads-up View toolbar.

509) Right-click the **back circular face** of the Boss-Stop feature as illustrated.

510) Click **Sketch** ✏ from the Context toolbar. The Sketch toolbar is displayed.

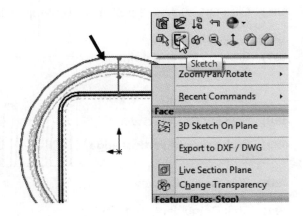

💡 The book is design to expose the user to various methods in creating sketches and features. In the next section, create a slot **without using** the Slot Sketch tool.

511) Click the **Centerline** ⁝ Sketch tool.

512) Sketch a **vertical centerline** collinear with the Right Plane, coincident to the Origin.

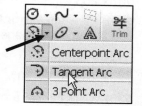

513) Sketch a **horizontal centerline**. The left end point of the centerline is coincident with the vertical centerline on the Boss-Stop feature. Do not select existing feature geometry.

514) **Zoom in** on the top of the Boss-Stop.

515) Click the **Line** ＼ Sketch tool.

516) Sketch a **line** above the horizontal centerline as illustrated.

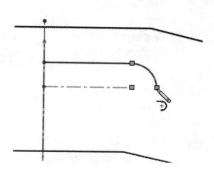

517) Click the **Tangent Arc** Sketch tool.

518) Sketch a **90° arc**.

519) Right-click **Select** to exit the Tangent Arc tool.

Add an Equal relation.

520) Click the **horizontal centerline**.

521) Hold the **Ctrl** key down.

522) Click the **horizontal line**.

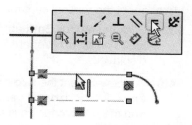

523) Release the **Ctrl** key.

524) Right-click **Make Equal** = from the Context toolbar.

525) Click **OK** ✔ from the Properties PropertyManager.

Add a Horizontal relation.

526) Click the **right end point** of the tangent arc.

527) Hold the **Ctrl** key down.

528) Click the **arc center point**.

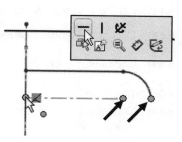

529) Click the left **end point** of the centerline.

530) Release the **Ctrl** key.

531) Right-click **Make Horizontal** ▬ from the Context toolbar.

532) Click **OK** ✔ from the Properties PropertyManager.

Mirror about the horizontal centerline.

533) Click the **Mirror Entities** ⚠ Sketch tool. The Mirror PropertyManager is displayed.

534) Click the **horizontal** line.

535) Click the **90° arc**. The selected entities are displayed in the Entities to mirror box.

536) Click inside the **Mirror about** box.

537) Click the **horizontal centerline**.

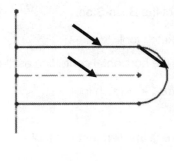

538) Click **OK** ✅ from the Mirror PropertyManager.

Mirror about the vertical centerline.

539) Click the **Mirror Entities** ⚠ Sketch tool. The Mirror PropertyManager is displayed.

540) Window-Select the **two horizontal lines**, the **horizontal centerline** and the **90° arc** for Entities to mirror. The selected entities are displayed in the Entities to mirror box.

541) Click inside the **Mirror about** box.

542) Click the **vertical centerline**.

543) Click **OK** ✅ from the Mirror PropertyManager.

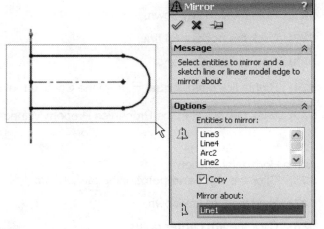

Add dimensions.

544) Click the **Smart Dimension** ✏ Sketch tool.

545) Enter **1.000**in, **[25.4]** between the arc center points.

546) Enter **.100**in, **[2.54]** for Radius.

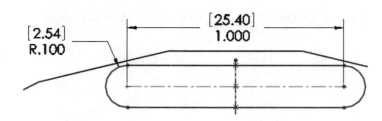

Add a Pierce relation.

547) Right-click **Select**.

548) Click **Isometric view** from the Heads-up View toolbar.

549) Click the **top midpoint** of the Sketch profile.

550) Hold the **Ctrl** key down.

551) Click the **line** from the Handle Path.

552) Release the **Ctrl** key.

553) Click **Pierce** from the Add Relations box. The sketch is fully defined.

554) Click **OK** from the Properties PropertyManager.

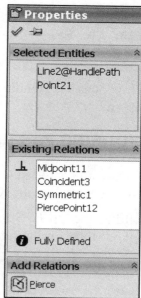

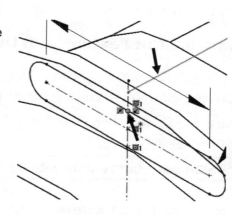

Close and rename the sketch.

555) Right-click **Exit Sketch**.

556) Rename **Sketch8** to **HandleProfile**.

557) If needed, **Hide** ThreadPlane and **Hide** BatterlyLoftPlane.

Insert the Swept feature.

558) Click the **Swept Boss/Base** feature tool. The Sweep PropertyManager is displayed. HandleProfile is the Sweep profile.

559) Click inside the **Profile** box.

560) Click **HandleProfile** from the fly-out FeatureManager.

561) Click the **HandlePath** from the fly-out FeatureManager.

562) Click **OK** from the Sweep PropertyManager. Sweep2 is displayed in the FeatureManager.

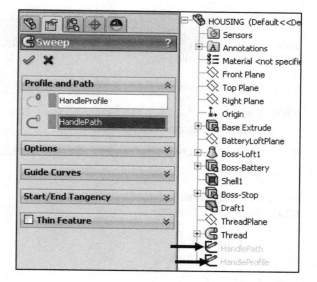

Fit the profile to the Graphics window.

563) Press the **f** key.

564) Click **Shaded With Edges** from the Heads-up View toolbar.

565) Rename **Sweep2** to **Handle**.

566) Click **Save**.

How does the Handle Swept feature interact with other parts in the FLASHLIGHT assembly? Answer: The Handle requires an Extruded Cut to insert the SWITCH.

HOUSING Part-Extruded Cut Feature with Up To Surface

Create an Extruded Cut in the Handle for the SWITCH. Utilize the top face of the Handle for the Sketch plane. Create a circular sketch centered on the Handle.

Utilize the Up To Surface End Condition in Direction 1. Select the inside surface of the HOUSING for the reference surface.

Activity: HOUSING Part-Extruded Cut Feature with Up To Surface

Select the Sketch plane.

567) Right-click the **top face** of the Handle. Handle is highlighted in the FeatureManager. This is your Sketch plane.

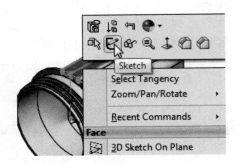

Create the sketch.

568) Click **Sketch** ✏ from the Context toolbar. The Sketch toolbar is displayed.

569) Click **Top view** 🔲 from the Heads-up View toolbar.

570) Click **Circle** ⊙ from the Sketch toolbar. The Circle PropertyManager is displayed.

571) Sketch a **circle** on the Handle near the front as illustrated.

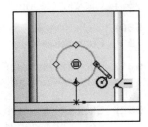

Deselect the circle sketch tool.

572) Right-click **Select**.

Add a Vertical relation.

573) Click the **Origin**.

574) Hold the **Ctrl** key down.

575) Click the **centerpoint** of the circle. The Properties PropertyManager is displayed. The selected entities are displayed in the Selected Entities box.

576) Release the **Ctrl** key.

577) Click **Vertical** Ⅰ.

578) Click **OK** ✔ from the Properties PropertyManager.

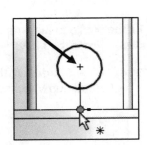

Add dimensions.

579) Click the **Smart Dimension** 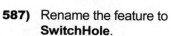 Sketch tool.

580) Enter **.510**in, [**12.95**] for diameter.

581) Enter **.450**in, [**11.43**] for the distance from the Origin.

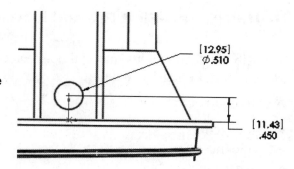

Insert an Extruded Cut feature.

582) **Rotate** the model to view the inside Shell1.

583) Click the **Extruded Cut**  feature tool. The Cut-Extrude PropertyManager is displayed.

584) Select the **Up To Surface** End Condition in Direction 1.

585) Click the **top inside face** of the Shell1 as illustrated.

586) Click **OK** ✔ from the Cut-Extrude PropertyManager. The Cut-Extrude1 feature is displayed in the FeatureManager.

587) Rename the feature to **SwitchHole**.

588) Click **Isometric view** from the Heads-up View toolbar.

589) Click **Save** 💾.

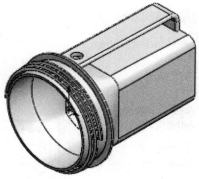

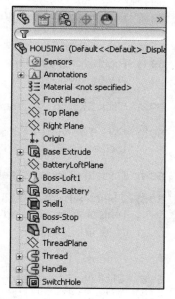

HOUSING Part-First Rib and Linear Pattern Feature

The Rib ⬗ feature adds material between contours of existing geometry. Use Ribs to add structural integrity to a part.

A Rib requires:

- A sketch
- Thickness
- Extrusion direction

The first Rib profile is sketched on the Top Plane. A 1° draft angle is required for manufacturing. Determine the Rib thickness by the manufacturing process and the material.

💡 Rule of thumb states that the Rib thickness is ½ the part wall thickness. The Rib thickness dimension is .100 inches [2.54mm] for illustration purposes.

The HOUSING requires multiple Ribs to support the BATTERY. A Linear Pattern feature creates multiple instances of a feature along a straight line. Create the Linear Pattern feature in two directions along the same vertical edge of the HOUSING.

Activity: HOUSING Part-First Rib and Linear Pattern Feature

Display all hidden lines.
590) Click **Hidden Lines Visible** ⬚ from the Heads-up View toolbar.

Create the sketch.
591) Right-click **Top Plane** from the FeatureManager. This is your Sketch plane.

592) Click **Sketch** ✎ from the Context toolbar.

593) Click **Top view** ⬗ from the Heads-up View toolbar.

594) Click the **Line** ＼ Sketch tool.

595) Sketch a **horizontal line** as illustrated. The endpoints are located on either side of the Handle.

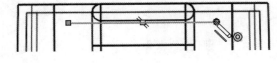

Add a dimension.
596) Click the **Smart Dimension** ◇ Sketch tool.

597) Click the **inner back edge**.

598) Click the **horizontal line**.

599) Click a **position** to the right off the profile.

600) Enter **.175**in, [**4.45**].

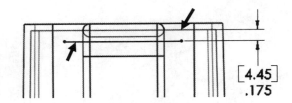

[4.45]
.175

Insert the Rib feature.

601) Click the **Rib**  feature tool. The Rib PropertyManager is displayed.

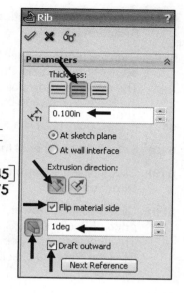

602) Click the **Both Sides** button.

603) Enter **.100**in, **[2.54]** for Rib Thickness.

604) Click the **Parallel to Sketch** button. The Rib direction arrow points to the back. Flip the material side if required. Select the Flip material side check box if the direction arrow does not point towards the back.

605) Click the **Draft On/Off** box.

606) Enter **1deg** for Draft Angle.

607) Click **Front view** from the Heads-up View toolbar.

608) Click the **back inside face** of the HOUSING for the Body.

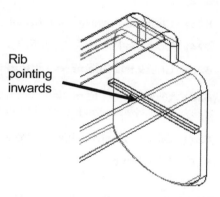

Rib pointing inwards

609) Click **OK** from the Rib PropertyManager. Rib1 is displayed in the FeatureManager.

610) Click **Isometric view** from the Heads-up View toolbar.

611) Click **Save** .

Existing geometry defines the Rib boundaries. The Rib does not penetrate through the wall.

Insert the Linear Pattern feature.

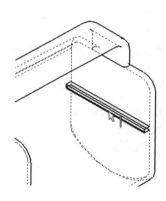

612) **Zoom to Area** on Rib1.

613) Click **Rib1** from the FeatureManager.

614) Click the **Linear Pattern** feature tool. Rib1 is displayed in the Features to Pattern box.

615) Click inside the **Direction 1 Pattern Direction** box.

616) Click the **hidden upper back vertical edge** of Shell1 in the Graphics window. The direction arrow points upward. Click the Reverse direction button if required.

617) Enter **.500**in, [**12.7**] for Spacing.

618) Enter **3** for Number of Instances.

619) Click inside the **Direction 2 Pattern Direction** box.

620) Click the hidden **lower back vertical edge** of Shell1 in the Graphics window. The direction arrow points downward. Click the Reverse direction button if required.

621) Enter **.500**in, [**12.7**] for Spacing.

622) Enter **3** for Number of Instances.

623) Click the **Pattern seed only** box.

624) Drag the Linear Pattern **Scroll bar** downward to display the Options box.

625) Check the **Geometry pattern** box. Accept the default values.

626) Click **OK** ✅ from the Linear Pattern PropertyManager. LPattern1 is displayed in the FeatureManager.

627) Click **Isometric view** from the Heads-up View toolbar.

628) Click **Save**.

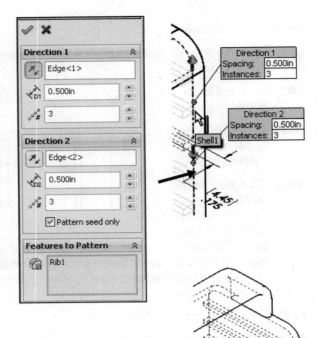

💡 Utilize the Geometry pattern option to efficiently create and rebuild patterns. Know when to check the Geometry pattern.

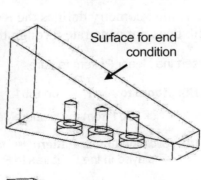

Check Geometry pattern. You require an exact copy of the seed feature. Each instance is an exact copy of the faces and edges of the original feature. End conditions are not calculated. This option saves rebuild time.

Uncheck Geometry pattern. You require the end condition to vary. Each instance will have a different end condition. Each instance is offset from the selected surface by the same amount.

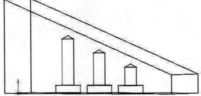

Suppress Patterns when not required. Patterns contain repetitive geometry that takes time to rebuild. Pattern features also clutter the part during the model creation process. Suppress patterns as you continue to create more complex features in the part. Unsuppress a feature to restore the display and load into memory for future calculations. Hide features to improve clarity. Show feature to display hidden features.

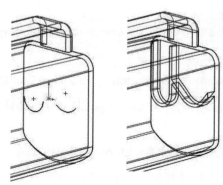

Rib sketches are not required to be fully defined. The Linear Rib option blends sketched geometry into existing contours of the model.

Example: Create an offset reference plane from the inside back face of the HOUSING.

Sketch two under defined arcs. Insert a Rib feature with the Linear option. The Rib extends to the Shell walls.

HOUSING Part-Second Rib Feature

The Second Rib feature supports and centers the BATTERY. The Rib is sketched on a reference plane created through a point on the Handle and parallel with the Right Plane. The Rib sketch references the Origin and existing geometry in the HOUSING. Utilize an Intersection and Coincident relation to define the sketch.

Activity: HOUSING Part-Second Rib Feature

Insert a Reference plane for the second Rib feature. Create a Parallel Plane at Point.

629) Click **Wireframe** from the Heads-up View toolbar

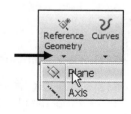

630) **Zoom to Area** on the back right side of the Handle.

631) Click **Plane** from the Features toolbar. The Plane PropertyManager is displayed.

632) Click **Right Plane** from the fly-out FeatureManager.

633) Click the **vertex** (point) at the back right of the handle as illustrated.

634) Click **OK** from the Plane PropertyManager. Plane2 is displayed in the FeatureManager.

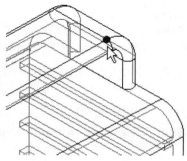

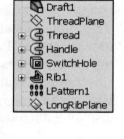

635) Rename **Plane2** to **LongRibPlane**.

Fit to the Graphics window.
636) Press the **f** key.

637) Click **Save** .

Create the second Rib.
638) Right-click **LongRibPlane** for the Sketch plane.

Create the sketch.
639) Click **Sketch** from the Context toolbar.

640) Click **Right view** from the Heads-up View toolbar.

641) Click the **Line** Sketch tool.

642) Sketch a **horizontal line**. Do not select the edges of the Shell1 feature.

Deselect the Line Sketch tool.
643) Right-click **Select**.

Add a Coincident relation.
644) Click the **left end point** of the horizontal sketch line.

645) Hold the **Ctrl** key down. Click the **BatteryLoftPlane** from the fly-out FeatureManager.

646) Release the **Ctrl** key.

647) Click **Coincident**.

648) Click **OK** from the Properties PropertyManager.

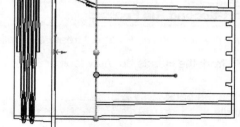

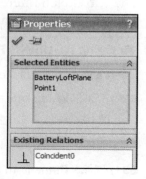

Add a dimension.
649) Click the **Smart Dimension** Sketch tool.

650) Click the **horizontal line**.

651) Click the **Origin**.

652) Click a **position** for the vertical linear dimension text.

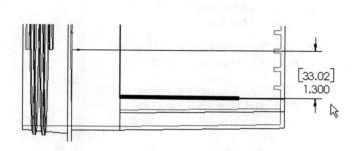

653) Enter **1.300**in, [**33.02**].

☀ When the sketch and reference geometry become complex, create dimensions by selecting Reference planes and the Origin in the FeatureManager.

☀ Dimension the Rib from the Origin, not from an edge or surface for design flexibility. The Origin remains constant. Modify edges and surfaces with the Fillet feature.

Sketch an arc.

654) **Zoom to Area** ⌕ on the horizontal Sketch line.

655) Click the **Tangent Arc** ⌓ Sketch tool.

656) Click the **left end** point of the horizontal line.

657) Click the **intersection** of the Shell1 and Boss Stop features. The sketch is displayed in black and is fully defined. If needed add an Intersection relation between the endpoint of the Tangent Arc, the left vertical Boss-Stop edge and the Shell1 Silhouette edge of the lower horizontal inside wall.

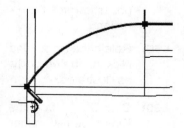

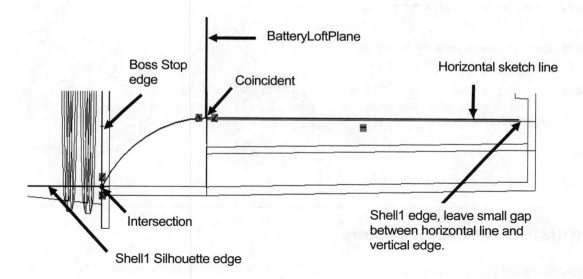

Insert the Rib feature.

658) Click the **Rib** 🠄 feature tool. The Rib PropertyManager is displayed.

659) Click the **Both Sides** box.

660) Enter **.075**in, **[1.91]** for Rib Thickness.

661) Click the **Draft On/Off** box.

662) Enter **1**deg for Angle.

663) Click the **Draft outward** box.

664) Click the **Flip material side** box if required. The direction arrow points towards the bottom.

665) Rotate the model and click the **inside body** as illustrated.

666) Click **OK** ✔ from the Rib PropertyManager. Rib2 is displayed in the FeatureManager.

667) **Hide** LongRibPlane from the FeatureManager.

668) Click **Trimetric view** 🔲 from the Heads-up View toolbar.

669) Click **Shaded With Edges** 🔲 from the Heads-up View toolbar. View the created Rib feature.

HOUSING Part-Mirror Feature

An additional Rib is required to support the BATTERY. Reuse features with the Mirror feature to create a Rib symmetric about the Right Plane.

The Mirror feature requires:

- Mirror Face or Plane reference

- Features or Faces to Mirror

Utilize the Mirror feature. Select the Right Plane for the Mirror Plane. Select the second Rib for the Features to Mirror.

Activity: HOUSING Part-Mirror Feature

Insert the Mirror feature.

670) Click the **Mirror**  feature tool. The Mirror PropertyManager is displayed.

671) Click inside the **Mirror Face/Plane** box.

672) Click **Right Plane** from the fly-out FeatureManager.

673) Click **Rib2** for Features to Mirror from the fly-out FeatureManager.

674) Click **OK** ✔ from the Mirror PropertyManager. Mirror1 is displayed in the FeatureManager.

675) Click **Trimetric view** from the Heads-Up View toolbar.

676) Click **Save** .

Close all parts.

677) Click **Window**, **Close All** from the Menu bar.

The parts for the FLASHLIGHT are complete! Review the HOUSING before moving on to the FLASHLIGHT assembly.

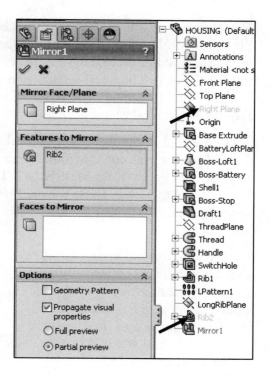

Additional information on Extrude Boss/Base, Extrude Cut, Swept, Loft, Helix/Spiral, Rib, Mirror, and Reference Planes are found in SolidWorks Help Topics.

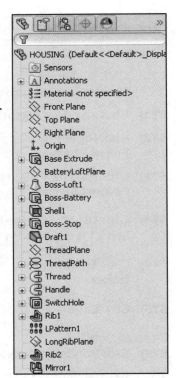

Review of the HOUSING Part

The HOUSING utilized the Extruded Boss/Base feature with the Draft Angle option. The Lofted Boss feature was created to blend the circular face of the LENS with the rectangular face of the BATTERY. The Shell feature removed material with a constant wall thickness. The Draft feature utilized the front face as the Neutral plane.

You created a Thread similar to the LENSCAP Thread. The Thread profile was copied from the LENSCAP and inserted into the Top Plane of the HOUSING. The Extruded Cut feature was utilized to create a hole for the Switch. The Rib features were utilized in a Linear Pattern and Mirror feature.

Each feature has additional options that are applied to create different geometry. The Offset From Surface option creates an Extruded Cut on the curved surface of the HOUSING and LENSCAP. The Reverse offset and Translate surface options produce a cut depth constant throughout the curved surface.

Utilize Tools, Sketch Entities, Text to create the text profile on an Offset Plane.

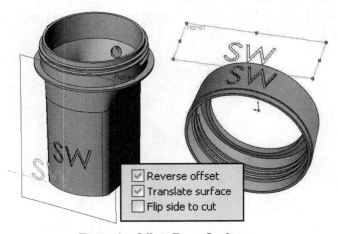

Example: Offset From Surface

Chapter Summary

You created four parts for the FLASHLIGHT assembly: O-RING, SWITCH, LENSCAP, and HOUSING. The FLASHLIGHT parts contain over a 100 features, reference planes, sketches, and components. You organized the features in each part.

The O-RING part utilized a Swept Base feature. The SWITCH part utilized a Loft Base feature. The simple Swept feature required two sketches: a path and a profile. A complex Swept feature requires multiple sketches and Guide Curves. The Loft feature required two or more sketches created on different planes.

The LENSCAP and HOUSING part utilized a variety of features. You applied design intent to reuse geometry through Geometric relationships, Symmetry, and patterns. Review the chapter exercises before moving on to the next chapter.

Chapter Terminology

Circular Pattern: A feature used to create instances of a seed feature rotated about an axis of revolution. You utilized a Circular Pattern to create Extruded Cuts around the LENSCAP.

CommandManager: The CommandManager is a context-sensitive toolbar that dynamically updates based on the toolbar you want to access. By default, it has toolbars embedded in it based on the document type. When you click a tab below the Command Manager, it updates to show that toolbar. For example, if you click the **Sketches** tab, the Sketch toolbar is displayed.

Draft: A feature used to add a specified draft angle to a face. You utilized a Draft feature with the Neutral Plane option.

Extruded Thin Cut: A feature used to remove material by extruding an open profile.

Helix/Spiral Curve: A Helix is a curve with pitch. The Helix is created about an axis. You utilized a Helix Curve to create a thread for the LENSCAP.

Linear Pattern: A feature used to create instances of a seed feature in a rectangular array, along one or two edges. The Linear Pattern was utilized to create multiple instances of the HOUSING Rib1 feature.

Loft: A feature used to blend two or more profiles on separate Planes. A Loft Boss adds material. A Loft Cut removes material. The HOUSING part utilized a Loft Boss feature to transition a circular profile of the LENSCAP to a square profile of the BATTERY.

Mirror: The feature used to create a symmetric feature about a Mirror Plane. The Mirror feature created a second Rib, symmetric about the Right Plane.

Revolved Cut Thin: A feature used to remove material by rotating a sketched profile around a centerline.

Rib: A feature used to add material between contours of existing geometry. Use Ribs to add structural integrity to a part.

Suppressed: A feature or component not loaded into memory. A feature is suppressed or unsuppressed. A component is suppressed or resolved. Suppress features and components to improve model rebuild time.

Swept: A Swept Boss/Base feature adds material. A Swept Cut removes material. A Swept requires a profile sketch and a path sketch. A Swept feature moves a profile along a path.

Questions

1. Identify the function of the following features:

   - Swept Boss/Base

   - Revolved Cut Thin

   - Loft Boss/Base

   - Rib

   - Circular Pattern

   - Linear Pattern

2. Describe a Suppressed feature.

3. Why would you suppress a feature?

4. The Rib features require a sketch, thickness and a _____ direction.

5. What is a Pierce Geometric relation?

6. Describe how to create a thread using the Swept feature. Provide an example.

7. Explain how to create a Linear Pattern feature. Provide an example.

8. Identify two advantages of utilizing Convert Entities in a sketch to obtain the profile.

9. How is symmetry built into a sketch? Provide an example.

10. How is symmetry built into a feature? Provide an example.

11. Define a Guide Curve. Identify the features that utilize Guide Curves.

12. Describe a Draft feature.

13. Identify the differences between a Draft feature and the Draft Angle option in the Extruded Boss/Base feature.

14. Describe the differences between a Circular Pattern feature and a Linear Pattern feature.

15. Identify the advantages of the Convert Entities tool.

16. True or False. A Loft feature can only be inserted as the first feature in a part. Explain your answer.

Exercises

Exercise 6.1: QUATTRO-SEAL-O-RING Part

Create the QUATTRO-SEAL-O-RING part as a single Swept feature.

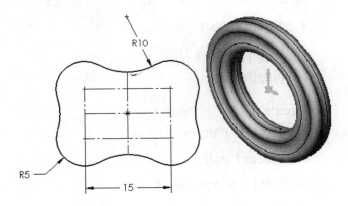

- Create a 100mm diameter circle on the Front plane for the path, Sketch1.

- Create the symmetric cross section on the Top Plane for the profile, Sketch2.

Exercise 6.2: HOOK Part

Create the HOOK part. Create the HOOK with a Swept feature. View the illustrated FeatureManager. Note: Not all dimensions are provided. Your HOOK Part will vary.

A Swept feature adds material by moving a profile along a path. A simple Swept feature requires two sketches. The first sketch is called the path. The second sketch is called the profile. The profile and path are sketched on perpendicular planes. Remove all Tangent edges in the final model.

The Swept feature uses:

- A path sketched on the Right Plane.

- A profile sketched on the Top Plane.

- Utilize a Dome feature to create a spherical feature on a circular face. A Swept Cut feature removes material.

- Utilize a Swept Cut feature to create the Thread for the HOOK part.

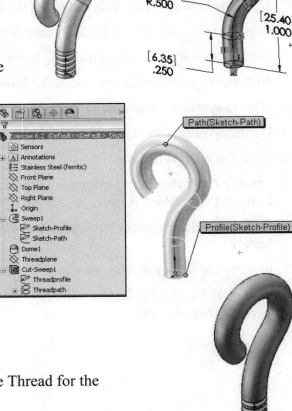

- Create the Swept Path.

- Create the Swept Profile.

- Insert a Swept feature.

- Use the Dome feature, .050in, [1.27].

- Create the Threads. Sketch a circle ⌀.020in, [.51mm] on the Right Plane for the

    Thread profile.

- Use Helix Curve feature for the path. Pitch 0.050in, Revolution 4.0, Starting Angle 0.0deg. Chamfer the bottom face.

- Create the Thread Profile.

- Insert The Swept Cut featue.

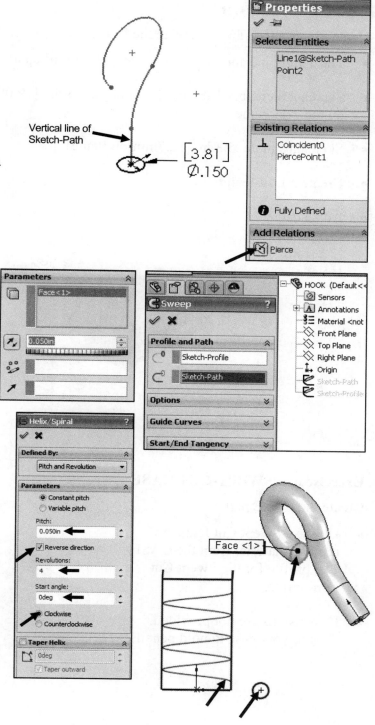

Exercise 6.3: WEIGHT Part

Create the WEIGHT part. Utilize the Loft Base Feature.

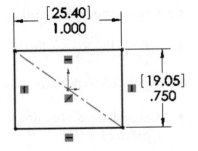

- The Top Plane and Plane1 are 0.5in, [12.7mm] apart.

- Sketch a rectangle 1.000in, [25.4mm] x .750in, [19.05] on the Top Plane.

- Sketch a square .500in, [12.7mm]on Plane1.

- Create a Loft feature.

- Add a centered ⌀.150in, [3.81mm] Thru Hole.

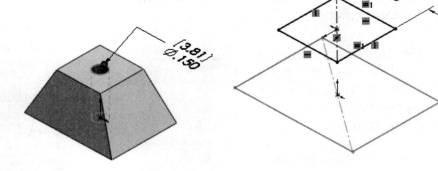

Exercise 6.4: SWEPT-CUT CASE

Create the CASE part.

- Utilize a Swept Cut feature to remove material from the CASE. The profile for the Swept Cut is a semi-circle.

- Dimensions are not provided. Design your case to hold pencils.

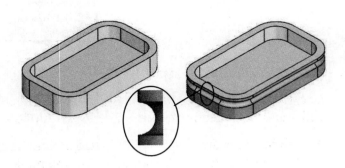

Exercise 6.5: Hole Wizard, Rib and Linear Pattern Features.

Create the part from the illustrated A-ANSI Third
Angle drawing: Front, Top, Right and Isometric views.

- Apply 6061 Alloy material.

- Calculate the volume of the part and locate the
  Center of mass.

- Think about the steps that you would take to build
  the model! **Note: ANSI standard states,
  "Dimensioning to hidden lines should be avoided
  wherever possible. However, sometimes it is
  necessary as below.**

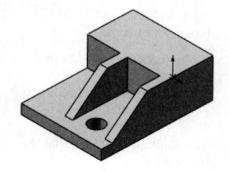

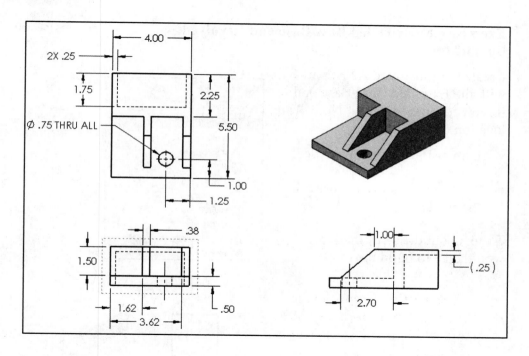

Exercise 6.6: Shell feature.

Create the illustrated part with the Extruded Boss/Base,
Fillet and Shell features. Note: The location of the Origin.

- Dimensions are not provided. Design your case to hold a
  bar of soap. Apply ABS material to the model. Think
  about the steps that you would take to build the model.

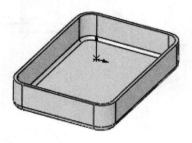

Exercise 6.7: Revolved Base, Hole Wizard, and Circular Pattern features.

Create the illustrated ANSI part with the Revolved Base, Hole Wizard, and Circular Pattern features. Note: The location of the Origin.

- Dimensions are not provided.

- Apply PBT General Purpose Plastic material to the model.

Think about the steps that you would take to build the model.

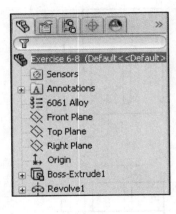

Exercise 6.8: Extruded Boss/Base and Revolved Boss features.

Create the illustrated ANSI part with the Extruded Boss/Base and Revolved Boss features. Note: The location of the Origin.

- Dimensions are not provided.

- Apply 6061 Alloy material to the model.

Think about the steps that you would take to build the model.

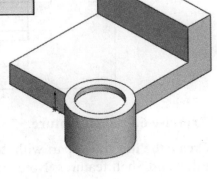

Exercise 6.9: Gem® Paper clip

Create a simple paper clip. You see this common item every day. Create an ANSI - IPS model. Apply material to the model.

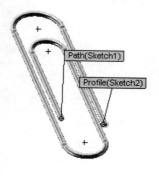

Think about where you would start. What is your Base Sketch?

What are the dimensions? Measure or estimate all needed dimensions from a small gem paper clip. You are the designer.

Note: The paper clip uses a circle as the profile and (lines and arcs) as the path.

Exercise 6.10: Anvil Spring

Create a Variable Pitch Helix. Create an Anvil Spring with 6 coils, two active as illustrated.

Create an ANSI - IPS model.

Sketch a circle, Coincident to the Origin on the Top plane with a .235in dimension.

Create a Variable Pitch Helix/Spiral. Enter the following information as illustrated in the Region Parameters table. The spring has 6 coils. Coils 1,2,5 & 6 are the closed ends of the spring. The spring will have a diameter of .020in. The pitch needs to be slightly larger than the wire. Enter .021in for Pitch. Enter .080in for the free state of the two active coils.

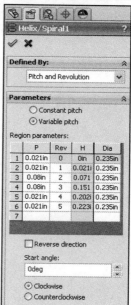

	P	Rev	H	Dia
1	0.021in	0	0in	0.235in
2	0.021in	1	0.021i	0.235in
3	0.08in	2	0.071	0.235in
4	0.08in	3	0.151	0.235in
5	0.021in	4	0.202i	0.235in
6	0.021in	5	0.223i	0.235in
7				

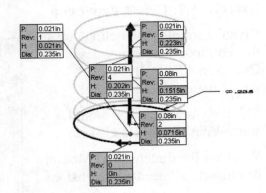

Second, create the profile (circle .021in) for the spring and add a Pierce relation (do not select the endpoint of the path).

Third, create the Sweep feature (path & profile).

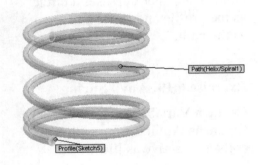

Exercise 6.11: Water Bottle

Create the container as illustrated.
Create an ANSI - IPS model.

You see this common item every day.

Apply material to the model.

Think about where you would start.

What is your Base Sketch?

What is your Base Feature?

What are the dimensions?

View the sample FeatureManager.
Your FeatureManager can (should) be
different. This is just ONE way to
create this part. You are the designer.
Be creative. Estimate any needed
dimension.

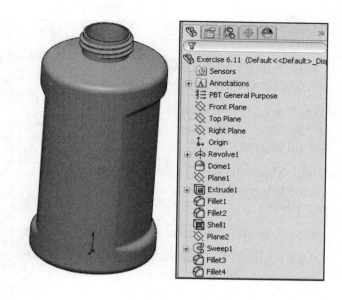

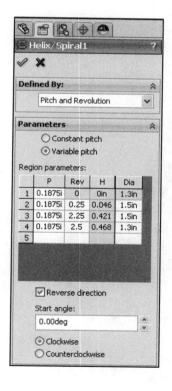

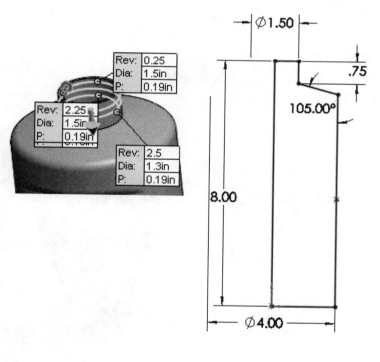

Exercise 6.12: Traditional Ice cream cone

Create a traditional Ice Cream Cone as illustrated. Create an ANSI - IPS model.

This is a common item that you see all of the time. Think about where you would start.

Think about the design features that create this model. Why does the cone use ribs?

Ribs are used for structural integrity.

Use a standard Cake Ice Cream Cone and measure all dimensions (approximately). Do your best.

View the sample FeatureManager. Your FeatureManager can (should) be different. This is just ONE way to create this part. You are the designer. Be creative.

Below are sample models from my Freshman Engineering class.

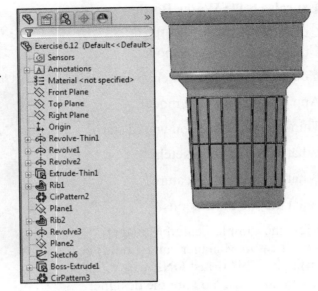

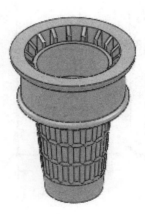

Below are sample models from my Freshman Engineering (Cont:).

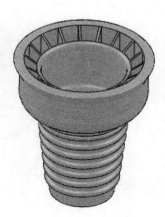

Notes:

Chapter 7
Assembly Modeling

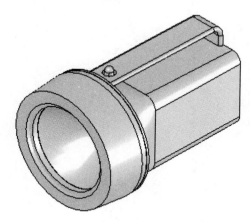

Below are the desired outcomes and usage competencies based on the completion of Chapter 7.

Desired Outcomes:	Usage Competencies:

• Create four assemblies in this project:
 	
   ○ LENSANDBULB assembly
 	
   ○ CAPANDLENS assembly
 	
   ○ BATTERYANDPLATE assembly
 	
   ○ FLASHLIGHT assembly
	• Develop an understanding of Assembly modeling techniques.
	
	• Combine the LENSANDBULB assembly, CAPANDLENS assembly, BATTERYANDPLATE assembly, HOUSING part, and SWITCH part to create the FLASHLIGHT assembly.
	
	• Ability to use the following tools: Insert Component, Hide/Show, Suppress/UnSuppress, Mate, Move Component, Rotate Component, Exploded View, and Interference Detection.

• Create an Inch and Metric Assembly Template.
 	
   ○ ASM-IN-ANSI
 	
   ○ ASM-MM-ISO
	• Ability to apply Document Properties and to create Custom Assembly Templates.

Notes:

Chapter 7 - Assembly Modeling

Chapter Overview

Create four assemblies in this chapter:

1. LENSANDBULB assembly

2. CAPANDLENS assembly

3. BATTERYANDPLATE assembly

4. FLASHLIGHT assembly

Create an inch and metric Assembly Template.

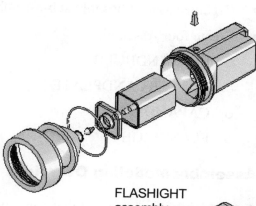

- ASM-IN-ANSI

- ASM-MM-ISO

FLASHIGHT
assembly

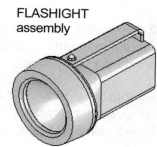

Develop an understanding of assembly modeling techniques. Combine the LENSANDBULB assembly, CAPANDLENS assembly, BATTERYANDPLATE assembly, HOUSING part, and SWITCH part to create the FLASHLIGHT assembly.

Review Standard mate types. Create the following Standard mates:

- Coincident

- Concentric

- Distance

Utilize the following tools: Insert Component, Hide/Show, Suppress/UnSuppress, Mate, Move Component, Rotate Component, Exploded View, and Interference Detection.

After completing the activities in this chapter, you will be able to:

- Create two Assembly Templates: ASM-IN-ANSI and ASM-MM-ISO.

- Apply the following Standard mates: Coincident, Concentric, and Distance.

- Utilize the following tools: Insert Component, Hide/Show Component, Mate, Move Component, Rotate Component, Interference Detection and Suppress/UnSuppress.

- Export a .STL file of the HOUSING part.

- Develop an eDrawing for the FLASHLIGHT assembly.

- Create an Exploded view of the FLASHLIGHT assembly.

- Animate a Collapse view and an Exploded view.

- Organize assemblies into sub-assemblies.

- Create four Assemblies:

   o LENSANDBULB

   o BATTERYANDPLATE

   o CAPANDLENS

   o FLASHLIGHT

Assembly Modeling Overview

An assembly is a document that contains two or more parts. An assembly inserted into another assembly is called a sub-assembly. A part or assembly inserted into an assembly is called a component.

Establishing the correct component relationship in an assembly requires forethought on component interaction. Mates are Geometric relationships that align and fit components in an assembly. Mates remove degrees of freedom from a component.

In dynamics, motion of an object is described in linear and rotational terms. Components possess linear motion along the x, y and z-axes and rotational motion around the x, y, and z-axes.

In an assembly, each component has 6 degrees of freedom: 3 translational (linear) and 3 rotational. Mates remove degrees of freedom. All components are rigid bodies.

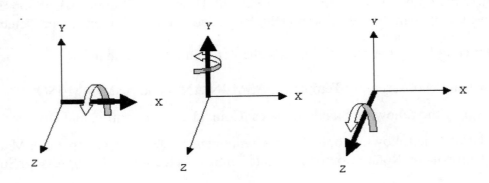

The components do not flex or deform. Components are assembled in this Project with Standard mate types. There are three Mate types displayed in the Mate PropertyManager, they are: *Standard*, *Advanced* and *Mechanical*.

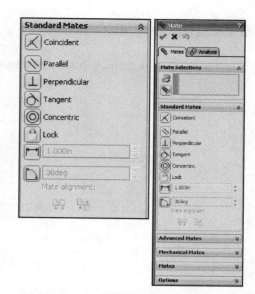

The Standard mate types are Coincident, Parallel, Perpendicular, Tangent, Concentric, Lock, Distance and Angle.

The Advanced mate types are Symmetric, Width, Path Mate, Linear/Linear Coupler, Distance (limit) and Angle.

The Mechanical mate types are Cam, Gear, Rack Pinion, Screw and Universal Joint.

Mates require geometry from two different components. Selected geometry includes Planar Faces, Cylindrical faces, Linear edges, Circular/Arc edges, Vertices, Axes, Temporary axes, Planes, Points and Origins.

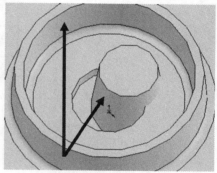

Mates reflect the physical behavior of a component in an assembly. Example: Utilize a Concentric mate between the BATTERY Extruded Boss (Terminal) cylindrical face and the BATTERYPLATE Extruded Boss (Holder) face.

The FLASHLIGHT assembly consists of the following components:

FLASHLIGHT Components:	
BATTERY	BATTERYPLATE
LENS	BULB
O-RING	SWITCH
LENSCAP	HOUSING

How do you organize these components into the FLASHLIGHT assembly? Answer: Create an assembly component layout diagram to determine which components to group into a sub-assembly.

FLASHLIGHT Assembly

Plan the sub-assembly component layout diagram.

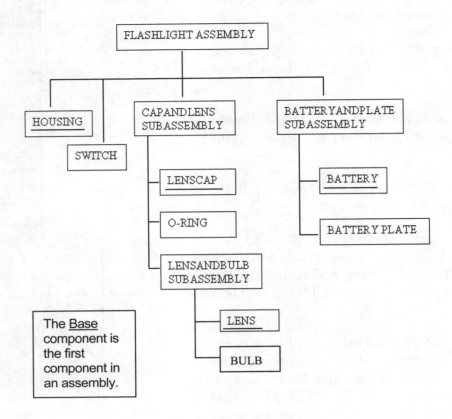

Assembly Layout Structure

The FLASHLIGHT assembly steps are as follows:

- Create the LENSANDBULB sub-assembly from the LENS and BULB component. The LENS is the Base component.

- Create the BATTERYANDPLATE sub-assembly from the BATTERY and BATTERYPLATE component.

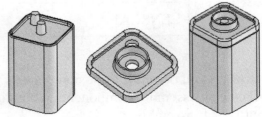

- Create the CAPANDLENS sub-assembly from the LENSCAP, O-RING and LENSANDBULB sub-assembly. The LENSCAP is the Base component.

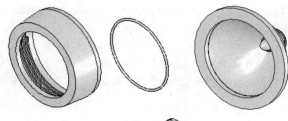

- Create the FLASHLIGHT assembly. The HOUSING is the Base component. Insert the SWITCH, CAPANDLENS and BATTERYANDPLATE component.

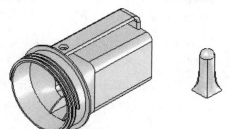

- Modify the dimensions to complete the FLASHLIGHT assembly.

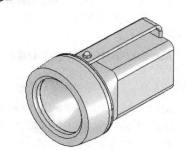

Assembly Techniques

Assembly modeling requires time and practice. Below are helpful hints and techniques to address the Bottom-up design modeling approach.

- Create an assembly layout structure. The layout structure will organize the sub-assemblies and components and save time.

- Insert sub-assemblies and components as lightweight components. Lightweight components save on file size, rebuild time, and overall complexity.

- Set Lightweight components in the Options, Performance section.

- Use the Zoom and Rotate commands to select the correct geometry in the mate process. Zoom in to select the correct face, edge, plane, point, etc.

- Improve display. Apply various colors to features and components.

- Mate with Reference planes when addressing complex geometry. Example: The O-RING does not contain a flat surface or edge.

- Activate the Temporary Axes and the required Planes from the Menu bar toolbar.

- Select Reference planes from the fly-out FeatureManager. Expand the component in the FeatureManager to view the planes and features.

Example: Select the Right Plane of the LENS and the Right Plane of the BULB to be collinear. Do not select the Right Plane of the HOUSING if you want to create a reference between the LENS and the BULB.

Remove display complexity. Hide components and features. Suppress components and features when not required.

- Apply the Move Component and Rotate Component tools if needed before mating. Position the component in the correct orientation.

- Remove unwanted entries. Use Right-click Clear Selections or Right-click Delete from the Assembly Mate Selections text box.

- Verify the position of the mated components. Use Top, Front, Right, and Section views.

- Use caution when you view the color blue in an assembly. Blue indicates that a part is being edited in the context of the assembly.

- Avoid unwanted references. Verify your geometry selections with the PropertyManager.

Assembly Template

An Assembly Document Template is the foundation of the assembly. The FLASHLIGHT assembly and its sub-assemblies require the Assembly Document Template. Utilize the default Assembly Template. Modify the Dimensioning Standard and Units. Create an Assembly Document Template using inch units, ASM-IN-ANSI. Create an Assembly Document Template using millimeter units, ASM-MM-ISO. Save the Templates in the MY-TEMPLATES folder.

Activity: Assembly Templates-ASM-IN-ANSI

Create an Assembly Template.

1) Click **New** ☐ from the Menu bar.

2) Double-click **Assembly** from the Templates tab. The Begin Assembly PropertyManager is displayed.

3) Click **Cancel** ✖ from the Begin Assembly PropertyManager.

Set the Assembly Document Template options.

4) Click **Options** 🗐 , **Document Properties** tab from the Menu bar.

Set units and precision.

5) Select **ANSI** for Overall drafting standard.

6) Click **Units**.

7) Select **IPS, (inch, pound, second)** for Unit system.

8) Select **.123** in the Length units Decimals drop-down box.

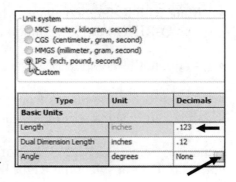

9)   Select **None** for Angular units in the Decimals drop-down box.

10) Click **OK** from the Document Properties - Units dialog box.

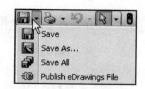

Save the assembly template.

11) Click **Save As** from the drop-down Menu bar.

12) Select the **Assembly Template (*asmdot)** from the Save As type box.

13) Select the **SOLIDWORKS-MODELS 2012/MY-TEMPLATES** folder.

14) Enter **ASM-IN-ANSI** in the File name box.

15) Click **Save**.

Activity: Assembly Templates-ASM-MM-ISO

Create an ISO assembly template.

16) Click **New** ⬜ from the Menu bar.

17) Double-click **Assembly** from the Templates tab.

18) Click **Cancel** ✖ from the Begin Assembly PropertyManager.

Set the Assembly Document Template options.

19) Click **Options** , **Document Properties** tab from the Menu bar.

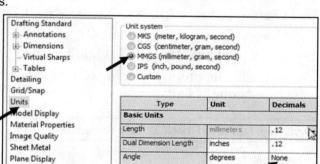

20) Select **ISO** for Overall drafting standard.

21) Click **Units**.

22) Select **MMGS, (millimeter, gram, second)** for Unit system.

23) Select **.12** in the Length units Decimals box.

24) Select **None** in the Angular units Decimals box.

25) Click **OK** from the Document Properties - Units dialog box.

Save the assembly template.

26) Click **Save As** from the drop-down Menu bar.

27) Select the **Assembly Template (*asmdot)** from the Save As type box.

28) Select the **SOLIDWORKS-MODELS 2012/MY-TEMPLATES** folder.

29) Enter **ASM-MM-ISO** in the File name box.

30) Click **Save**.

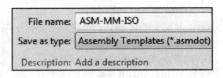

LENSANDBULB Sub-assembly

Create the LENSANDBULB sub-assembly. The LENS is the Base component. LENSANDBULB sub-assembly mates the BULB component to the LENS component. The Right Plane of the LENS and the Right Plane of the BULB are Coincident.

The Top Plane of the LENS and the Top Plane of the BULB are Coincident. The inside Counterbore face of the LENS and the back face of the BULB utilize a Distance mate. The LENS name is added to the LENSANDBULB assembly FeatureManager with the symbol (f). The symbol (f) represents a fixed component. A fixed component cannot move and is locked to the assembly Origin.

The Fixed component state can be removed to create a component that is free to move or rotate. To remove the fixed state, Right-click on the component name in the FeatureManager. Click Float. The component is free to move and rotate.

Suppress the Lens Shield feature to view all surfaces during the mate process. Utilize Open Part to open the LENS from inside the LENSANDBULB assembly.

Utilize the Suppress ↓⊟ Suppress tool from the FeatureManager to Suppress a component. Utilize the UnSuppress ↑⊟ tool to restore the component. Note: The Mates of a suppressed component are also suppressed.

Activity: LENSANDBULB Sub-assembly

Close all documents.
31) Click **Windows**, **Close All** from the Menu bar.

Create the LENSANDBULB sub-assembly.
32) Click **New** ⬜ from the Menu bar.

33) Click the **MY-TEMPLATES** tab.

34) Double-**click ASM-IN-ANSI, [ASM-MM-ISO]**.

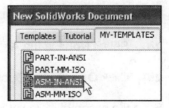

Insert the LENS.
35) Click the **Browse** button.

36)   Select **Part** for file type in the PROJECTS folder.

37)   Double-click **LENS**.

38)   Click **OK** ✔ from the Begin Assembly PropertyManager. The LENS is fixed to the Origin.

39)   Click **Save**.

40)   Enter **LENSANDBULB** for File name in the PROJECTS folder.

41)   Enter **LENS AND BULB ASSEMBLY** for Description.

42)   Click **Save**.

Insert the BULB.

43)   Click **Insert Components** 🖐 from the Assembly toolbar.

44)   Click the **Browse** button.

45)   Double-click **BULB** from the PROJECTS folder.

46)   Click a **position** in front of the LENS as illustrated.

Fit the model to the Graphics window.
47)   Press the **f** key.

Move the BULB.
48)   Click and drag the **BULB** in the Graphics window.

Save the LENSANDBULB.
49)   Click **Save**. View the Assembly FeatureManager.

Suppress the LensShield feature.
50)   **Expand** LENS in the FeatureManager.

51)   **Expand** the Features folder.

52)   Right-click **LensShield** in the FeatureManager.

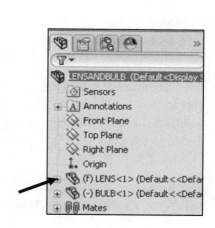

53)   Click **Feature Properties**.

54)   Check the **Suppressed** box.

55)   Click **OK** from the Feature Properties dialog box.

Insert a Coincident mate.

56) Click the **Mate** ✎ Assembly tool. The Mate PropertyManager is displayed.

57) Click **Right Plane** of the LENS from the fly-out FeatureManager.

58) **Expand** BULB in the fly-out FeatureManager.

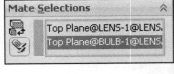

59) **Expand** the Features folder.

60) Click **Right Plane** of the BULB. Coincident mate is selected by default.

61) Click **OK** ✔ from the Mate dialog box. Coincident1 is created.

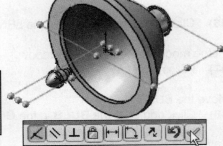

Insert the second Coincident mate.

62) Click **Top Plane** of the LENS from the fly-out FeatureManager.

63) Click **Top Plane** of the BULB from the fly-out FeatureManager. Coincident mate is selected by default.

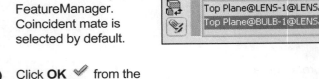

64) Click **OK** ✔ from the Mate dialog box. Coincident2 is created.

🔆 Select face geometry efficiently. Position the mouse pointer in the middle of the face. Do not position the mouse pointer near the edge of the face. Zoom in on geometry. Utilize the Face Selection Filter for narrow faces.

Activate the Face Selection Filter.

65) Click **View**, **Toolbars**, **Selection Filter**. The Selection Filter toolbar is displayed.

66) Click **Filter Faces**. The Filter icon is displayed ⬉▽.

Only faces are selected until the Face Selection Filter is deactivated. Select Clear All Filters 🐦 from the Selection Filter toolbar to deactivate all filters.

Insert a Coincident mate.

67)   **Zoom to Area** 🔍 and **Rotate** 🔄 on the CBORE.

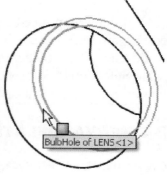

BulbHole of LENS<1>

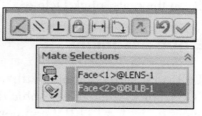

68)   Click the **BulbHole face** of the LENS in the Graphics window as illustrated.

69)   Click the **bottom back flat face, Rovolve1** of the BULB. The Coincident Mate is selected by default.

Revolve1 of BULB<1>

70)   Click **OK** ✔ from the Mate dialog box. Coincident3 is created. Close the Mate PropertyManager.

71)   Click **OK** ✔ from the Mate PropertyManager. The LENSANDBULB is fully defined.

Clear the Face filter.

72)   Click **Clear All Filters** 🐦 from the Selection Filter toolbar.

Display the Mate types.

73)   **Expand** the Mates folder in the FeatureManager. View the inserted mates. Note: The Mates under each sub-component in the FeatureManager.

74)   Click **Right view** 🔲 from the Heads-up View toolbar.

75)   Click **Wireframe** 🔲 from the Heads-up View toolbar.

Save the LENSANDBULB.

76)   Click **Isometric view** 🔲 from the Heads-up View toolbar.

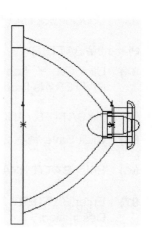

77)   Click **Shaded With Edges** 🔲 from the Heads-up View toolbar.

78)   Click **Save**. View the results in the Graphics window.

If the wrong face or edge is selected, click the face or edge again to remove it from the Mate Selections text box. Right-click Clear Selections to remove all geometry from the Mate Selections text box. To delete a mate from the FeatureManager, right-click on the mate, click Delete.

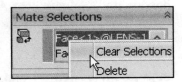

BATTERYANDPLATE Sub-assembly

Create the BATTERYANDPLATE sub-assembly. Utilize two Coincident Mates and one Concentric Mate to assemble the BATTERYPLATE component to the BATTERY component.

Note: Utilize the Selection Filter required. Select planes from the FeatureManager when the Selection Filters are activated.

Activity: BATTERYANDPLATE Sub-assembly

Create the BATTERYANDPLATE sub-assembly.

79) Click **New** ⬜ from the Menu bar. The New SolidWorks Document dialog box is displayed.

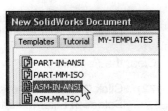

80) Click the **MY-TEMPLATES** tab.

81) Double-click **ASM-IN-ANSI**. The Begin Assembly PropertyManager is displayed.

Insert the BATTERY part.
82) Click **Browse** from the Begin Assembly PropertyManager.

83) Double-click **BATTERY** from the PROJECTS folder.

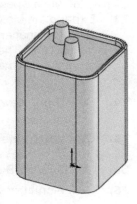

Place the BATTERY.
84) Click **OK** ✔ from the Begin Assembly PropertyManager. The BATTERY is fixed to the Origin.

Save the BATTERYANDPLATE sub-assembly.
85) Click **Save** 💾. Select the **PROJECTS** folder.

86) Enter **BATTERYANDPLATE** for File name.

87) Enter **BATTERY AND PLATE FOR 6-VOLT FLASHLIGHT** for Description.

88) Click **Save**. The BATTERYANDPLATE FeatureManager is displayed.

Insert the BATTERYPLATE part.

89) Click **Insert Components**  from the Assembly toolbar.

90) Click the **Browse** button.

91) Double-click **BATTERYPLATE** from the PROJECTS folder.

92) Click a **position** above the BATTERY as illustrated.

Insert a Coincident mate.

93) Click the **Mate** Assembly tool. The PropertyManager is displayed.

Mate

94) Click the **outside bottom face** of the BATTERYPLATE.

95) Click the **top narrow flat face** of the BATTERY Base Extrude feature as illustrated. Coincident mate ⟨ is selected by default.

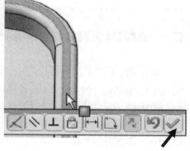

96) Click **OK** from the Mate dialog box. Coincident1 is created.

Insert a Coincident mate.

97) Click **Right Plane** of the BATTERY from the fly-out FeatureManager.

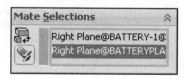

98) Click **Right Plane** of the BATTERYPLATE from the fly-out FeatureManager. Coincident mate ⟨ is selected by default.

99) Click **OK** from the Mate dialog box. Coincident2 is created.

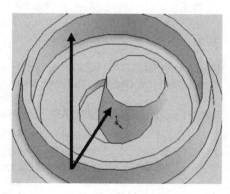

Insert a Concentric mate.

100) Click the center Terminal feature **cylindrical face** of the BATTERY as illustrated.

101) Click the Holder feature **cylindrical face** of the BATTERYPLATE. Concentric mate is selected by default.

102) Click **OK** from the Mate dialog box. Concentric1 is created. Click **OK** from the Mate PropertyManager. Note: If required, deactivate view Sketches from the Menu bar.

103) Expand the Mates folder. View the created mates.

Save the BATTERYANDPLATE.

104) Click **Isometric view**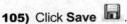
from the Heads-up View
toolbar.

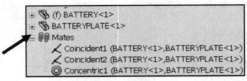

105) Click **Save** .

💡 Tangent Edges and Origins are displayed for educational
purposes.

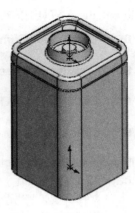

CAPANDLENS Sub-assembly

Create the CAPANDLENS sub-assembly. Utilize two Coincident mates and one Distance
mate to assemble the O-RING to the LENSCAP. Utilize three Coincident mates to
assemble the LENSANDBULB sub-assembly to the LENSCAP component.

Caution: Select the correct reference. Expand the LENSCAP and O-RING. Click the
Right Plane within the LENSCAP. Click the Right Plane within the O-RING.

Activity: CAPANDLENS Sub-assembly

Create the CAPANDLENS sub-assembly.
106) Click **New** from the Menu bar.

107) Click the **MY-TEMPLATES** tab.

108) Double-click **ASM-IN-ANSI**.

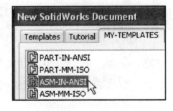

Insert the LENSCAP sub-assembly.
109) Click the **Browse** button.

110) Double-click **LENSCAP** from the PROJECTS folder.

111) Click **OK** ✓ from the Begin Assembly PropertyManager. The
LENSCAP is fixed to the Origin.

Save the CAPANDLENS assembly.
112) Click **Save**. Select the **PROJECTS** folder.

113) Enter **CAPANDLENS** for File name.

114) Enter **LENSCAP AND LENS** for Description.

115) Click **Save**. The CAPANDLENS FeatureManager is displayed.

Insert the O-RING part.

116) Click **Insert Components**  from the Assembly toolbar.

117) Click the **Browse** button.

118) Double-click **O-RING** from the PROJECTS folder.

119) Click a **position** behind the LENSCAP as illustrated.

Insert the LENSANDBULB assembly.

120) Click **Insert Components**  from the Assembly toolbar.

121) Click the **Browse** button.

122) Select **Assembly** for file type in the PROJECTS folder.

123) Double-click **LENSANDBULB**.

124) Click a **position** behind the O-RING as illustrated.

125) Click **Isometric view** .

Move and hide components.

126) Click and drag the **O-RING** and **LENSANDBULB** as illustrated in the Graphics window.

127) Right-click **LENSANDBULB** in the FeatureManager. Click **Hide components** from the Context toolbar.

Insert three mates between the LENSCAP and O-RING.

128) Click the **Mate** Assembly tool. The Mate PropertyManager is displayed.

Insert a Coincident mate.

129) Click **Right Plane** of the LENSCAP in the fly-out FeatureManager.

130) Click **Right Plane** of the O-RING in the fly-out FeatureManager. Coincident mate is selected by default.

131) Click **OK** from the Mate dialog box. Coincident1 is created.

Insert a second Coincident mate.

132) Click **Top Plane** of the LENSCAP in the fly-out FeatureManager. Click **Top Plane** of the O-RING in the fly-out FeatureManager. Coincident mate is selected by default. Click **OK** from the Mate dialog box. Coincident2 is created.

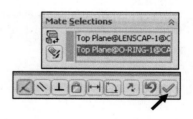

Insert a Distance mate.

133) Click the Shell1 **back inside face** of the LENSCAP as illustrated.

134) Click **Front Plane** of the O-RING in the fly-out FeatureManager.

135) Click **Distance**.

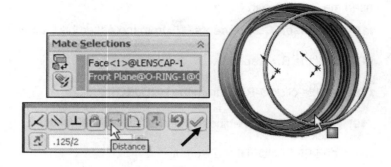

136) Enter **.125/2**in, [**3.175/2mm**].

137) Click **OK** ✔ from the Mate dialog box.

138) Click **OK** ✔ from the Mate PropertyManager.

139) Click **Isometric view** from the Heads-up View toolbar.

140) **Expand** the Mates folder. View the created mates. Note: View the created mates under each sub-component.

141) Click **Save**.

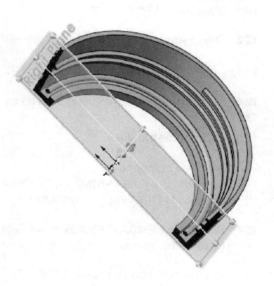

How is the Distance mate, .0625in, [1.588] calculated? Answer:

O-RING Radius (.1250in/2) = .0625in.

O-RING Radius [3.175mm/2] = [1.588mm].

🔆 Utilize a Section view to locate internal geometry for mating and verify position of components.

🔆 Build flexibility into the mate. A Distance mate offers additional flexibility over a Coincident mate. You can modify the value of a Distance mate.

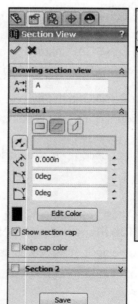

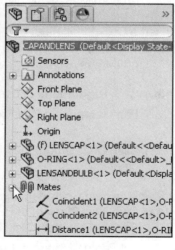

Show the LENSANDBULB.
142) Right-click **LENSANDBULB** in the FeatureManager.

143) Click **Show components** from the Contexts toolbar.

Fit the model to the Graphics window.
144) Press the **f** key.

145) Click **Mate** ✎ from the Assembly toolbar. The Mate
PropertyManager is displayed.

Insert a Coincident mate.
146) Click **Right Plane** of the LENSCAP in the fly-out
FeatureManager.

147) Click **Right Plane** of the LENSANDBULB in the fly-out
FeatureManager. Coincident mate ✕ is selected by default.

148) Click **OK** ✅ from the Mate dialog box.

Insert a Coincident Mate.
149) Click **Top Plane** of the LENSCAP in the fly-out
FeatureManager.

150) Click **Top Plane** of the LENSANDBULB in
the fly-out FeatureManager. Coincident mate
✕ is selected by default.

151) Click **OK** ✅ from the Mate dialog box.

Insert a Coincident Mate.
152) Click the flat inside **narrow back face** of the
LENSCAP.

153) Click the **front flat face** of the
LENSANDBULB. Coincident ✕ is selected
by default.

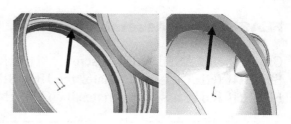

154) Click **OK** ✅ from the Mate dialog box.

155) Click **OK** ✅ from the Mate
PropertyManager. View the created mates.

156) **Expand** the Mates folder. View the created
mates.

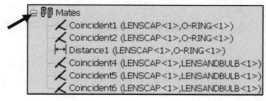

Confirm the location of the O-RING.

157) Click **Right Plane** of the CAPANDLENS from the FeatureManager.

158) Click **Section view** 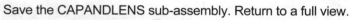 from the Heads-up View toolbar.

159) Click **Isometric view** from the Heads-up View toolbar. **Expand** the Section 2 box.

160) Click **inside** the Reference Section Plane box. Click **Top Plane** of the CAPANDLENS from the fly-out FeatureManager.

161) Click **OK** from the Section View PropertyManager.

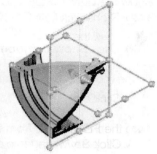

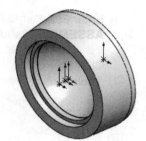

Save the CAPANDLENS sub-assembly. Return to a full view.

162) Click **Section view** from the Heads-up View toolbar.

163) Click **Save** .

The LENSANDBULB, BATTERYANDPLATE, and CAPANDLENS sub-assemblies are complete. The components in each assembly are fully defined. No minus (-) sign or red error flags exist in the **FeatureManager**. Insert the sub-assemblies into the final FLASHLIGHT assembly.

FLASHLIGHT Assembly

Create the FLASHLIGHT assembly. The HOUSING is the Base component. The FLASHLIGHT assembly mates the HOUSING to the SWITCH component. The FLASHLIGHT assembly mates the CAPANDLENS and BATTERYANDPLATE.

Activity: FLASHLIGHT Assembly

Create the FLASHLIGHT assembly.

164) Click **New** from the Menu bar. Click the **MY-TEMPLATES** tab. Double-click **ASM-IN-ANSI**. The Begin Assembly PropertyManager is displayed.

Insert the HOUSING and SWITCH.

165) Click the **Browse** button. Select **Parts** for file type.

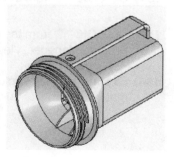

166) Double-click **HOUSING** from the PROJECTS folder.

167) Click **OK** from the Begin Assembly PropertyManager. The HOUSING is fixed to the Origin.

168) Click **Insert Components** from the Assembly toolbar.

169) Click the **Browse** button.

170) Double-click **SWITCH** from the PROJECTS folder.

171) Click a **position** in front of the HOUSING as illustrated.

Save the FLASHLIGHT assembly.
172) Click **Save**. Select the **PROJECTS** folder.

173) Enter **FLASHLIGHT** for File name. Enter **FLASHLIGHT ASSEMBLY** for Description.

174) Click **Save**. The FLASHLIGHT FeatureManager is displayed.

Insert a Coincident mate.
175) Click the **Mate** Assembly tool.

176) Click **Right Plane** of the HOUSING from the fly-out FeatureManager.

177) Click **Right Plane** of the SWITCH from the fly-out FeatureManager. Coincident mate is selected by default.

178) Click **OK** from the Mate dialog box.

Insert a Coincident mate.
179) Click **View**, check **Temporary Axes** from the Menu bar.

180) Click the **Temporary axis** inside the Switch Hole of the HOUSING. Click **Front Plane** of the SWITCH from the fly-out FeatureManager. Coincident mate is selected by default.

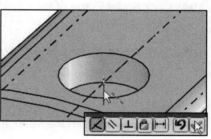

181) Click **OK** from the Mate dialog box.

Insert a Distance mate.
182) Click the **top face** of the Handle.

183) Click the **Vertex** on the Loft top face of the SWITCH.

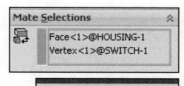

184) Click **Distance**. Enter **.100**in, [**2.54**]. Check the **Flip Direction** box.

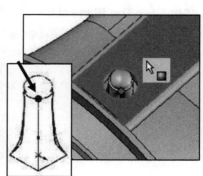

185) Click **OK** ✔ from the Mate dialog box.

186) Click **OK** ✔ from the Mate PropertyManager.

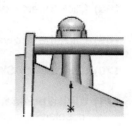

Insert the CAPANDLENS assembly.
187) Click **View**, un-check **Temporary Axis** from the Menu bar.

188) Click **View**, un-check **Origins** from the Menu bar.

189) Click **Insert Components** 🗗 from the Assembly toolbar.

190) Click the **Browse** button. Select **Assembly** for file type.

191) Double-click **CAPANDLENS** from the PROJECTS folder.

Place the sub-assembly.
192) Click a **position** in front of the HOUSING as illustrated.

Insert Mates between the HOUSING component and the
CAPANDLENS sub-assembly.
193) Click the **Mate** 🖉 Assembly tool. The Mate PropertyManager is
displayed.

Insert a Coincident mate.
194) Click **Right Plane** of the HOUSING from the fly-out
FeatureManager.

195) Click **Right Plane** of the CAPANDLENS from the fly-out
FeatureManager. Coincident mate ⟨ is selected by
default.

196) Click **OK** ✔ from the Mate dialog box.

Insert a Coincident mate.
197) Click **Top Plane** of the HOUSING from the fly-out
FeatureManager.

198) Click **Top Plane** of the CAPANDLENS from the fly-out
FeatureManager. Coincident mate ⟨ is selected by
default.

199) Click **OK** ✔ from the Mate dialog box.

Insert a Coincident mate.
200) Click the **front face** of the Boss-Stop on the HOUSING.

Rotate the view.
201) Press the **Left arrow key** to view the back face.

202) Click the **back face** of the CAPANDLENS. Coincident mate
is selected by default.

203) Click **OK** from the Mate dialog
box. Click **OK** from the Mate
PropertyManager.

Save the FLASHLIGHT assembly.
204) Click **Isometric view**.

205) Click **Save**.

Insert the BATTERYANDPLATE sub-assembly.
206) Click **Insert Components** from the Assembly toolbar.

207) Click the **Browse** button. Select **Assembly** for file type.

208) Double-click **BATTERYANDPLATE** from the PROJECTS
folder.

209) Click a **position** to the left of the
HOUSING as illustrated.

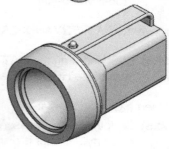

Rotate the part.
210) Click **BATTERYANDPLATE** in the
FeatureManager.

211) Click **Rotate Component** from the
Assembly toolbar.

212) Rotate the **BATTERYANDPLATE** until
it is approximately parallel with the
HOUSING. Click **OK** from the
Rotate Component PropertyManager.

Insert a Coincident mate.
213) Click the **Mate** Assembly tool. Click **Right Plane** of the
HOUSING from the fly-out FeatureManager.

214) Click **Front Plane** of the BATTERYANDPLATE from the fly-out
FeatureManager. Coincident is selected by default.

215) Click **OK** from the Mate dialog box.

216) Move the **BATTERYANDYPLATE** in front of the HOUSING.

Insert a Coincident mate.
217) Click **Top Plane** of the HOUSING in the fly-out FeatureManager.

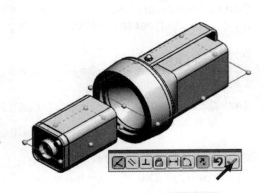

218) Click **Right Plane** of the BATTERYANDPLATE in the fly-out FeatureManager. Coincident mate ⊼ is selected by default.

219) Click **OK** ✔ from the Mate dialog box.

220) Click **OK** ✔ from the Mate PropertyManager.

Display the Section view.
221) Click **Right Plane** in the FLASHLIGHT Assembly FeatureManager.

222) Click **Section view** ▨ from the Heads-up View toolbar.

223) Click **OK** ✔ from the Section View PropertyManager.

Move the BATTERYANDPLATE in front of the HOUSING.
224) Click and drag the **BATTERYANDPLATE** in front of the HOUSING as illustrated.

Insert a Coincident mate.
225) Click the **Mate** ◈ Assembly tool. The Mate PropertyManager is displayed.

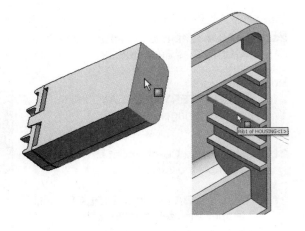

226) Click the **back center Rib1 face** of the HOUSING.

227) Click the **bottom face** of the BATTERYANDPLATE. Coincident ⊼ is selected by default.

228) Click **OK** ✔ from the Mate dialog box.

229) Click **Isometric view** ◻ from the Heads-up View toolbar.

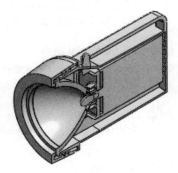

230) Click **OK** ✔ from the Mate PropertyManager.

Display the Full view.

231) Click **Section view**  from the Heads-up View toolbar.

Save the FLASHLIGHT.

232) Click **Save** 💾.

🔍　Additional information on Assembly, Move Component, Rotate Component, and Mates is available in SolidWorks Help Topics.

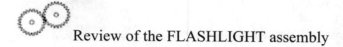　Review of the FLASHLIGHT assembly

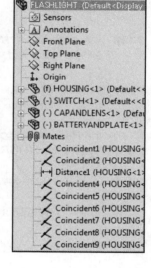

The FLASHLIGHT assembly consisted of the HOUSING part, SWITCH part, CAPANDLENS sub-assembly and BATTERYANDPLATE sub-assembly.

The CAPANDLENS sub-assembly contained the BULBANDLENS sub-assembly, the O-RING and the LENSCAP part. The BATTERANDPLATE sub-assembly contained the BATTERY and BATTERYPLATE part.

You inserted eight Coincident mates and a Distance mate. Through the Assembly Layout illustration you simplified the number of components into a series of smaller assemblies. You also enhanced your modeling techniques and skills.

You still have a few more areas to address. One of the biggest design issues in assembly modeling is interference. Let's investigate the FLASHLIGHT assembly.

💡　Clearance Verification checks the minimum distance between components and reports any value that fails to meet your input value of the minimum clearance. View SolidWorks Help for additional information.

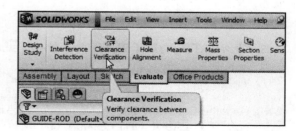

Addressing Interference Issues

There is an interference issue between the FLASHLIGHT components. Address the design issue. Adjust Rib2 on the HOUSING. Test with the Interference Check command. The FLASHLIGHT assembly is illustrated in inches.

Activity: Addressing Interference Issues

Check for interference.

233) Click the **Interference Detection** tool from the Evaluate tab in the CommandManager.

234) Delete **FLASHLIGHT.SLDASM** from the Selected Components box.

235) Click **BATTERYANDPLATE** from the fly-out FeatureManager.

236) Click **HOUSING** from the fly-out FeatureManager.

237) Click the **Calculate** button. The interference is displayed in red in the Graphics window.

238) Click each **Interference** in the Results box to view the interference in red with Rib2 of the HOUSING.

239) Click **OK** from the Interference Detection PropertyManager.

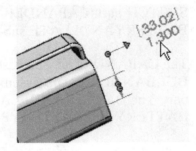

Modify the Rib2 dimension to address the interference issue.
240) **Expand** the HOUSING in the FeatureManager.

241) Double-click on the **Rib2** feature.

242) Double click **1.300**in, [33.02]. Enter **1.350**in, [34.29].

243) **Rebuild** the model. Click **OK** .

Recheck for Interference.

244) Click **Interference Detection** from the Evaluate tab. The Interference dialog box is displayed.

245) Delete **FLASHLIGHT.SLDASM** from the Selected Components box.

246) Click **BATTERYANDPLATE** from the FeatureManager.

247) Click **HOUSING** from the FeatureManager.

248) Click the **Calculate** button. No Interference is displayed in the Results box. The FLASHLIGHT design is complete.

249) Click **OK** ✅ from the Interference Detection PropertyManager.

Save the FLASHLIGHT.
250) Click **Save**.

Exploded View

The Exploded view  tool utilizes the Explode PropertyManager.

An Exploded view illustrates how to assemble the components in an assembly.

Create an Exploded view in this section with seven steps. Click and drag components in the Graphics window.

The Manipulator icon ⊥ indicates the direction to explode. Select an alternate component edge for the Explode direction. Drag the component in the Graphics window or enter an exact value in the Explode distance box. Manipulate the top-level components in the assembly.

Access the Explode View 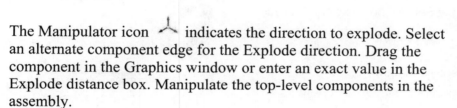 tool from the following:

- Click the **ConfigurationManager** tab.

- Right-click **Default**.

- Click **New Exploded View**. The Explode PropertyManager is displayed.

- Click the **Exploded View** 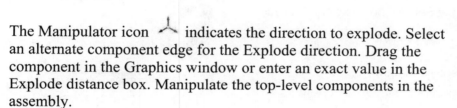 tool in the Assemble toolbar. The Explode PropertyManager is displayed.

- Select **Insert**, **Exploded View** from the Menu bar. The Explode PropertyManager is displayed.

Activity: FLASHLIGHT Assembly-Exploded View

Insert an Exploded view.

251) Click the **Exploded View**  tool from the
Assembly toolbar. The Explode
PropertyManager is displayed.

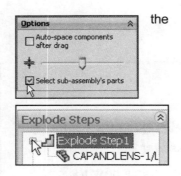

Create Explode Step1.
252) Check **Select sub-assembly's parts** in
the Options box.

253) Click **CAPANDLENS** from the Graphics
window.

254) Click and drag the **blue/yellow manipulator handle** to
the front.

255) Release the **mouse** button.

256) Click **Done** from the Settings box. Explode Step1 is
created.

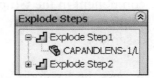

Fit the model to the Graphics window.
257) Press the **f** key.

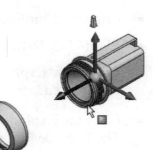

Create Explode Step2.
258) Click **SWITCH** from the fly-out FeatureManager.

259) Drag the **green/yellow manipulator handle** upward.

260) Click **Done** from the Settings box. Explode Step2 is created.

Create Explode Step3.
261) Click **LENS** in the Graphics window. Note: You can also select
LENS from the Fly-out FeatureManager.

262) Drag the **blue/yellow manipulator handle** to the front as
illustrated.

263) Click **Done** from the Settings box.

Create Explode Step4.
264) Click **O-RING** from the fly-out FeatureManager.

265) Drag the **blue/yellow manipulator handle** to the front of the LENS as illustrated.

266) Click **Done** from the Settings box.

Create Explode Step5.
267) Click **HOUSING** from the fly-out FeatureManager.

268) Drag the **blue/yellow manipulator handle** backwards to expose the BATTERYANDPLATE.

269) Click **Done** from the Settings box.

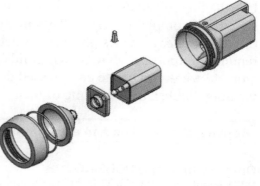

Create Explode Step6
270) Click **BATTERYPLATE** from the fly-out FeatureManager.

271) Drag the **blue/yellow manipulator handle** forward.

272) Click **Done** from the Settings box..

Create Explode Step7.
273) Click **BULB** from the fly-out FeatureManager.

274) Drag the **blue/yellow manipulator handle** to the back of the LENS as illustrated.

275) Click **Done**.

276) Click **OK** ✔ for the Explode PropertyManager.

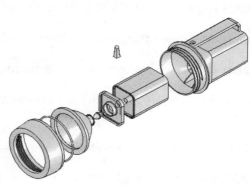

Fit the model to the Graphics window.
277) Press the **f** key.

Remove the Exploded State.
278) **Right-click** in the Graphics window.

279) Click **Collapse** from the Pop-up menu.

280) Click **Isometric view** ⬠ from the Heads-up view toolbar.

281) Click **Save** 💾.

Export Files and eDrawings

You receive a call from the sales department. They inform you that the customer increased the initial order by 200,000 units. However, the customer requires a prototype to verify the design in six days. What do you do? Answer: Contact a Rapid Prototype supplier. You export three SolidWorks files:

- HOUSING

- LENSCAP

- BATTERYPLATE

Use the Stereo Lithography (STL) format. Email the three files to a Rapid Prototype supplier. Example: Paperless Parts Inc. (www.paperlessparts.com). A Stereolithography (SLA) supplier provides physical models from 3D drawings. 2D drawings are not required. Export the HOUSING. SolidWorks eDrawings provides a facility for you to animate, view and create compressed documents to send to colleagues, customers and vendors. Publish an eDrawing of the FLASHLIGHT assembly.

Activity: Export Files and eDrawings

Open and Export the HOUSING.

282) Right-click **HOUSING** from the FeatureManager.

283) Click **Open Part** from the Context toolbar.

284) Click **Save As** from the drop-down Menu bar.

285) Select **STL (*.stl)** from the Save as type drop-down menu. The dialog box is displayed.

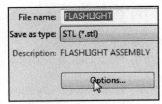

286) Click the **Options** button.

287) Click the **Binary** box from the Output as format box.

288) Click the **Course** box for Resolution.

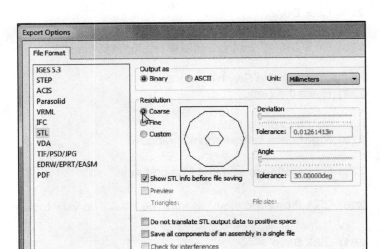

Create the binary STL file.
289) Click **OK** from the Export Options dialog box.

290) Click **Save** from the Save dialog box. A status report is provided.

291) Click **Yes**.

Publish an eDrawing and email the document to a colleague.

Create the eDrawing and animation.
292) Click **File**, **Publish to eDrawings** from the Menu bar.

293) Click the **Play** button.

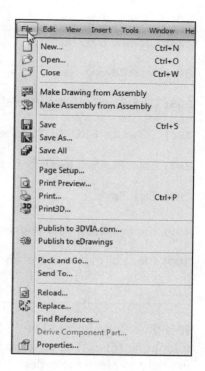

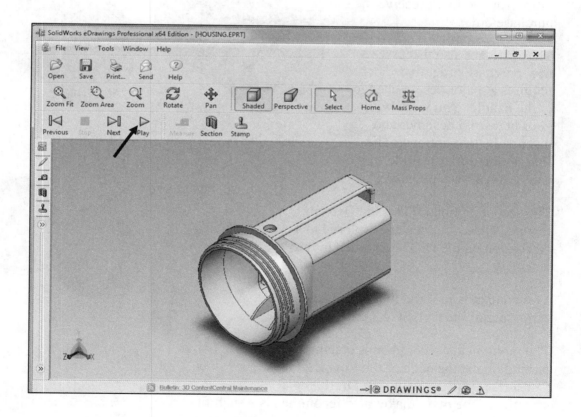

Stop the animation.
294) Click the **Stop** button.

Save the eDrawing.

295) Click **Save** Save from the eDrawing Main menu.

296) Select the **PROJECTS** folder.

297) Enter **FLASHLIGHT** for File name.

298) Click **Save**.

299) Close ☒ the eDrawing dialog box.

300) Close all models in the session.

It is time to go home. The
telephone rings. The customer
is ready to place the order.
Tomorrow you will receive the
purchase order.

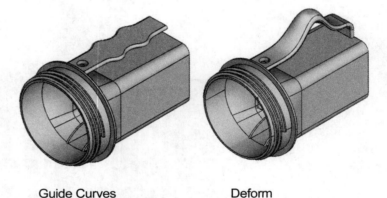

Guide Curves   Deform

The customer also discusses a
new purchase order that
requires a major design change
to the handle. You work with
your industrial designer and
discuss the two options. The
first option utilizes Guide
Curves on a Sweep feature.

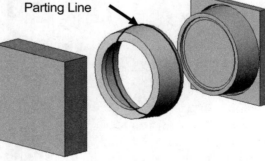

Parting Line

The second option utilizes the Deform
feature. These features are explored in a
discussion on the DVD contained in the
text.

You contact your mold maker and send an
eDrawing of the LENSCAP.

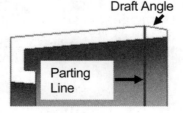

Draft Angle

Parting
Line

The mold maker recommends placing the
parting line at the edge of the Revolved
Cut surface and reversing the Draft Angle
direction. The mold maker also recommends a snap fit
versus a thread to reduce cost. The Core-Cavity mold
tooling is explored in the project exercises.

Additional information on Interference Detection, eDrawings, STL files (stereolithography), Guide Curves, Deform and Mold Tools are available in SolidWorks Help Topics.

Chapter Summary

The FLASHLIGHT contains over 100 features, Reference planes, sketches and components. You organized the features in each part. You developed an assembly layout structure to organize your components.

The O-RING utilized a Swept Base feature. The SWITCH utilized a Loft Base feature. The simple Swept feature requires two sketches: a path and a profile. A complex Swept requires multiple sketches and Guide Curves. The Loft feature requires two or more sketches created on different planes.

The LENSCAP and HOUSING utilized a variety of features. You applied design intent to reuse geometry through Geometric relationships, symmetry and patterns.

The assembly required an Assembly Template. You utilized the ASM-IN-ANSI Template to create the LENSANDBULB, CAPANDLENS, BATTERYANDPLATE and FLASHLIGHT assemblies.

You created an STL file of the Housing and an eDrawing of the FLASHLIGHT assembly to communicate with your vendor, mold maker and customer. Review the chapter exercises before moving on to the next chapter.

Chapter Terminology

Assembly Component Layout Diagram: A diagram used to plan the top-level assembly organization. Organize parts into smaller subassemblies. Create a flow chart or manual sketch to classify components.

Assembly Techniques: Methods utilized to create efficient and accurate assemblies. You can create assemblies using bottom-up design, Top-down design, or a combination of both methods.

Bottom-Up Assembly: You can create assemblies using bottom-up design, top-down design, or a combination of both methods. Bottom-up design is the traditional method. You first design and model part, then insert it into an assembly and apply mates to position the part. To modify the part, you must edit the individually part. The change is then reflected in the assembly.

Coincident mate: Positions selected faces, edges, and planes, (in combination with each other or combined with a single vertex) so they share the same infinite plane. Note: Position two vertices so they touch.

CommandManager: The CommandManager is a Context-sensitive toolbar that dynamically updates based on the toolbar you want to access. By default, it has toolbars embedded in it based on the document type. When you click a tab below the Command Manager, it updates to show that toolbar. For example, if you click the **Sketches** tab, the Sketch toolbar is displayed.

Concentric mate: Places the selections so that they share the same center line.

Distance mate: Places the selected items with the specified distance between them.

eDrawings: A compressed document used to animate and view SolidWorks documents.

Exploded View: You create exploded views by selecting and dragging parts in the graphics area, creating one or more explode steps.

Interference Detection: With Interference Detection, you can: Determine the interference between components, display the true volume of interference as a shaded volume, or change the display settings of the interfering and non-interfering components to see the interference better.

Mate References: Mate references specify one or more entities of a component to use for automatic mating. When you drag a component with a mate reference into an assembly, the SolidWorks software tries to find other combinations of the same mate reference name and mate type. If the name is the same, but the type does not match, the software does not add the mate.

Parallel mate: Places the selected items so they remain a constant distance apart from each other.

Perpendicular mate: Places the selected items at a 90° angle to each other.

Stereo Lithography (STL) format: STL format is the type of file format requested by Rapid Prototype manufacturers.

Suppress: In a part document, you can suppress any feature. When you suppress a feature, it is removed from the model (but not deleted). The feature disappears from the model view and is shown in gray in the FeatureManager design tree. In an assembly document, you can suppress features that belong to the assembly. These include mates, assembly feature holes and cuts, and component patterns.

Tangent mate: Places the selected items tangent to each other (at least one selection must be a cylindrical, conical, or spherical face).

Templates: Templates are part, drawing, and assembly documents that include user-defined parameters and are the basis for new documents.

Top-Down Assembly: Top-down design is also referred to as "In-Context design" in the SolidWorks Help. In Top-down design parts' shapes, sizes, and locations are designed in the assembly using the Edit Component tool.

Questions

1. True or False. A Part Template is the foundation for an assembly document. Explain you answer.

2. Describe the difference between the Distance Mate option and the Coincident Mate option. Provide an example.

3. Describe an assembly or sub-assembly. Are they the same?

4. Describe five proven assembly modeling techniques. Can you add a few more?

5. Explain how to determine an interference between components in an assembly. Provide an example.

6. Describe the Deform feature.

7. Describe Mates. Why are Mates important in assembling components?

8. In an assembly, each component has_____# degrees of freedom. Name them.

9. True or False. A fixed component cannot move and is locked to the Origin.

10. Describe a Section view.

11. Describe a Suppressed feature and component? Provide an example.

12. Identify the type a faces utilized for a Concentric mate.

13. List the Standard mate types. Where would you locate additional information on a Tangent Mate?

14. True or False. Only Planes are utilized for Mate References. Explain your answer.

15. True or False. Only faces are utilized for Mate References. Explain your answer.

16. Define the steps required to create an Exploded view.

17. List the names of the following icons from the Assembly toolbar.

A   B   C   D   E   F   G   H   I

A	B	C
D	E	F
G	H	I

Exercises

Exercise 7.1: Weight-Hook Assembly

Create an ANSI, IPS Weight-Hook assembly. The Weight-Hook assembly has two components: WEIGHT and HOOK.

- Create a new assembly document. Copy and insert the WEIGHT part from the Chapter 7 Homework folder on the DVD in the book. Note: Do not use the components directly from the folder in the DVD. Copy all parts to your computer.

- Fix the WEIGHT to the Origin as illustrated in the Assem1 FeatureManager.

- Insert the HOOK part from the Chapter 7 Homework folder into the assembly.

- Insert a Concentric mate between the inside top cylindrical face of the WEIGHT and the cylindrical face of the thread. Concentric is the default mate.

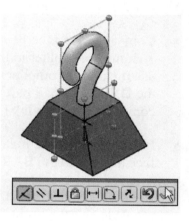

- Insert the first Coincident mate between the top edge of the circular hole of the WEIGHT and the top circular edge of Sweep1, above the thread.

- Coincident is the default mate. The HOOK can rotate in the WEIGHT.

- Fix the position of the HOOK. Insert the second Coincident mate between the Right Plane of the WEIGHT and the Right Plane of the HOOK. Coincident is the default mate.

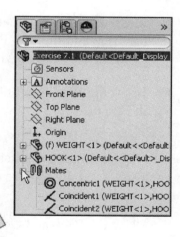

- Expand the Mates folder and view the created mates. Note: Tangent edges and origins are displayed for educational purposes.

Exercise 7.2: Weight-Link Assembly

Create an ANSI, IPS Weight-Link assembly. The Weight-Link assembly has two components and a sub-assembly: Axle component, FLATBAR component, and the Weight-Hook sub-assembly that you created in Exercise 7.1. Note: Tangent edges and origins are displayed for educational purposes.

- Create a new assembly document. Copy and insert the Axle part from the Chapter 7 Homework folder in the book DVD. Note: Do not use the parts / components directly from the folder on the DVD. Copy all parts and components to your computer from the DVD.

- Fix the Axle component to the Origin.

- Copy and insert the FLATBAR part from the Chapter 7 Homework folder in the book DVD. Note: Do not use the parts / components directly from the folder on the DVD. Copy all parts and components to your computer from the DVD.

- Insert a Concentric mate between the Axle cylindrical face and the FLATBAR inside face of the top circle.

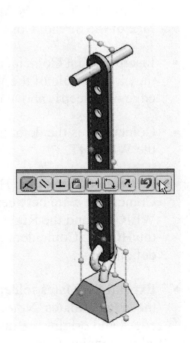

- Insert a Coincident mate between the Front Plane of the Axle and the Front Plane of the FLATBAR.

- Insert a Coincident mate between the Right Plane of the Axle and the Top Plane of the FLATBAR. Position the FLATBAR as illustrated.

- Insert the Weight-Hook sub-assembly that you created in Exercise 7.1.

🔅 Determine the static and dynamic behavior of mates in each sub-assembly before creating the top level assembly.

🔅 Use the Pack and Go option to save an assembly or drawing with references. The Pack and Go tool saves either to a folder or creates a zip file to e-mail. View SolidWorks help for additional information.

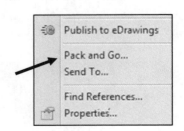

- Insert a Tangent mate between the inside bottom cylindrical face of the FLATBAR and the top circular face of the HOOK, in the Weight-Hook assembly. Tangent mate is selected by default.

- Insert a Coincident mate between the Front Plane of the FLATBAR and the Front Plane of the Weight-Hook sub-assembly. Coincident mate is selected by default. The Weight-Hook sub-assembly is free to move in the bottom circular hole of the FLATBAR.

Exercise 7.3: Binder Clip

- Create a simple Gem® binder clip. You see this item every day.

- Create an ANSI - IPS assembly. Create two components - Binder and Binder Clip.

- Apply material to each component and address all needed mates. Think about where you would start.

What is your Base Sketch for each component?

What are the dimensions? Measure (estimate) all dimensions (approximately) from a small or large Gem binder clip.

You are the designer.

View SolidWorks Help or Chapter 4 on the Swept Base feature to create the Binder Clip component.

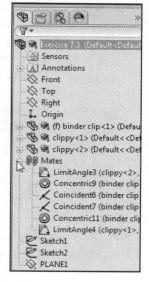

🔅 Determine the static and dynamic behavior of mates in each sub-assembly before creating the top level assembly.

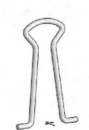

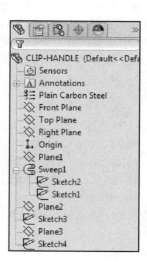

Exercise 7.4: Limit Mate (Advanced Mate Type)

- Open the assembly (Limit Mate) from the Chapter 7 Homework/Limit Mate folder located on the DVD in the book.

- Copy the assembly (components and sub-assemblies) to your working folder. Do not work directly from the DVD.

- Insert a Limit Mate to restrict the movement of the Slide Component - lower and upper movement. (Use the Measure tool to obtain max and min distances).

- Use SolidWorks Help for additional information.

A Limit Mate is an Advanced Mate type. Limit mates allow components to move within a range of values for distance and angle mates. You specify a starting distance or angle as well as a maximum and minimum value.

- Save the model and move the Slide to view the results in the Graphics window. Think about how you would use this mate type in other assemblies.

💡 Use the Pack and Go option to save an assembly or drawing with references. The Pack and Go tool saves either to a folder or creates a zip file to e-mail. View SolidWorks help for additional information.

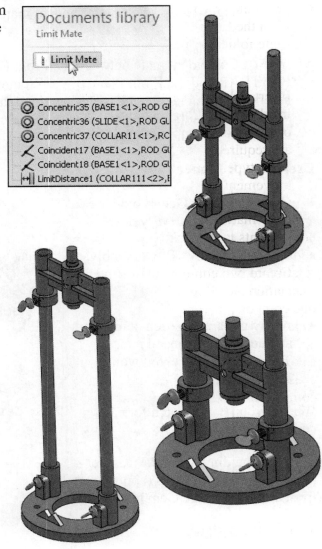

Exercise 7.5: Screw Mate (Mechanical Mate Type)

- Open the assembly (**Screw Mate)** from the Chapter 7 Homework/Screw Mate folder located on the DVD in the book.

- Insert a Screw mate between the inside Face of the Base and the Face of the vice and any other mates that are required. A Screw is a Mechanical Mate type. View the avi file for proper movement.

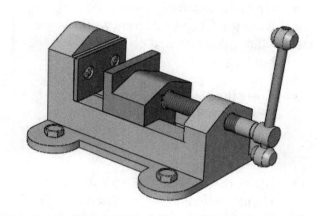

A Screw mate constrains two components to be concentric, and also adds a pitch relationship between the rotation of one component and the translation of the other. Translation of one component along the axis causes rotation of the other component according to the pitch relationship. Likewise, rotation of one component causes translation of the other component. Use SolidWorks Help if needed.

Note: Use the Select Other tool (See SolidWorks Help if needed) to select the proper inside faces and to create the Screw mate for the assembly.

- Rotate the handle and view the results. Think about how you would use this mate type in other assemblies..

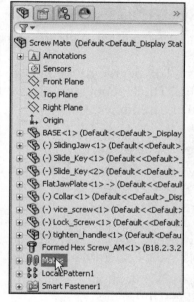

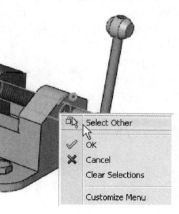

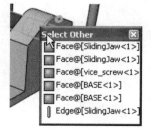

Exercise 7.6: Angle Mate

- Open the assembly (**Angle Mate**) from the Chapter 7 Homework/Angle Mate folder located on the DVD in the book.

- Copy the assembly (components and sub-assemblies) to your working folder. Do not work directly from the DVD.

- Move the Handle in the assembly. The Handle is free to rotate. Set the angle of the Handle. Insert an Angle mate (165 degrees) between the Handle and the Side of the valve using Planes. An Angle mate places the selected items at the specified angle to each other.

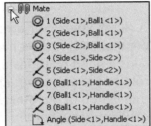

- The Handle has a 165-degree Angle mate to restrict flow through the valve. Think about how you would use this mate type in other assemblies.

Exercise 7.6A: Angle Mate (Cont:)

- Open the SolidWorks FlowXpress Tutorial under the Design Analysis folder. Follow the directions.

- Create two end caps (lids) for the ball value using the Top-down Assembly method. Note: The Reference - In-Content symbols in the FeatureManager.

- Modify the Appearance of the body to observe the changes - enhance visualization. Apply the Select-other tool to obtain access to hidden faces and edges

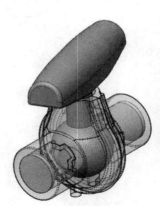

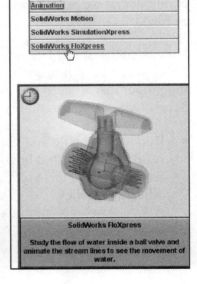

Exercise 7.7: Symmetric Mate (Advanced Mate)

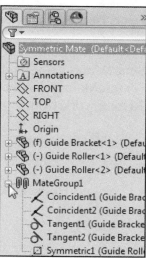

- Open the assembly (Symmetric Mate) from the Chapter 7 Homework/Symmetric Mate folder located on the DVD in the book.

- Copy the assembly (components and sub-assemblies) to your working folder. Do not work directly from the DVD.

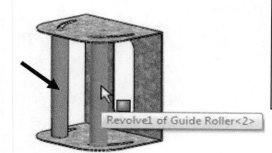

- Insert a Symmetric Mate for the Guild Rollers. Think about how you would use this mate type in other assemblies

Exercise 7.8: Gear Mate (Mechanical mate)

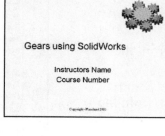

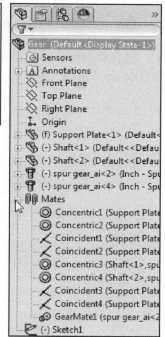

- View the ppt presentation on Gears located in the Chapter 7 Homework/Gear Mate folder on the DVD in the book.

- Create the Gear assembly as illustrated below. All needed information is provided in the ppt. Create all needed components and mates.

- Use the SolidWorks Toolbox. The SolidWorks Toolbox is an add-in.

- View the avi file in the Chapter 7 Homework/Gear Mate folder for proper movement.

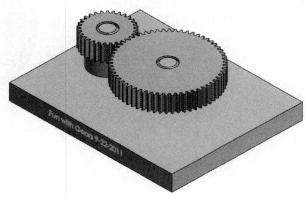

Exercise 7.9: Counter Weight Assembly

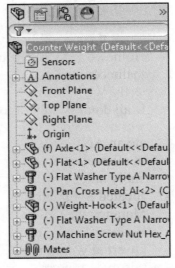

Create the Counter Weight assembly as illustrated using SmartMates and Standard mates from the Assembly FeatureManager. All components are supplied in the Chapter 7 Homework/Counter-Weight folder on the DVD.

Copy all components to your working folder. The Counter Weight consists of the following items:

- Weight-Hook sub-assembly

- Weight

- Eye Hook

- Axle component (f). Fixed to the origin.

- Flat component

- Flat Washer Type A (from the SolidWorks Toolbox)

- Pan Cross Head Screw (from the SolidWorks Toolbox)

- Flat Washer Type A (from the SolidWorks toolbox)

- Machine Screw Nut Hex (from the SolidWorks Toolbox)

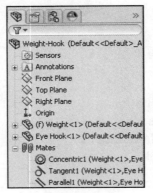

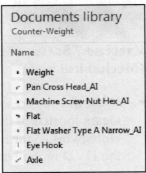

Apply SmartMates with the Flat Washer Type A Narrow_AI, Machine Screw Nut Hex_AI and the Pan Cross Head_AI components.

Use a Distance mate to fit the Axle in the middle of the Flat. Note a Symmetric mate could replace the Distance mate.

Think about the design of the assembly.

Apply all needed Lock mates.

The symbol (f) represents a fixed component. A fixed component cannot move and is locked to the assembly Origin.

Additional Term Project Exercises

Exercise 7.10: Butterfly Valve Assembly Project

Copy the components from the Chapter 7 Homework/ Butterfly Valve Term Project folder located on the DVD. View all components.

Create an ANSI IPS Butterfly Valve assembly document. Insert all needed components and mates to assemble the assembly and to simulate proper movement per the avi file located in the folder. You are the designer. Address all tolerancing and dimension modifications if needed. Use Standard, Advanced and Mechanical Mates. Create and insert any additional components if needed.

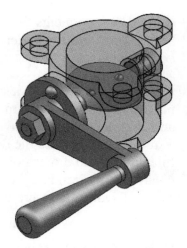

Create a C-ANSI Landscape - Third Angle Isometric Exploded Drawing document with Explode lines of the assembly using your knowledge of SolidWorks. Insert a BOM with Balloons. Insert all needed General notes in the Title Block.

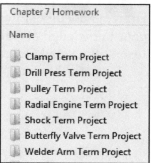

Chapter 7 Homework

Name

- Clamp Term Project
- Drill Press Term Project
- Pulley Term Project
- Radial Engine Term Project
- Shock Term Project
- Butterfly Valve Term Project
- Welder Arm Term Project

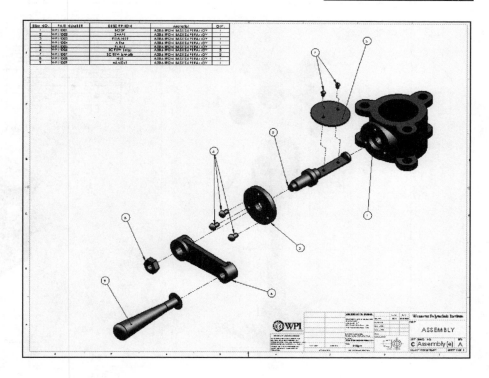

Exercise 7.11: Shock Assembly Project

Copy the components from the Chapter 7 Homework/ Shock Term Project folder located on the DVD. View all components.

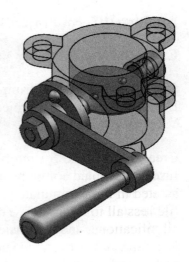

Create an ANSI IPS Shock assembly document. Insert all needed components and mates to assemble the assembly and to simulate proper movement per the avi file located in the folder. You are the designer. Address all tolerancing and dimension modifications if needed. Be creative. Use Standard, Advanced and Mechanical Mates. Create and insert any additional components if needed.

Create a C-ANSI Landscape - Third Angle Isometric Exploded Drawing document with Explode lines of the assembly using your knowledge of SolidWorks. Insert a BOM with Balloons. Insert all needed General notes in the Title Block.

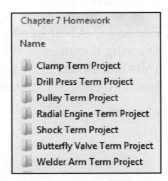

Chapter 7 Homework

Name

- Clamp Term Project
- Drill Press Term Project
- Pulley Term Project
- Radial Engine Term Project
- Shock Term Project
- Butterfly Valve Term Project
- Welder Arm Term Project

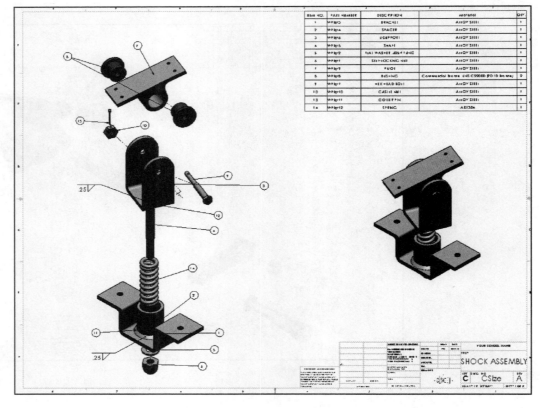

Exercise 7.12: Clamp Assembly Project

Copy the components from the Chapter 7
Homework/Clamp Term Project folder located
on the DVD. View all components.

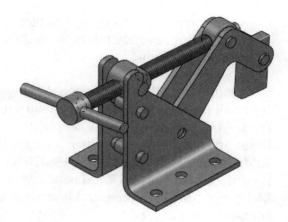

Create an ANSI IPS Clamp assembly
document. Insert all needed components and
mates to assemble the assembly and to
simulate proper movement per the avi file
located in the folder. You are the designer.
Address all tolerancing and dimension
modifications if needed. Be creative. Use
Standard, Advanced and Mechanical Mates.
Create and insert any additional components if needed.

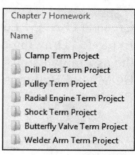

Create a C-ANSI Landscape - Third Angle Isometric Exploded
Drawing document with Explode lines of the assembly using your
knowledge of SolidWorks. Insert a BOM with Balloons. Insert all
needed General notes in the Title Block.

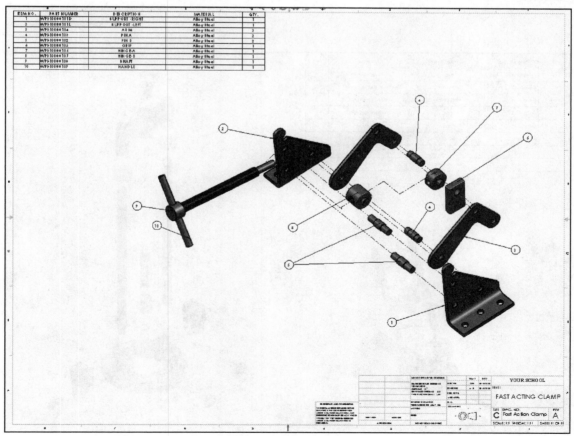

Exercise 7.13: **Drill Press Assembly Term Project**

Copy the components from the Chapter 7 Homework/
Drill Press Term Project folder located on the DVD.
View all components.

Create an ANSI IPS Drill Press assembly document.
Insert all needed components and mates to assemble
the assembly and to simulate proper movement per
the avi file located in the folder. You are the designer.
Address all tolerancing and dimension modifications
if needed. Be creative. Use Standard, Advanced and
Mechanical Mates. Create and insert any additional
components if needed.

Create a C-ANSI Landscape - Third Angle Isometric
Exploded Drawing document with Explode lines of
the assembly using your knowledge of SolidWorks. Insert a
BOM with Balloons. Insert all needed General notes in the Title
Block.

Chapter 7 Homework

Name

📁 Clamp Term Project
📁 Drill Press Term Project
📁 Pulley Term Project
📁 Radial Engine Term Project
📁 Shock Term Project
📁 Butterfly Valve Term Project
📁 Welder Arm Term Project

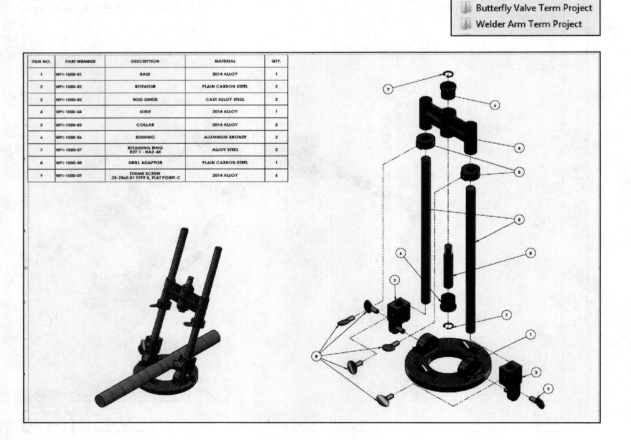

ITEM NO.	PART NUMBER	DESCRIPTION	MATERIAL	QTY.
1	WPI-1000-01	BASE	2014 ALLOY	1
2	WPI-1000-02	ROTATOR	PLAIN CARBON STEEL	2
3	WPI-1000-03	ROD GUIDE	CAST ALLOY STEEL	2
4	WPI-1000-04	SLIDE	2014 ALLOY	1
5	WPI-1000-05	COLLAR	2014 ALLOY	2
6	WPI-1000-06	BUSHING	ALUMINUM BRONZE	2
7	WPI-1000-07	RETAINING RING B27.1 - NA2-46	ALLOY STEEL	2
8	WPI-1000-08	DRILL ADAPTOR	PLAIN CARBON STEEL	1
9	WPI-1000-09	THUMB SCREW .25-20x0.51 TYPE B, FLAT POINT-C	2014 ALLOY	6

Exercise 7.14: Pulley Assembly Term Project

Copy the components from the Chapter 7 Homework/ Pulley Term Project folder located on the DVD. View all components.

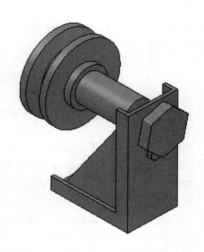

Create an ANSI Pulley assembly document. Insert all needed components and mates to assemble the assembly and to simulate proper movement per the avi file located in the folder. You are the designer. Address all tolerancing and dimension modifications if needed. Be creative. Use Standard, Advanced and Mechanical Mates. Create and insert any additional components if needed.

Create a C-ANSI Landscape - Third Angle Isometric Exploded Drawing document with Explode lines of the assembly using your knowledge of SolidWorks. Insert a BOM with Balloons. Insert all needed General notes in the Title Block.

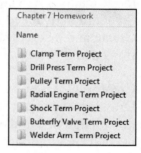

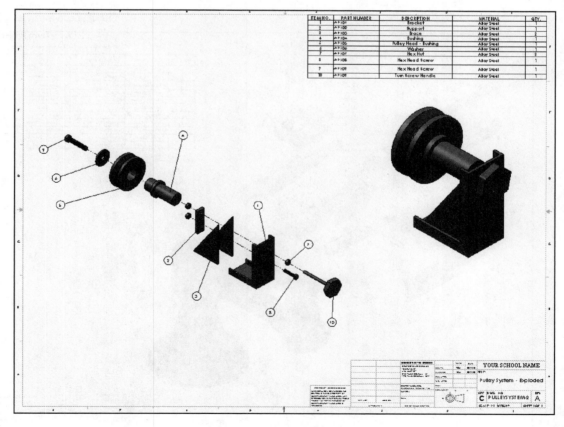

Exercise 7.15: Welder Arm Assembly Term Project

Copy the components from the Chapter 7 Homework/Welder Arm Term Project folder located on the DVD. View all components.

Create an ANSI Welder Arm assembly document. Insert all needed components and mates to assemble the assembly and to simulate proper movement per the avi file located in the folder. You are the designer. Address all tolerancing and dimension modifications if needed. Be creative. Use Standard, Advanced and Mechanical Mates. Create and insert any additional components if needed.

Create a C-ANSI Landscape - Third Angle Isometric Exploded Drawing document with Explode lines of the assembly using your knowledge of SolidWorks. Insert a BOM with Balloons. Insert all needed General notes in the Title Block.

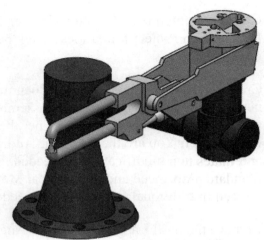

Chapter 7 Homework

Name

- Clamp Term Project
- Drill Press Term Project
- Pulley Term Project
- Radial Engine Term Project
- Shock Term Project
- Butterfly Valve Term Project
- Welder Arm Term Project

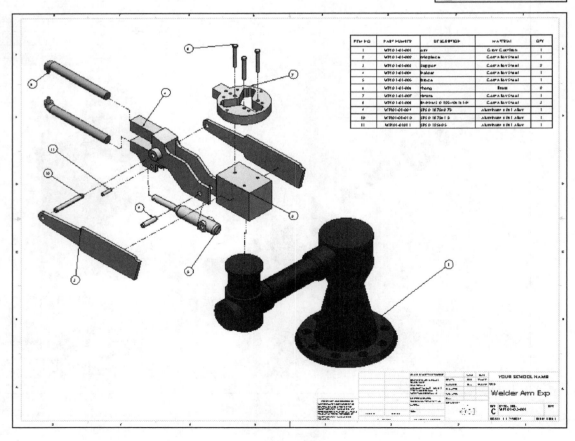

Exercise 7.16: Radial Engine Assembly Term Project

Copy the components from the Chapter 7 Homework/Radial Engine Term Project folder located on the DVD. View all components.

Create an ANSI Radial Engine assembly document. Insert all needed components and mates to assemble the assembly and to simulate proper movement per the avi file located in the folder.

You are the designer. Address all tolerancing and dimension modifications if needed. Be creative.

Use Standard, Advanced and Mechanical Mates.

Create and insert any additional components if needed.

Create a C-ANSI Landscape - Third Angle Isometric Exploded Drawing document with Explode lines of the assembly using your knowledge of SolidWorks.

Insert a BOM with Balloons. Insert all needed General notes in the Title Block.

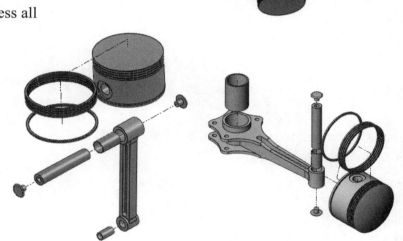

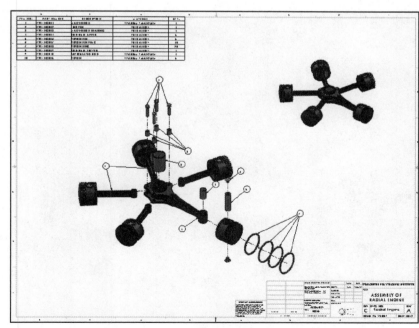

Notes:

Chapter 8

Fundamentals of Drawing

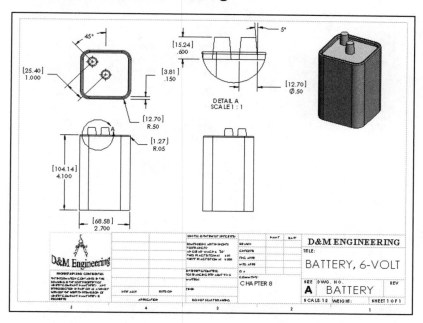

Below are the desired outcomes and usage competencies based on the completion of Chapter 8.

Desired Outcomes:	Usage Competencies:
• Custom Drawing and Sheet Template.	• Define Dimensioning standards, Units and Precision. • Create Title block information, and a Company logo.
• Three drawings: BATTERY, FLASHLIGHT, and O-RING-DESIGN-TABLE.	• Skill to create the following drawing views: Standard, Detail, Section and Exploded. • Proficiency to insert, and modify dimensions, BOM, Balloon text and Annotations.
• Design Table.	• Capability to create three configurations in a design table: Small, Medium and Large.

Notes:

Chapter 8 - Fundamentals of Drawing

Chapter Overview

Create three drawings in this chapter:

- BATTERY

- FLASHLIGHT

- O-RING-DESIGN-TABLE

The BATTERY part drawing contains the Front, Top, Right, Detail and Isometric views. Orient the views to fit the drawing sheet. Incorporate the BATTERY part dimensions into the drawing.

The FLASHLIGHT assembly drawing contains an Exploded view, a Bill of Materials and balloon text. The Balloon items correspond to the Item Number in the BOM. The numeric part number is a user-defined property in each part.

Insert a Design Table for the O-RING part. A Design Table is an Excel spreadsheet that contains parameters. Define the Sketch-path and Sketch-profile diameters of the Swept feature.

Create three configurations of the O-RING part:

- Small

- Medium

- Large

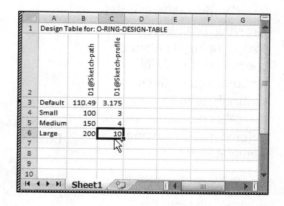

The O-RING-DESIGN-TABLE drawing contains three configurations of the O-RING. Utilize the View Palette from the Task Pane. Use the drawing view properties to control each part configuration. The three drawings utilize a custom Drawing Template and Sheet Format. The Drawing Template defines the dimensioning standard, units and precision. The Sheet Format contains the Title block information and a Company logo.

There are two major design modes used to develop a drawing: *Edit Sheet Format* and *Edit Sheet*. Work between the two drawing modes in this project. The Edit Sheet Format mode provides the ability to:

- Change the Title block size and text headings.

- Incorporate a Company logo.

- Add drawing, design, or company text.

The Edit Sheet mode provides the ability to:

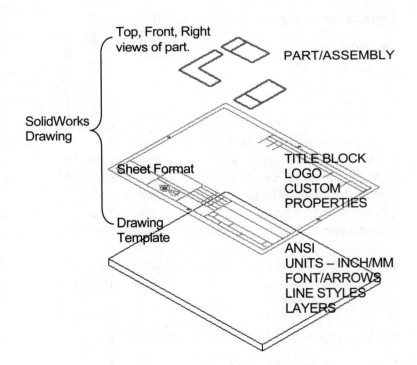

- Add or modify views.
- Add or modify dimensions.
- Add or modify text.

After completing the activities in this chapter, you will be able to:

- Utilize the View Layout tool and the Annotate toolbar for the following tools: Model View, Projected View, Detail View, Note, Model Items and Balloon.
- Create two Drawing Templates: A-IN-ANSI and A-MM-ISO.
- Insert, move and edit part and drawing dimensions.
- Develop a Design Table for the O-RING part and insert the correct configuration into a drawing.
- Insert an Exploded view with a Bill of Materials.
- Apply the Edit Sheet and Edit Sheet Format modes.

New Drawing and the Drawing Template

The foundation of a new SolidWorks drawing is the Drawing Template. Drawing size, drawing standards, company information, manufacturing, and or assembly requirements, units and other properties are defined in the Drawing Template.

The Sheet Format is incorporated into the Drawing Template. The Sheet Format contains the border, title block information, revision block information, company name, and or logo information, Custom Properties and SolidWorks Properties.

Custom Properties and SolidWorks Properties are shared values between documents.

Views from the part or assembly are inserted into the SolidWorks Drawing. Views are inserted in Third or First Angle projection. Notes and dimensions for millimeter drawings are provided in brackets [x] for this chapter.

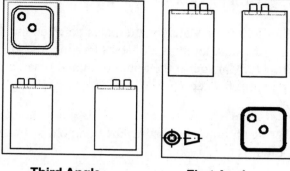

Third Angle   **First Angle**

Utilize an A size Drawing Template with Sheet Format for the BATTERY drawing and FLASHLIGHT assembly drawing. A copy of the default Drawing Template illustrated in this activity is contained in the SOLIDWORKS-MODELS 2012\MY-TEMPLATES folder on the DVD in the book. The default Drawing Templates contain predefined Title block Notes linked to Custom Properties and SolidWorks Properties.

For printers supporting millimeter paper sizes, utilize the Printer, Properties and Scale to Fit option.

Activity: New Drawing and the Drawing Template

Close all parts and drawings.

1) Click **Windows**, **Close All** from the Menu bar.

Create a new drawing.

2) Click **New** ⬜ from the Menu bar.

3) Double-click **Drawing** from the Templates tab. Uncheck the Only show standard formats box if needed.

4) Select **A (ANSI) Landscape**.

5) Click **OK** from the Sheet Format/Size dialog box.

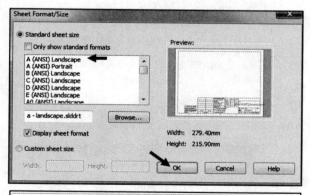

The A (ANSI) Landscape paper is displayed in the Graphics window. The sheet border defines the drawing size, 11″ × 8.5″ or (279.4mm × 215.9mm).

The View Layout toolbar is displayed in the CommandManager. Draw1 is the default drawing name. Sheet1 is the default first sheet name.

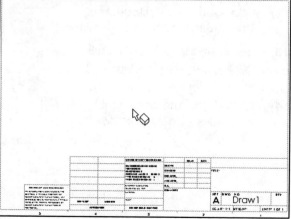

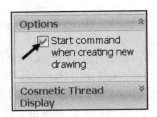

🔅 If the Start command when creating new drawing option is checked, the Model View PropertyManager is selected by default.

The CommandManager alternates between the View Layout, Sketch, and Annotate toolbars. A New Drawing invokes the Model View PropertyManager if the Start command when creating new drawing option is checked.

🔅 Right-click in the gray area and check the needed toolbars. The toolbar is displayed.

Set Sheet Properties and Document Properties for the Drawing Template. Sheet Properties control the Sheet Size, Sheet Scale and Type of Projection.

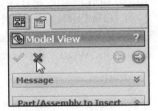

Document Properties control the display of dimensions, annotations and symbols in the drawing.

Exit model view.

6) Click **Cancel** ✖ from the Model View PropertyManager. The Draw1 FeatureManager is displayed. The Draw1 FeatureManager is displayed to the left of the Graphics window.

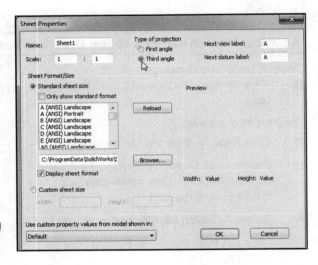

Set the Sheet Properties.

7) Right-click in the **Graphics window.**

8) Click **Properties**. The Sheet Properties dialog box is displayed.

9) Select Sheet Scale **1:1**.

10) Select **Third angle** for Type of projection.

11) Click **OK** from the Sheet Properties dialog box.

Set the Document Properties.

12) Click **Options** 🗒 , **Document Properties** tab from the Menu bar.

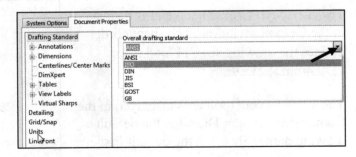

13) Select **ANSI**, **[ISO]** for Overall drafting standard from the drop-down menu.

14) Click **Units**. The Document Properties - Units dialog box is displayed.

15)   Select **IPS**, **[MMGS]** for Unit system.

16)   Select **.123**, **[.12]** for Length units Decimal places.

17)   Select **None** for Angular units Decimal places.

18)   Click **OK** from the Document Properties - Units dialog box.

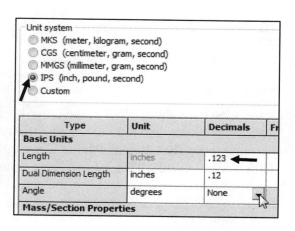

Detailing options provide the ability to address: dimensioning standards, text style, center marks, extension lines, arrow styles, tolerance and precision.

There are numerous text styles and sizes available in SolidWorks.

Save the Drawing.

19)   Click **Save** 💾 .

20)   Select the **MY-TEMPLATES** folder. Accept the defaults.

21)   Click **Save**. The Draw1 FeatureManager is displayed.

Title Block

The Title block contains text fields linked to System Properties and Custom Properties. System Properties are determined from the SolidWorks documents. Custom Property values are assigned to named variables. Save time. Utilize System Properties and define Custom Properties in your Sheet Formats.

System Properties and Custom Properties for Title Block:			
System Properties Linked to fields in default Sheet Formats:	**Custom Properties of drawings linked to fields in default Sheet Formats:**		**Custom Properties of parts and assemblies linked to fields in default Sheet Formats:**
SW-File Name (in DWG. NO. field):	CompanyName:	EngineeringApproval:	Description (in TITLE field):
SW-Sheet Scale:	CheckedBy:	EngAppDate:	Weight:
SW-Current Sheet:	CheckedDate:	ManufacturingApproval:	Material:
SW-Total Sheets:	DrawnBy:	MfgAppDate:	Finish:
	DrawnDate:	QAApproval:	Revision:
	EngineeringApproval:	QAAppDate:	

The drawing document contains two modes:

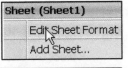

- *Edit Sheet Format*

- *Edit Sheet*

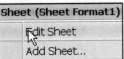

Insert views and dimensions in the Edit Sheet mode. Modify the Sheet Format text, lines, and the Title block information in the Edit Sheet Format mode. The CompanyName Custom Property is located in the Title block above the TITLE box. There is no value defined for CompanyName. A small text box indicates an empty field. Define a value for the Custom Property CompanyName. Example: D&M ENGINEERING. The Tolerance block is located in the Title block.

The Tolerance block provides information to the manufacturer on the minimum and maximum variation for each dimension on the drawing.

If a specific tolerance or note is provided on the drawing, the specific tolerance or note will override the information in the Tolerance block. General tolerance values are based on the design requirements and the manufacturing process. Modify the Tolerance block in the Sheet Format for ASME Y14.5 machined parts. Delete unnecessary text. The FRACTIONAL text refers to inches. The BEND text refers to sheet metal parts.

Activity: Title Block

Invoke the Edit Sheet Format Mode.

22) Right-click **Edit Sheet Format** from the Pop-up menu in the Graphics window. The Title block lines are displayed in blue.

23) **Zoom in** on the Sheet Format Title block.

Define COMPANYNAME Custom Property.

24) Click a **center position** above the TITLE box as illustrated. The Note text $PRP:"COMPANYNAME" is displayed. The Note PropertyManager is displayed.

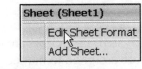

25) Click the **Link to Property** icon as illustrated from the Text Format box. The Link to Property dialog box is displayed.

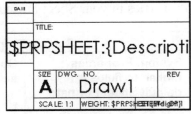

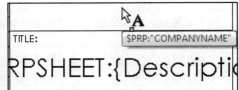

26) Click the **File Properties** button. The Summary Information dialog box is displayed.

27) Click the **Custom** tab from the Summary Information dialog box.

28) Select **CompanyName** from the Property Name box.

29) Click inside the **Value / Text Expression** box.

30) Enter: **D&M ENGINEERING** or your **company or school** name.

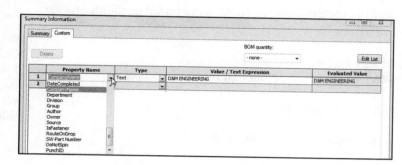

31) Click inside the **Evaluated Value** box. The CompanyName is displayed.

32) Click **OK** from the Summary Information dialog box.

33) Click **OK** from the Link to Property dialog box. D&M ENGINEERING is displayed.

Modify the font size.
34) Uncheck the **Use document's font** box in the Note PropertyManager.

35) Click the **Font** button. The Choose Font dialog box is displayed.

Select a new font size.
36) Click **Bold**.

37) Click **Points**.

38) Select **18**.

39) Click **OK** from the Choose Font dialog box. The text is displayed in the Title block.

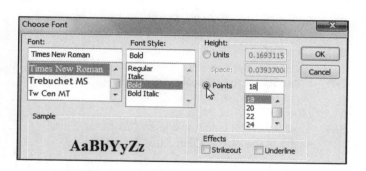

40) Click **OK** ✔ from the Note PropertyManager.

Modify the Tolerance Note in the text box.
41) **Zoom in** on the Tolerance Note in the text box.

42) Double-click the text **INTERPRET GEOMETRIC TOLERANCING PER:** The Note PropertyManager is displayed.

INTERPRET GEOMETRIC
TOLERANCING PER:

43) Enter **ASME Y14.5** as illustrated.

44) Click **OK** ✔ from the Note PropertyManager.

INTERPRET GEOMETRIC
TOLERANCING PER: ASME Y14.5

💡 Click outside the Note text box to end the Note, or Click OK ✔ from the Note PropertyManager.

Tolerance values are different for inch and millimeter templates. Enter the Tolerance values for the inch template. Enter the Tolerance values for the millimeter template. Save drawing templates with unique filenames.

Modify the Note text.

45) Right-click the **Note text block**.

46) Click **Edit Text**.

47) Delete the line **FRACTIONAL +-**.

48) Delete the text **BEND +-**.

Enter ANGULAR tolerance.

49) Click a **position** at the end of the line.

50) Enter **0**.

51) Click the **Add Symbol** icon from the Text Format box.

52) Select **Degree** from the Modifying Symbols library.

53) Click **OK**.

54) Enter **30'** for minutes of a degree.

Modify the DECIMAL LINES.

55) Modify the first Decimal line as illustrated. Enter **+- .01** at the end of ONE PLACE DECIMAL.

56) Modify the second Decimal line as illustrated. Enter **+- .005** at the end of TWO PLACE DECIMAL.

57) Click **OK** from the Note PropertyManager.

Fit the sheet to the Graphics window.

58) Press the **f** key.

DIMENSIONS ARE IN INCHES
TOLERANCES:
FRACTIONAL ±
ANGULAR: MACH ±   BEND ±
TWO PLACE DECIMAL   ±
THREE PLACE DECIMAL ±

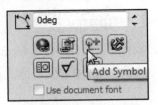

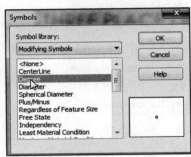

UNLESS OTHERWISE SPECIFIED:

DIMENSIONS ARE IN INCHES
TOLERANCES:
ANGULAR: MACH ± 0°30'
ONE PLACE DECIMAL ±.01
TWO PLACE DECIMAL ±.005

The Add Symbol ⌀⁺ icon is accessible through the Note PropertyManager. The ± symbol is located in the Modify Symbols list. The ± symbol is displayed as <MOD-PM>. The degree symbol ° is displayed as <MOD-DEG>. Select icon symbols or enter values from the keyboard.

Interpretation of tolerances is as follows for dimensions:

- The angular dimension 110° is machined between 109.5° and 110.5°.

- The dimension 2.04 is machined between 2.03 and 2.05.

Additional Custom Properties and Notes are added later in this project.

Company Logo and Save Sheet Format

A Company logo is normally located in the Title block of the drawing. You can create your own Company logo or copy and paste an existing picture.

☀ The COMPASS.jpeg file is enclosed on the DVD in the book. Copy all folders and files from the DVD to your hard drive. Do not work directly from the DVD. Insert the provided Company logo in the Edit Sheet Format mode. If you have your own Logo, use it for the drawing and skip the process of copying it from the DVD.

Activity: Insert the Company Logo and Save the Sheet Format

Insert a Company Logo.

59) Copy the folder **MY-SHEETFORMATS** from the DVD to the working folder on your hard drive.

60) Click **Insert**, **Picture** from the Menu bar. The Open dialog box is displayed.

61) Select **MY-SHEETFORMATS\COMPASS.jpeg**.

62) Click **Open**. The Sketch Picture PropertyManager is displayed.

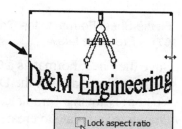

63) Drag the picture handles to size the **picture** to the left side of the Title block. If needed, uncheck the lock aspect ratio box.

64) Click **OK** ✔ from the Sketch Picture PropertyManager.

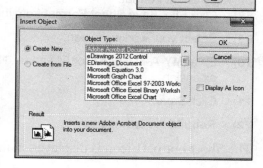

☀ Text can be added to create a custom logo. You can insert a picture or an object.

Return to the Edit Sheet mode.

65) Right-click in the **Graphics window**.

66) Click **Edit Sheet**. The Title block is displayed in black.

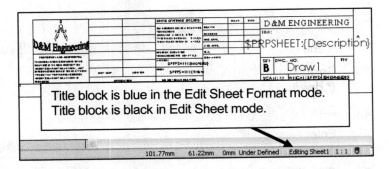

Title block is blue in the Edit Sheet Format mode.
Title block is black in Edit Sheet mode.

101.77mm   61.22mm   0mm   Under Defined   Editing Sheet1   1:1

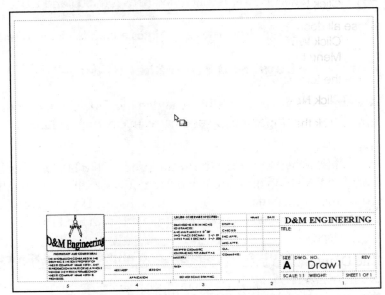

Fit the Sheet Format to the Graphics window.

67) Press the **f** key.

Save the Sheet Format as a Custom Sheet Format. Combine the Sheet Format with the Drawing Template to create a Custom Drawing Template. Use the Custom Sheet Format and Drawing Template to create drawings in this chapter.

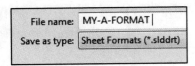

File name: MY-A-FORMAT

Save as type: Sheet Formats (*.slddrt)

Save the Sheet Format.

68) Click **File**, **Save Sheet Format** from the Menu bar.

69) **Select** your working folder.

70) Enter **MY-A-FORMAT** for File name. Note: .slddrt file extension.

71) Click **Save**.

72) **Rebuild** the drawing.

The Sheet Format1 icon is displayed in the FeatureManager.

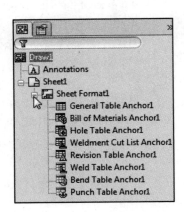

Save the Drawing Template.

73) Click **Save As** from the drop-down Menu bar.

74) Select **Drawing Templates (*.drwdot)** for Save as type.

75) Select the **MY-TEMPLATES** folder.

76) Enter **A-IN-ANSI**, **[A-MM-ISO]** for File name.

77) Click **Save**.

Close all documents.

78) Click **Window**, **Close All** from the Menu bar.

Verify the template.

79) Click **New** ⬜ from the Menu bar.

80) Click the **MY-TEMPLATES** tab.

81) Double-click the **A-IN-ANSI**, **[A-MM-ISO]** Drawing Template. The Model View PropertyManager is displayed.

82) Click **Cancel** ✖ from the Model View PropertyManager. The Draw2 FeatureManager is displayed.

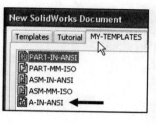

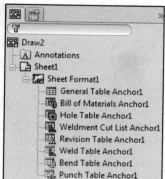

The Draw2-Sheet1 drawing is displayed in the Graphics window. The Model View PropertyManager is selected by default. The Draw# - Sheet1 is a sequential number created in the current SolidWorks session.

☀ Your default system document templates may be different if you are a new user of SolidWorks 2012 vs. an existing user who has upgraded from a previous version.

Utilize the A-IN-ANSI, [A- MM-ISO] Drawing Template for the BATTERY drawing.

Utilize descriptive filenames for the Drawing Template that contains the size, dimension standard, and units.

File Locations is a System Option. The option is active only for the current session of SolidWorks in some network environments.

Combine custom Drawing Templates and Sheet Formats to match your company's drawing standards. Save the empty Drawing Template and Sheet Format separately to reuse information.

Additional details on Drawing Templates, Sheet Format and Custom Properties are available in SolidWorks Help Topics. Keywords: Documents (templates, properties), Sheet Formats (new, new drawings, note text), Properties (drawing sheets) and Customize Drawing Sheet Formats.

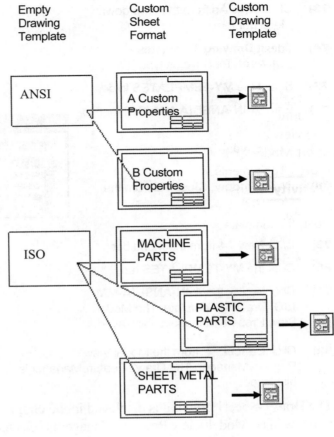

Review Drawing Templates

A custom Drawing Template was created from the default Drawing Template. Sheet Properties and Document Properties controlled the Sheet size, Sheet scale, units and dimension display.

The Sheet Format contained a Title block and Custom Property information. A Company Logo was inserted and you modified the Title block. The Save Sheet Format option was utilized to save the MY-A-FORMAT.slddrt Sheet Format.

The Save As option was utilized to save the A-IN-ANSI, [A-MM-ISO].drwdot Drawing Template. The Sheet Format was saved in the MY-SHEETFORMATS folder. The Drawing Template was saved in the MY-TEMPLATES folder.

BATTERY Drawing

A drawing contains part views, geometric dimensioning and tolerances, notes, and other related design information. When a part is modified, the drawing automatically updates. When a driving dimension in the drawing is modified, the part is automatically updated.

Create the BATTERY drawing from the BATTERY part. Utilize the Model View feature in the Drawings toolbar. The Front view is the first view inserted into the drawing. Note: The Top view and Right view are Projected views. Insert dimensions into the drawing with the Insert Model Items feature.

Activity: Insert Views - BATTERY Drawing

Insert Views.

83) Click **Model View** from the View Layout tab.

84) Click **Browse** from the Model View PropertyManager.

85) Double-click the **BATTERY** part from the PROJECTS folder.

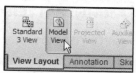

Insert the Front, Top, and Right view.

86) Check the **Create multiple views** box.

87) Click ***Top** and ***Right** view from the Orientation box. Note: *Front is activated by default.

88) Click **OK** from the Model View PropertyManager. Three views are displayed on Sheet1. If required, deactivate the Origins.

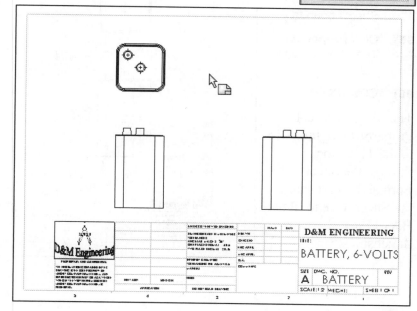

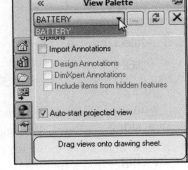

🔆 A part can't be inserted into a drawing when the Edit Sheet Format mode is selected.

🔆 Move parent and child views independently by dragging their view boundary. Hold the Shift key down and select multiple views to move as a group.

🔆 Click the View Palette icon in the Task Pane. Click the drop down arrow to view an open part or click the Browse button to locate a part. Click and drag the desired view/views into the active drawing sheet.

Activity: BATTERY Drawing-Insert a View

Insert an Isometric view.

89) Click **Model View** 🔲 from the View Layout tab in the CommandManager. The Model View PropertyManager is displayed.

90) Click **Next** ➡ from the Model View PropertyManager.

91) Click ***Isometric** from the Orientation box. The Isometric view is placed on the mouse pointer.

92) Click a **position** in the upper right corner of the Graphics window on Sheet1.

93) Click **OK** ✅ from the Drawing View4 PropertyManager.

View the Sheet Scale.

94) Right-click a **position** inside the Graphics window, Sheet1 boundary.

95) Click **Properties**. The Sheet Scale is 1:2.

96) Click **OK**.

The SW-Sheet scale Property 1:2 is linked to the Title block through the Sheet Format. Later, change the Sheet scale to fit the BATTERY dimensions.

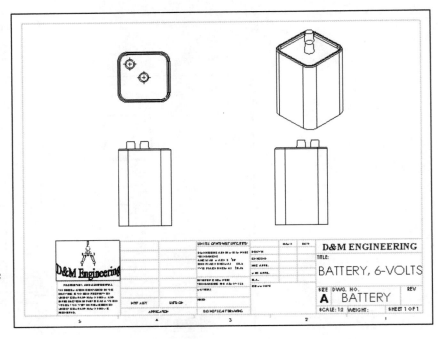

Save the Drawing.

97)   Click **Save As** from the drop-down Menu bar.

98)   Select the **PROJECTS** folder.

99)   Click **Save**. BATTERY is the default drawing file name.

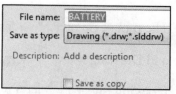

Each drawing has a unique file name. Drawing file names end with a .slddrw suffix in SolidWorks. Part file names end with a .sldprt suffix.

A drawing or part file can have the same prefix. A drawing or part file cannot have the same suffix. Example: Drawing file name: BATTERY.slddrw. Part file name: BATTERY.sldprt.

Text in the Title block is linked to the Filename and description created in the part. The DWG. NO. text box utilizes the Property, $PRP: "SW-File Name" passed from the BATTERY part to the BATTERY drawing. The TITLE text box utilizes the Property, $PRPSHEET: "Description".

The filename BATTERY is displayed in the DWG. NO. box. The Description BATTERY, 6-VOLT is displayed in the Title box.

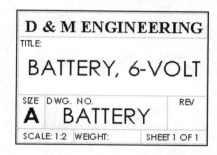

 Tangent Edges and Origin are displayed for educational purposes.

The BATTERY drawing contains three Principle Views (Standard Views): Front, Top, Right and an Isometric view. You created the views with the Model View tool. Drawing views can be inserted with the following processes:

- Utilize the Model View tool.

- Click and drag a part into the drawing or select the Standard 3 Orientation view option.

- Predefine views in a custom Drawing Template.

- Drag a hyperlink through Internet Explorer.

- Utilize the View Palette located in the Task Pane to the right of the Graphics window.

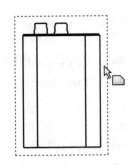

The mouse pointer provides feedback in both the Drawing Sheet and Drawing View modes. The mouse pointer displays the Drawing Sheet icon when the Sheet properties and commands are executed.

The mouse pointer displays the Drawing View icon when the View properties and commands are executed.

View the mouse pointer for feedback to select Sheet, View, Component and Edge properties in the Drawing.

- Sheet Properties display properties of the selected sheet. Right-click in the sheet boundary .

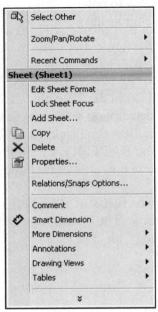

- View Properties display properties of the selected view. Right-click on the view boundary .

- Component Properties display properties of the selected component. Right-click on the face of the component .

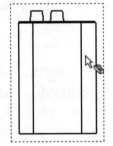

- Edge Properties display properties of the geometry. Right-click on an edge .

The Drawing views are complete. Move the views to allow for ample spacing for the dimensions and notes. Zoom in on narrow view boundary boxes if required.

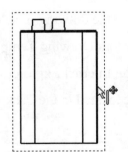

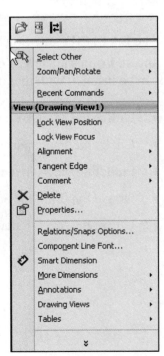

Detail View

A Detail view enlarges an area of an existing view. You need to specify location, name, shape and scale. Create a Detail view named A with a circle and a 1:1 scale.

A Detail circle specifies the area of an existing view to enlarge. The circle contains the same letter as the view name.

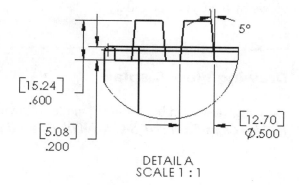

DETAIL A
SCALE 1 : 1

Activity: BATTERY Drawing-Detail View

Add a Detail view to the drawing.

100) Click the **Detail View** 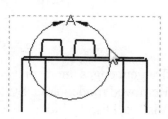 tool from the View Layout toolbar. The circle sketch tool is selected.

Sketch a circle with the center point located between the two terminals in Drawing View1 (Front).
101) Sketch the circle as illustrated.

Position the Detail View.
102) Click a **position** to the right of the Drawing View3 (Top).

103) Click the **Use custom scale** box.

104) Enter **1:1** in the Custom Scale box.

105) Click **OK** ✔ from the Drawing View A PropertyManager.

106) Rebuild the model.

Fit the Drawing to the Graphics window.
107) Press the **f** key.

108) Click **Save** 💾.

Center marks are displayed be default. Center marks and centerlines are controlled by the Tools, Options, Document Properties, Auto insert on view creation option.

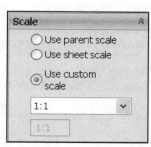

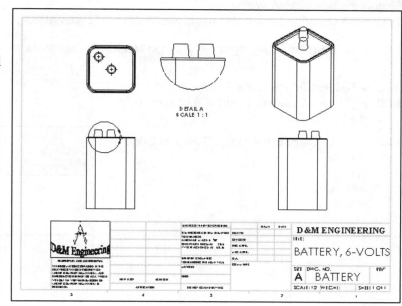

The Drawing view is complete. Allow for spacing of the dimensions and notes. Move the views by their view boundary.

Drawing View Display

Drawing views can be displayed in the following modes: *Wireframe, Hidden Lines Visible, Hidden Lines Removed, Shaded With Edges and Shaded mode.*

   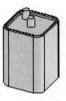

Tangent edges are displayed either in: Visible, With Font or Removed mode. Note: System default is Tangent Edges Visible. Display hidden lines, profile lines and tangent edges in various view modes.

Activity: BATTERY Drawing-View Display

Display hidden lines in the Detail View.
109) Click **Detail View A** from the FeatureManager.

110) Click **Wire frame** ⊞ from the Heads-up View toolbar.

Display the Isometric view Shaded.
111) Click **Drawing View4,** (Isometric view) from the FeatureManager.

112) Click **Shaded With Edges** ⬛ from the Heads-up View toolbar.

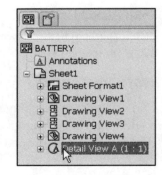

Display Tangent Edges in the Front view.
113) Right-click **Drawing View1** (Front view) from the FeatureManager.

114) Click **Tangent Edge**, check **Tangent Edge Visible**.

Insert Model Items and Move Dimensions

Dimensions created for each feature in the part are inserted into the drawing. Dimensions are inserted by sheet, view or feature. Select inside the sheet boundary to insert the dimensions into Sheet1. Move the dimensions off the profile lines in each view.

Illustrations for this project are provided in inches and millimeters. The Detailing option and Dual Dimensions Display produces both inches and millimeters for each dimension. The primary units are set to inches. The secondary units are set to millimeters.

Inches
or
Millimeters

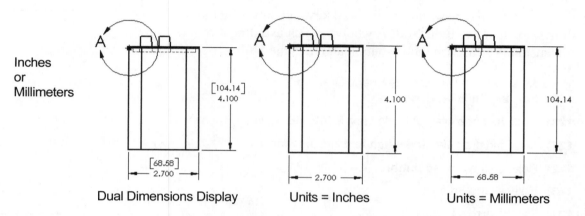

Dual Dimensions Display   Units = Inches   Units = Millimeters

The drawing dimension location is dependent on: Feature dimension creation and Selected drawing views.

Move dimensions within the same view. Use the mouse pointer to drag dimensions and leader lines to new locations. Move dimensions to a different view. Utilize the Shift key to drag a dimension to another view.

Leader lines reference the size of the profile. A gap must exist between the profile lines and the leader lines. Shorten the leader lines to maintain a drawing standard. Use the Arrow buttons in the PropertyManager to flip dimension arrows.

 Import dimensions and tolerances you created using the DimXpert tool for parts into a drawing. If the DimXpert dimensions are not displayed in the drawing after import, follow this procedure: 1.) In the FeatureManager design tree, right-click the **Annotations** folder. 2.) Select **Display Annotations** and **Show DimXpert Annotations**.

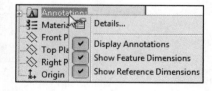

 DimXpert dimensions and tolerances are magenta-color by default.

Activity: BATTERY Drawing-Insert Model Items and Move Dimensions

Insert dimensions into the Sheet.

115) Click inside the **sheet boundary**.

116) Click the **Model Items** ✎ tool from the Annotation toolbar. The Model Items PropertyManager is displayed.

117) Select **Entire model** from the Source/Destination box.

118) Click **OK** ✔ from the Model Items PropertyManager. Note: Change the Sheet scale to address the dimensions.

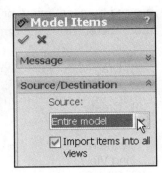

💡 Dimensions may appear in different view. To move a dimension from a view, utilize the Shift key and only drag the dimension text. Hold the Shift key down. Click and drag the dimension. Release the mouse button and then the Shift key.

Move dimensions in Drawing View1 and Drawing View3.

119) Hold the **Shift** key down.

120) Click the horizontal dimension text **2.700**, **[68.58]** in the Top view.

121) Click and drag the **dimension text** into the Front view.

122) Release the **mouse button**.

123) Release the **Shift** key.

124) Flip the **arrow head** to the inside.

125) Repeat the above procedure for the **other dimensions** as illustrated.

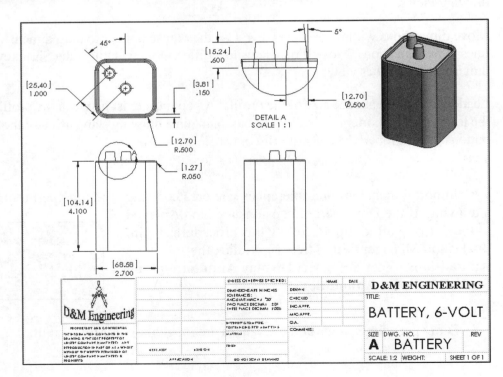

126) Click the vertical dimension text **4.100**, **[104.14]**.

127) Drag the **text** to the right of the BATTERY.

128) Move the **leader line endpoints** if required. The end points of the leader line are displayed in blue.

129) Drag each **square blue control end point** to the left until it is off the profile line.

130) Create a **gap** between the profile line and the leader lines.

Flip Arrows

68.58

2.700

Modify the radius text in Drawing View1 and Drawing View3.

131) Click the **R.050**, [Ø**1.27**] text on Drawing View1. The Dimension PropertyManager is displayed.

132) Select **.12**, 2 places in the Tolerance/Precision box. R.05 is displayed.

133) Click the **R.500**, [Ø**12.70**] text on Drawing View3.

134) Select **.12**, 2 place in the Tolerance/Precision box. R.50 is displayed.

135) Click **OK** ✓ from the Dimension PropertyManager.

Tolerance/Precision

None

.12

Primary Value

D1@Top Face Fillet

0.050in

Override value:

Save the drawing.

136) Click **Save** 💾 .

Shorten the Bent leader length.

137) Click **Options**, **Document Properties** tab from the Menu bar.

138) Click **Dimensions**.

139) Enter **.25**in, **[6.35]** for Bent leader length.

140) Click **OK**.

Bent leaders

Leader length: 0.25in

Leading zeroes: Standard

Trailing zeroes: Smart

Activity: BATTERY Drawing-Insert a Note

Insert a Note.

141) Click the **Note** A tool from the Annotation toolbar. The Note PropertyManager is displayed.

142) Click a **position** in the COMMENTS box.

143) Enter **CHAPTER 8**.

144) Click **OK** ✓ from the Note PropertyManager.

Save the BATTERY drawing.

145) Click **Save** 💾 .

There are hundreds of Document Property options. Where do you go for additional information on these Properties? Answer: Select the Help button in the lower right hand corner of the Document Property dialog box.

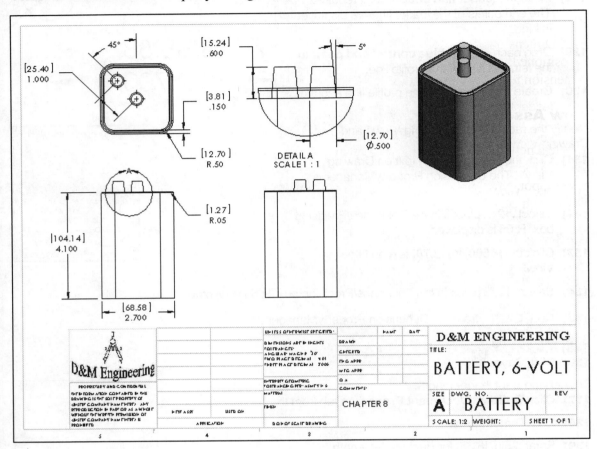

Additional details on Drawing Views, New Drawing, Details, Dimensions and Annotations are available in SolidWorks Help Topics.

 Tangent Edges and Origin are displayed for educational purposes.

 Review BATTERY Drawing

You created a new drawing, BATTERY with the A-IN-ANSI, [A-MM-ISO] Drawing Template. The BATTERY drawing utilized the BATTERY part in the Model View PropertyManager. The Model View PropertyManager provides the ability to insert new views with the View Orientation option.

You selected Front, Top, Right and Isometric to position the BATTERY views. The Detail View tool inserted a Detail view of the BATTERY. You moved the views by dragging the blue view boundary. You inserted the dimensions and annotations to detail the BATTERY drawing.

You inserted part dimensions and annotations into the drawing with the Insert Model Items tool. Dimensions were moved to new positions. Leader lines and dimension text were repositioned. Annotations were edited to reflect the drawing standard. You modified the dimension text by inserting additional text.

New Assembly Drawing and Exploded View

Create a new drawing named FLASHLIGHT. Insert the FLASHLIGHT assembly Isometric view. Modify the view properties to display the Exploded view. The Bill of Materials reflects the components of the FLASHLIGHT assembly. Create a drawing with a Bill of Materials. Label each component with Balloon text.

Activity: New Assembly Drawing and Exploded View

Close all parts and drawings.
146) Click **Windows, Close All** from the Menu bar.

Create a new drawing.
147) Click **New** ⬜ from the Menu bar.

148) Click the **MY-TEMPLATES** tab.

149) Double-click the **A-IN-ANSI, [A-MM-ISO]** Drawing Template.

Insert the FLASHLIGHT assembly.
150) Click **Browse** from the Model View PropertyManager.

151) Select **Assembly** for file type.

152) Double-click **FLASHLIGHT** from the PROJECTS folder.

153) Select ***Isometric** for the Orientation box.

154) Click **Shaded With Edges** from the Display Style box.

155) Click a **position** on the right side of the drawing.

156) Click **OK** ✔ from the Drawing View1 PropertyManager.

157) If needed, **deactivate** the Origins.

☼ Save parts and assemblies to preview picture thumbnails in the Open dialog box.

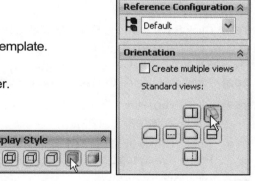

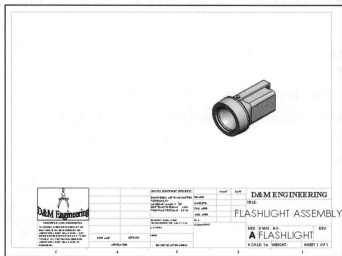

Edit the Title text.

158) Right-click a **position** in the sheet boundary.

159) Click **Edit Sheet Format**.

160) Double-click the title text **FLASHLIGHT ASSEMBLY**. The Note PropertyManager and the Formatting dialog box are displayed.

161) Enter **.19**in for Text Height.

162) Click **OK** ✓ from the Note PropertyManager.

163) Right-click a **position** in the Sheet boundary.

164) Click **Edit Sheet**. The FLASHLIGHT ASSEMBLY text is sized to the Title box.

Save the drawing.

165) Click **Save** 🖫. The Save As dialog box is displayed.

166) Select the **PROJECTS** folder. FLASHLIGHT is the default File name.

167) Click **Save** from the Save As dialog box.

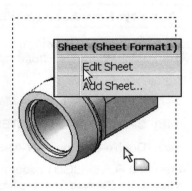

Display the Exploded view of the assembly.

168) Right-click inside the **Isometric view**.

169) Click **Properties**. The Drawing View Properties dialog box is displayed.

170) Check **Show in exploded state** from the Drawing View Properties dialog box.

171) Click **OK**. The Isometric view is displayed in an Exploded state.

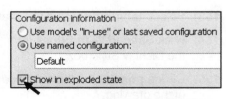

Modify the view scale.

172) Click inside the **Isometric view**.

173) Check the **Use custom scale** box.

174) Select **User Defined** from the drop-down menu.

175) Enter **1:4** in the Scale box as illustrated.

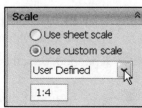

176) Click **OK** ✔ from the Model View1 PropertyManager. View the Exploded view.

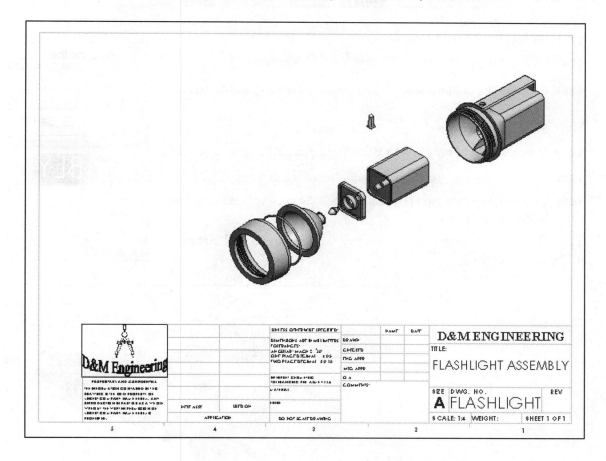

Bill of Materials and Balloons

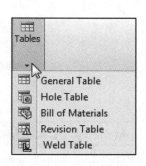

Apply the Bill of Materials 🗐 tool from the Consolidated Tables drop-down menu. A Bill of Materials (BOM) is a table inserted into a drawing to keep a record of the parts used in an assembly. The default BOM template contains the Item Number, Quantity, Part No. and Description.

The default Item number is determined by the order in which the component is inserted into the assembly. Quantity is the number of instances of a part or assembly. Part No. is determined by the following: file name, default and the User Defined option, Part Number used by the Bill of Materials. Description is determined by the description entered when the document is saved.

Activity: FLASHLIGHT Drawing-Bill of Materials

Insert a Bill of Materials.

177) Click inside the **Isometric view**.

178) Click the **Annotation** tab from the CommandManager.

179) Click the **Bill of Materials** 🗂 tool from the Consolidated Table drop-down menu.

180) Select **bom-standard** for the Table Template.

181) Check the **Parts only** box for BOM Type. Accept the default settings.

182) Click **OK** ✔ from the Bill of Materials PropertyManager.

183) Click a **position** in the upper left corner of Sheet1 as illustrated. The BOM is displayed.

The Bill of Materials requires additional work that you will complete in the next section.

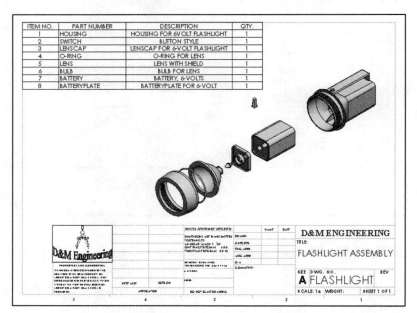

ITEM NO.	PART NUMBER	DESCRIPTION	QTY.
1	HOUSING	HOUSING FOR 6 VOLT FLASHLIGHT	1
2	SWITCH	BUTTON STYLE	1
3	LENSCAP	LENSCAP FOR 6-VOLT FLASHLIGHT	1
4	O-RING	O-RING FOR LENS	1
5	LENS	LENS WITH SHIELD	1
6	BULB	BULB FOR LENS	1
7	BATTERY	BATTERY, 6-VOLTS	1
8	BATTERYPLATE	BATTERYPLATE FOR 6-VOLT	1

Activity: FLASHLIGHT Drawing-Balloons

Label each component.

184) Click inside the **Isometric view** boundary.

185) Click the **Auto Balloon** 🎈 tool from the Annotation toolbar. The Auto Balloon PropertyManager is displayed. Accept the defaults. Note: new of 2012 is the insert magnetic lines option. See SolidWorks help for additional information.

186) Click **OK** ✔ from the Auto Balloon
PropertyManager.

187) Click and drag the **balloons** inside the
view boundary.

Insert a balloon for the O-RING.

188) Click the **Balloon** 🎈 tool from the
Annotatiton toolbar.

189) Click the **O-RING** in the Graphics
window.

190) **Position** the Balloon in the drawing.

191) Click **OK** ✔ from the Balloon
PropertyManager.

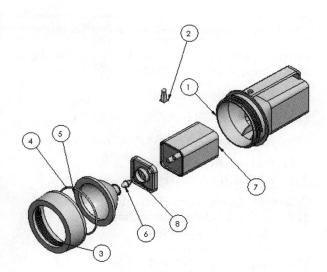

The Balloon note displays an arrowhead
on a selected edge and a filled dot on a selected face when the Drawing Standard is set to
ANSI. The Balloon note displays a "?" if no edge or face is selected.

Part Numbers

Use the following prefix codes to categories created parts and drawings. The part names and
part numbers are as follows:

Category:	Prefix:	Part Name:	Part Number:
Molded Parts	44-	BATTERYPLATE	44-A26
		LENSCAP	44-A27
		HOUSING	44-A28
Purchased Parts	B99-	BATTERY	B99-B01
	99-	LENS	99-B02
		O-RING	99-B03
		SWITCH	99-B04
		BULB	99-B05
Assemblies	10-	FLASHLIGHT	10-F123

The Bill of Materials requires editing. The current part file name determines the PART
NUMBER parameter values. The Configuration Properties controls the display of the PART
NUMBER in the Bill of Materials. Redefine the PART NUMBER for each part.

Activity: FLASHLIGHT Drawing-ConfigurationManager

Open the BATTERYPLATE part from the drawing.
192) **Expand** Drawing View1.

193) **Expand** FLASHLIGHT.

194) **Expand** BATTERYANDPLATE.

195) Right-click **BATTTERYPLATE**.

196) Click **Open Part**. The
BATTERPLATE part is displayed.

Display the Configuration Properties.
197) Click the **ConfigurationManager**
tab.

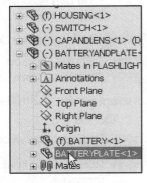

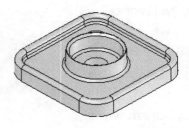

198) Right-click the **Default**
[BATTTERYPLATE] configuration.

199) Click **Properties**.

200) Select **User Specified Name** from the drop-down menu.

201) Enter **44-A26**.

202) Click **OK** ✔ from the Configuration Properties PropertyManager.
The ConfigurationManager displays the Default [44-A26]
configuration.

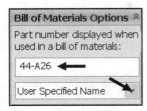

203) Click the **FeatureManager** 🖱 tab.

Return to the FLASHLIGHT drawing.
204) Click **Window, FLASHLIGHT-Sheet1** from the Menu bar.

Activity: FLASHLIGHT Drawing-Update the Bill of Materials

Update the Bill of Materials.
205) **Rebuild** the model.

Enter the BATTERY PART NUMBER.
206) Right-click **BATTERY** part from the FeatureManager.

207) Click **Open Part**. The BATTERY part is displayed.

208) Click the BATTERY **ConfigurationManager** 🖱 tab.

209) Right-click **Default [BATTERY]** from the ConfigurationManager.

210) Click **Properties**.

211) Select **User Specified Name** from the drop-down menu.

212) Enter **B99-B01**.

213) Click **OK** ✔ from the Configuration Properties PropertyManager.

214) Click the **FeatureManager** 🖱 tab.

Return to the FLASHLIGHT drawing.

215) Click **Window, FLASHLIGHT-Sheet1** from the Menu bar.

Update the Bill of Materials.

216) Rebuild the model.

Resize the PART NUMBER column.

217) Drag the **vertical line** between the PART NUMBER and the DESCRIPTION column to the left.

218) Drag the **vertical line** between the DESCRIPTION and the QTY to the left.

Complete the PART NUMBER column. The User Specified Name for the remaining PART numbers is left as an exercise.

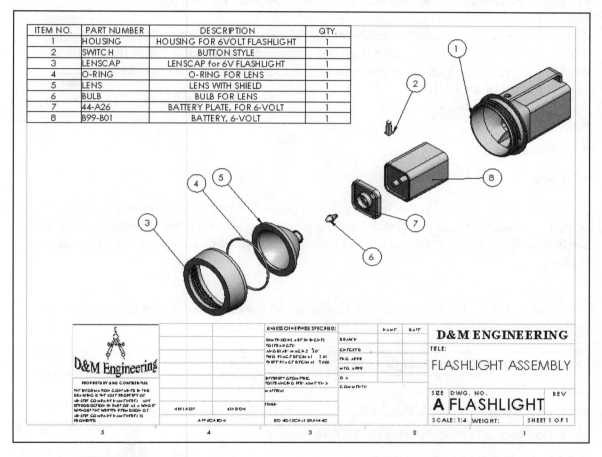

ITEM NO.	PART NUMBER	DESCRIPTION	QTY.
1	HOUSING	HOUSING FOR 6VOLT FLASHLIGHT	1
2	SWITCH	BUTTON STYLE	1
3	LENSCAP	LENSCAP for 6V FLASHLIGHT	1
4	O-RING	O-RING FOR LENS	1
5	LENS	LENS WITH SHIELD	1
6	BULB	BULB FOR LENS	1
7	44-A26	BATTERY PLATE, FOR 6-VOLT	1
8	B99-B01	BATTERY, 6-VOLT	1

Save the FLASHLIGHT drawing.

219) Click **Save** 💾.

Close All documents.

220) Click **Windows, Close All** from the Menu bar.

Additional details on Exploded View, Notes, Properties, Bill of Materials and Balloons, are available in SolidWorks Help Topics. Keywords: Exploded, Notes, Properties (configurations), Bill of Materials, Balloons and Auto Balloon.

Review the FLASHLIGHT Drawing

The FLASHLIGHT drawing contained an Exploded view. The Exploded view was created in the FLASHLIGHT assembly. The Bill of Materials listed the Item Number, Part Number, Description and Quantity of components in the assembly. Balloons were inserted to label the top level components in the FLASHLIGHT assembly. You developed Properties in the part to modify the Part Number utilized in the Bill of Materials.

Design Tables and O-RING-DESIGN-TABLE Drawing

A design table is a spreadsheet used to create multiple configurations in a part or assembly. The design table controls the dimensions and parameters in the part. Utilize the design table to modify the overall path diameter and profile diameter of the O-RING. Create three configurations of the O-RING:

- Small
- Medium
- Large

Save the O-RING part with the Save As Copy option. Create a new drawing for the O-RING using the View Palette in the Task Pane. The part configurations utilized in the drawing are controlled through the Properties of the view. Insert the three O-RING configurations in the drawing.

The O-RING contains two dimension names in the design tables. Parts contain hundreds of dimensions and values. Rename dimension names for clarity.

Activity: O-RING Part-Design Table

Open the O-RING part.

221) Click **Open** from the Menu bar.

222) Double-click **O-RING** from the PROJECTS folder.

Save a copy of the O-RING.

223) Click **Save As** from the drop-down Menu bar.

224) Check the **Save as copy** box.

225) Enter **O-RING-DESIGN-TABLE** for File name.

226) Enter **O-RING WITH DESIGN TABLE** for Description.

227) Click **Save**.

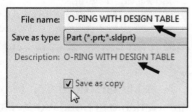

Open the O-RING-DESIGN-TABLE part.

228) Click **Open** from the Menu bar.

229) Double-click **O-RING-DESIGN-TABLE** from the PROJECTS folder.

Modify the Primary Units.

230) Click **Options**, **Document Properties** tab.

231) Click **Units**. ANSI is the default Overall Drafting standard.

232) Select **MMGS**.

233) Select **.12** for Length units Decimal.

234) Click **OK** from the Document Properties - Units dialog box.

235) Double-click the **face** of the O-RING. The two diameter dimensions are displayed in millimeters. The O-RING is displayed in blue.

Insert a Design Table.

236) Click **Insert**, **Design Table** from the Menu bar. The Auto-create option is selected. Accept the defaults.

237) Click **OK** from the Design Table PropertyManager.

238) Hold the **Ctrl** key down.

239) Select **D1@Sketch-path** and **D1@Sketch-profile**.

240) Release the **Ctrl** key.

241) Click **OK** from the Dimensions dialog box.

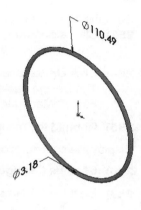

Note: The dimension variable name will be different if sketches or features were deleted. The input dimension names and default values are automatically entered into the Design Table. The value Default is entered in Cell A3. The values for the O-RING are entered in Cells B3 through C6. The sketch-path diameter is controlled in Column B. The sketch-profile diameter is controlled in Column C.

Enter the three configuration names.
242) Click **Cell A4**. Enter **Small**.

243) Click **Cell A5**. Enter **Medium**.

244) Click **Cell A6**. Enter **Large**.

Enter the dimension values for the Small configuration.
245) Click **Cell B4**. Enter **100**.

246) Click **Cell C4**. Enter **3**.

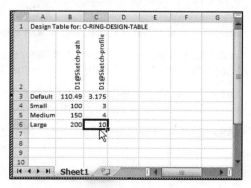

Enter the dimension values for the Medium configuration.
247) Click **Cell B5**. Enter **150**.

248) Click **Cell C5**. Enter **4**.

Enter the dimension values for the Large configuration.
249) Click **Cell B6**.

250) Enter **200**.

251) Click **Cell C6**.

252) Enter **10**.

Build the three configurations.
253) Click a **position** outside the EXCEL Design Table in the Graphics window.

254) Click **OK** to generate the configurations. The Design Table icon is displayed in the FeatureManager.

255) **Rebuild** the model.

Display the configurations.
256) Click the **ConfigurationManager** tab.

257) Double-click **Small**.

258) Double-click **Medium**.

259) Double-click **Large**.

260) Double-click **Default**.

Return to the FeatureManager.
261) Click the O-RING-DESIGN-TABLE **FeatureManager** tab.

Activity: Create the O-RING-DESIGN-TABLE Drawing

Create a new drawing.
262) Click **New** from the Menu bar.

263) Click the **MY-TEMPLATES** tab.

264) Double-click the **A-IN-ANSI, [A-ISO-MM]** Drawing Template. The Model View PropertyManager is displayed.

265) Click **Cancel** from the Model View PropertyManager

Insert an Isometric view using the View Palette.
266) Click the **View Palette** tab from the Task Pane.

267) Select **O-RING-DESIGN-TABLE** from the drop-down menu. The available views are displayed.

268) Click and drag the **Isometric view** to the left side of the drawing.

269) Select **1:2** for Scale.

270) Click **OK** ✔ from the Drawing View1 PropertyManager.

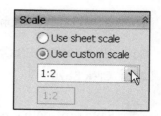

Deactivate the Origin.
271) Click **View**, uncheck **Origins** from the Menu bar.

272) Click **Save** 🖫 . Accept the defaults.

273) Click **Save**. Note: Adjust the font size in the drawing for the Title box and DWG. No.

Activity: O-RING Drawing-Design Table

Copy the Isometric view.
274) Click inside the **Isometric view** boundary.

275) Click **Edit**, **Copy** from the Menu bar.

Paste the Isometric view.
276) Click a **position** to the right of the Isometric view.

277) Click **Edit**, **Paste** from the Menu bar.

Display the medium configuration.
278) Right-click inside the second **Isometric view** boundary.

279) Click **Properties**.

280) Select **Medium** from the Use named configuration list.

281) Click **OK** from the Drawing View Properties dialog box.

282) Click **OK** ✔ from the Drawing View2 PropertyManager.

283) **Rebuild** the model.

284) **Modify** the Title box text.

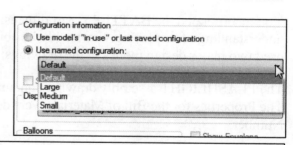

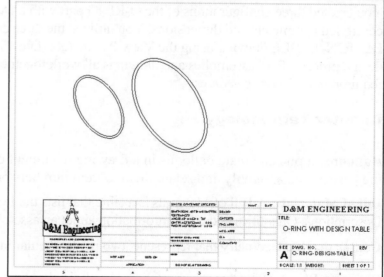

The Large configuration is left as an exercise.

Save the O-RING-DESIGN-TABLE drawing.

285) Click **Save** 💾 .

💡 Save time with repetitive dimensioning in configurations. Insert dimensions into the first view. Copy the view. The view and the dimensions are copied to the new view.

🔍 Additional details on Design Tables and Configurations are available in SolidWorks Help Topics.

Chapter Summary

You produced three drawings: BATTERY, FLASHLIGHT and O-RING-DESIGN-TABLE. The drawings contained Standard views, Detail view and Isometric views. The drawings utilized a custom Sheet Format and custom Drawing Template. The Sheet Format contained the Company logo and Title block information.

You incorporated the BATTERY part dimensions into the drawing. You obtained an understanding of displaying views with the ability to insert, add and modify dimensions. You used two major design modes in the drawings: *Edit Sheet Format and Edit Sheet*.

The FLASHLIGHT assembly drawing contained an Exploded view and Bill of Materials. The Properties for the Bill of Materials were developed in each part with a user defined Part Number.

You created three configurations of the O-RING part with a Design Table. A Design Table controlled parameters and dimensions. You utilized the three configurations in the O-RING-DESIGN-TABLE drawing using the View Palette tool. Drawings are an integral part of the design process. Part, assemblies and drawings all work together to fulfill the design requirements of your customer.

Chapter Terminology

Balloons: You can create balloons in a drawing document or in a note. The balloons label the parts in the assembly and relate them to item numbers on the bill of materials (BOM).

Bill of Materials: A Bill of Materials is a table that lists the Item Numbers, Part Numbers, Descriptions, Quantities and other information about an assembly.

Center marks: Geometry that represents two perpendicular intersecting centerlines.

Drawing Sheets: The "paper sheets" used to hold the views, dimensions and annotations and create the drawing.

Design Table: A Design Table is a table used to create multiple configurations in a part or assembly. The Design Table controls the dimensions and parameters in the part or assembly.

Detached Drawing: A drawing format that allows opening and working in a drawing without loading the corresponding models into memory. The models are loaded on an as-needed basis.

Drawing Template: The foundation of a SolidWorks drawing is the Drawing Template. Drawing size, drawing standards, company information, manufacturing and or assembly requirements, units and other properties are defined in the Drawing Template. In this chapter the Drawing Template contained the drawing Size and Document Properties.

Detail view: A view inserted into a drawing that enlarges an area of an existing view.

Edit Sheet Format Mode: Provides the ability to: Change the title block size and text headings, incorporate a company logo and add a drawing, design or company text. Remember: A part cannot be inserted into a drawing when the Edit Sheet Format mode is selected.

Edit Sheet Mode: Provides the ability to: Add or modify views, dimensions or text.

First Angle: Standard 3 Views are in either third angle or first angle projection. In first angle projection, the front view is displayed at the upper left and the other two views are the top and left views.

Insert Model Items: The tool utilized to insert part dimensions and annotations into drawing views.

Leader lines: Dimension entity that references the size of the profile. A gap must exist between the profile lines and the leader lines for proper drafting practice.

Model View: The tool utilized to insert named views into a drawing.

Multiple Drawing Sheets: The drawing can have multiple sheets, if required. To create an additional sheet, use Add Sheet. The size and format of the new sheet is copied from the original but can be edited and changed.

Note: Annotation tool used to add text with leaders or as a stand-alone text string.

Sheet Format: A document applied to the Drawing Template. The Sheet Format contains the border, title block information, revision block information, company name and or logo information, Custom Properties and SolidWorks Properties.

Third Angle: Standard 3 Views are in either third angle or first angle projection. In Third Angle projection, the default front view from the part or assembly is displayed at the lower left, and the other two views are the top and right views.

Title block: The area in the Sheet Format containing vital part or assembly information. Each company has a unique version of a title block.

View Palette: The View Palette is located in the Task Pane. The View Palette provides the ability to insert drawing views. It contains images of standard views, annotation views, section views, and flat patterns (sheet metal parts) of the selected model. You can drag views onto an active drawing sheet to create a drawing view.

Questions

1. Identify the differences between a Drawing Template and a Sheet Format. Provide an example.

2. Identify the command to save the Sheet Format.

3. Identify the command to save the Drawing Template.

4. Describe a Bill of Materials. Provide an example.

5. Name the two major design modes used to develop a drawing in SolidWorks.

6. Name seven components that are commonly found in a title block.

7. Describe the procedure to insert an Isometric view into a drawing.

8. In SolidWorks, drawing file names end with a _____ suffix.

9. True or False. Most engineering drawings us the following font: Time New Roman - All small letters. Explain your answer.

10. Describe Leader lines. Provide an example.

11. Describe a Note on a drawing. Provide an example.

12. Explain the procedure to add an Exploded view from an assembly to a drawing.

13. Explain the procedure on labeling components in an Exploded view on an assembly drawing. Provide an example.

14. Describe the procedure to create a Design Table.

15. True or False. You cannot display different configurations in the same drawing. Explain your answer.

16. True or False. The Part Number is only entered in the Bill of Materials. Explain your answer.

17. There are hundreds of options in the Document Properties, Drawings and Annotations toolbars. How would you locate additional information on these options and tools?

18. Describe the View Palette.

Exercises

Exercise 8.1: FLATBAR - 3 HOLE Drawing

Note: Dimensions are enlarged for clarity. Utilize inch, millimeter, or dual dimensioning.

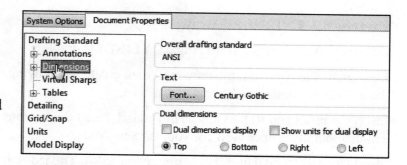

Create the ANSI-IPS Third Angle Projection FLATBAR - 3HOLE drawing. First create the part from the drawing - then create the drawing. Use the default A-Landscape Sheet Format/Size.

Insert a Shaded Isometric view. No Tangent Edges displayed.

Insert a Front and Top view. Insert dimensions. Insert 3X - EQ. SP. Insert the Company and Third Angle Projection icon. Add a Parametric Linked Note for MATERIAL THICKNESS.

Hide the Thickness dimension in the Top view. Insert needed Centerlines.

Insert Custom Properties for Material (2014 Alloy), DRAWNBY, DRAWNDATE, COMPANYNAME, etc.

Third Angle Projection icon is located in the Chapter 8 Homework folder on the DVD in the book.

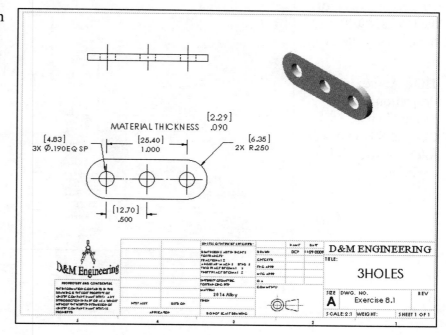

Exercise 8.2: **CYLINDER Drawing**

Create the ANSI - IPS - Third Angle CYLINDER drawing.

First create the part from the drawing - then create the drawing. Use the default A-Landscape Sheet Format/Size.

Insert the Front and Right view as illustrated. Insert dimensions. Think about the proper view for your dimensions!

Insert Company and Third Angle projection icons. The icons are available in the homework folder.

Insert needed Centerlines and Center Marks.

Insert Custom Properties: Material, Description, DrawnBy, DrawnDate, CompanyName, etc. Note: Material is AISI 1020.

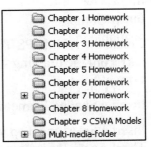

Utilize the Mass Properties tool from the Evaluate toolbar to calculate the volume and mass of the CYLINDER part. Set decimal places to 4.

Third Angle Projection icon is located in the Chapter 8 Homework folder on the DVD in the book.

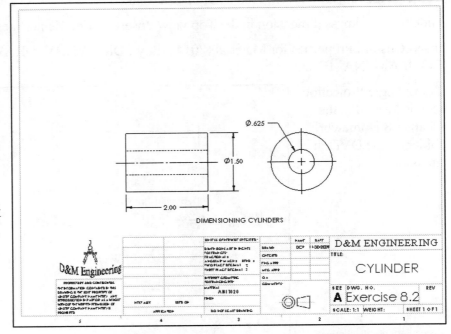

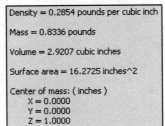

Density = 0.2854 pounds per cubic inch

Mass = 0.8336 pounds

Volume = 2.9207 cubic inches

Surface area = 16.2725 inches^2

Center of mass: (inches)
    X = 0.0000
    Y = 0.0000
    Z = 1.0000

Exercise 8.3: PRESSURE PLATE Drawing

Create the ANSI - IPS - Third Angle PRESSURE PLATE drawing.

First create the part from the drawing - then create the drawing. Use the default A-Landscape Sheet Format/Size.

Insert the Front and Right view as illustrated. Insert dimensions. Think about the proper view for your dimensions!

Insert Company and Third Angle projection icons. The icons are available in the homework folder.

Insert needed Centerlines and Center Marks.

Insert Custom Properties: Material, Description, DrawnBy, DrawnDate, CompanyName, etc. Note: Material is 1060 Alloy.

Third Angle Projection icon is located in the Chapter 8 Homework folder on the DVD in the book.

📁 Chapter 1 Homework
📁 Chapter 2 Homework
📁 Chapter 3 Homework
📁 Chapter 4 Homework
📁 Chapter 5 Homework
📁 Chapter 6 Homework
⊞ 📁 Chapter 7 Homework
📁 Chapter 8 Homework
📁 Chapter 9 CSWA Models
⊞ 📁 Multi-media-folder

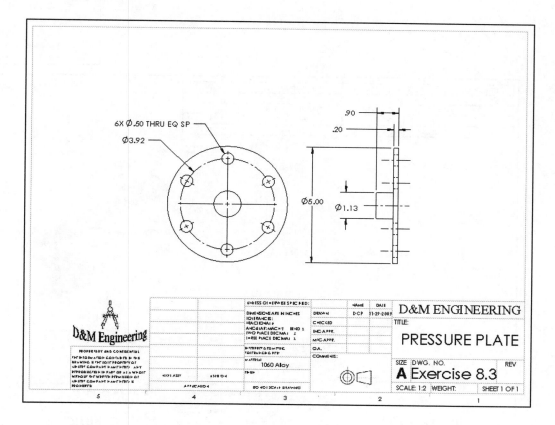

Exercise 8.4: LINKS Assembly Drawing

Create the LINK assembly. Utilize three different FLATBAR configurations and a SHAFT-COLLAR.

Create the LINK assembly drawing as illustrated. Use the default A-Landscape Sheet Format/Size.

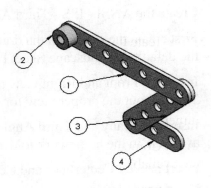

Insert Company and Third Angle projection icons. The icons are available in the homework folder. Remove all Tangent Edges.

Insert Custom Properties: Description, DrawnBy, DrawnDate, CompanyName, etc.

Insert a Bill of Materials as illustrated with Balloons.

Third Angle Projection icon is located in the Chapter 8 Homework folder on the DVD in the book.

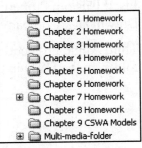

- Chapter 1 Homework
- Chapter 2 Homework
- Chapter 3 Homework
- Chapter 4 Homework
- Chapter 5 Homework
- Chapter 6 Homework
- Chapter 7 Homework
- Chapter 8 Homework
- Chapter 9 CSWA Models
- Multi-media-folder

ITEM NO.	PART NUMBER	DESCRIPTION	QTY.
1	GIDS-SC-10009-7	7 HOLES	1
2	GIDS-SC-10012-3-16	SHAFT-COLLAR	1
3	GIDS-SC-10009-5	5 HOLES	1
4	GIDS-SC-10009-3	3 HOLES	1

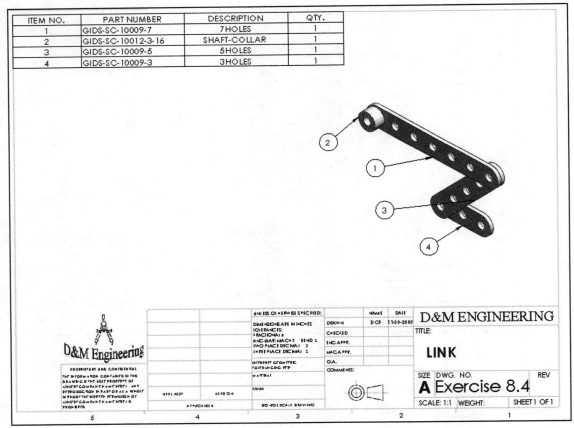

Exercise 8.5: PLATE-1 Drawing

Create the ANSI - MMGS - Third Angle PLATE-1 drawing.

First create the part from the drawing - then create the drawing. Use the default A-Landscape Sheet Format/Size.

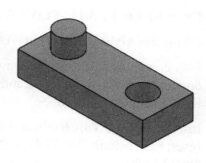

Insert the Front and Right view as illustrated. Insert dimensions. Think about the proper view for your dimensions!

Insert Company and Third Angle projection icons. The icons are available in the homework folder.

Insert needed Centerlines and Center Marks.

Insert Custom Properties: Material, Description, DrawnBy, DrawnDate, CompanyName, etc. Note: Material is 1060 Alloy.

Third Angle Projection icon is located in the Chapter 8 Homework folder on the DVD in the book.

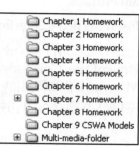

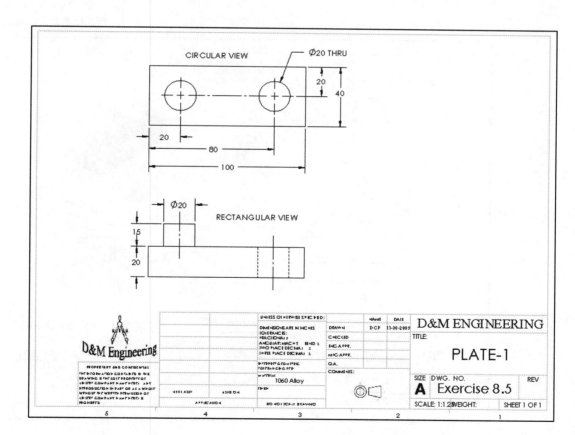

Exercise 8.6: FLATE-PLATE Drawing

Create the ANSI - IPS - Third Angle PLATE-1 drawing.

First create the part from the drawing - then create the
drawing. Use the default A-Landscape Sheet Format/Size.
Remove all Tangent Edges.

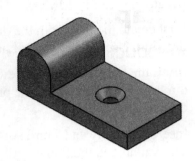

Insert the Front, Top, Right and Isometric view as illustrated.
Insert dimensions. Think about the proper view for your
dimensions!

Insert needed Centerlines and Center Marks.

Insert Custom Properties: Material, Description, DrawnBy,
DrawnDate, CompanyName, Hole Annotation, etc. Note:
Material is 1060 Alloy. Third Angle Projection icon is located
in the Chapter 8 Homework folder on the DVD in the book.

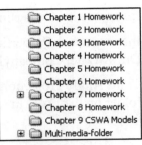

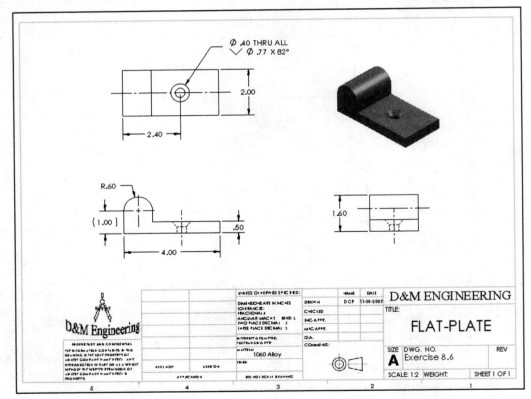

Chapter 9

Introduction to the Certified SolidWorks Associate Exam

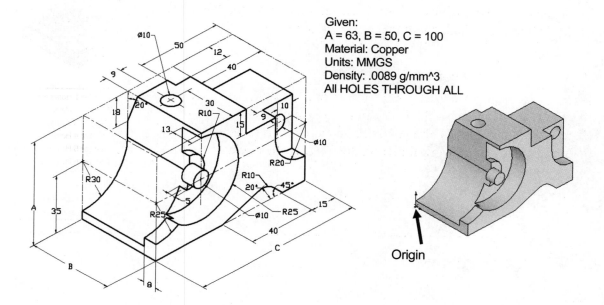

Given:
A = 63, B = 50, C = 100
Material: Copper
Units: MMGS
Density: .0089 g/mm^3
All HOLES THROUGH ALL

Origin

Below are the desired outcomes and usage competencies based on the completion of Chapter 9.

Desired Outcomes:	Usage Competencies:
• Procedure and process knowledge • Exam categories: o Drafting Competencies, Basic Part Creation and Modification, Intermediate Part Creation and Modification, Advanced Part Creation and Modification, and Assembly Creation and Modification	• Familiarity of the CSWA exam. • Comprehension of the skill sets to past the CSWA exam. • Awareness of the question types. • Capability to locate additional CSWA exam information.

Notes:

Chapter 9 - Certified SolidWorks Associate CSWA Exam

Chapter Objective

Provide a basic introduction into the curriculum and categories of the Certified SolidWorks Associate CSWA exam. Awareness to the exam procedure, process, and required model knowledge needed to take and past the CSWA exam. The five exam categories are:

- Drafting Competencies

- Basic Part Creation and Modification

- Intermediate Part Creation and Modification

- Advanced Part Creation and Modification

- Assembly Creation and Modification

Introduction

DS SolidWorks Corp. offers various stages of certification. Each stage representing increasing levels of expertise in 3D CAD design: Certified SolidWorks Associate CSWA, Certified SolidWorks Professional CSWP and Certified SolidWorks Expert CSWE along with specialty fields in Simulation, Sheet Metal, and Surfacing.

The CSWA Certification indicates a foundation in and apprentice knowledge of 3D CAD design and engineering practices and principles. The main requirement for obtaining the CSWA certification is to take and pass the on-line proctored 180 minute exam (minimum of 165 out of 240 points). The new CSWA exam consists of fourteen questions in five categories.

Intended Audience

The intended audience for the CSWA exam is anyone with a minimum of 6 - 9 months of SolidWorks experience and basic knowledge of engineering fundamentals and practices. SolidWorks recommends that you review their SolidWorks Tutorials on Parts, Assemblies, and Drawings as a prerequisite and have at least 45 hours of classroom time learning SolidWorks or using SolidWorks with basic engineering design principles and practices.

To prepare for the CSWA exam, it is recommended that you first perform the following:

- Take a CSWA exam preparation class or review a text book written for the CSWA exam.

- Complete the SolidWorks Tutorials

- Practice creating models from the isometric working drawings sections of any Technical Drawing or Engineering Drawing Documentation text books.

- Complete the sample CSWA exam in a timed environment, available at www.solidworks.com.

Additional references to help you prepare are as follows:

- **SolidWorks Users Guide**, SolidWorks Corporation, 2012.

- ***Official Certified SolidWorks® Associate Examination Guide, Version 3; 2011, 2010, 2009**.

- **Engineering Drawing and Design**, Jensen & Helsel, Glencoe, 1990.

- **Drawing and Detailing with SolidWorks**, Planchard & Planchard, SDC Pub., Mission, KS 2010.

*For detailed exam information see the Official Certified SolidWorks Associate Examination Guide book. The primary goal of this book is not only to help you pass the CSWA exam, but also to ensure that you understand and comprehend the concepts and implementation details of the CSWA process.

CSWA Exam Content

The CSWA exam is divided into five key categories. Questions on the timed exam are provided in a random manor. The following information provides general guidelines for the content likely to be included on the exam. However, other related topics may also appear on any specific delivery of the exam. In order to better reflect the contents of the exam and for clarity purposes, the guidelines below may change at any time without notice.

- *Drafting Competencies*: (Three questions - multiple choice - 5 points each).

  - Questions on general drawing views: Projected, Section, Break, Crop, Detail, Alternate Position, etc.

- *Basic Part Creation and Modification*: (Two questions - one multiple choice / one single answer - 15 points each).

  - Sketch Planes:

    - Front, Top, Right

  - 2D Sketching:

    - Geometric Relations and Dimensioning

  - Extruded Boss/Base Feature

  - Extruded Cut feature

  - Modification of Basic part

In the *Basic Part Creation and Modification* category there is a dimension modification.

- *Intermediate Part Creation and Modification*: (Two questions - one multiple choice / one single answer - 15 points each).

  - Sketch Planes:

    - Front, Top, Right

  - 2D Sketching:

    - Geometric Relations and Dimensioning

A00006: Drafting Competencies - To create drawing view 'B' it is necessary to select drawing view 'A' and insert which SolidWorks view type?

B22001: Basic Part (Hydraulic Cylinder Half) - Step 1
Build this part in SolidWorks.
(Save part after each question in a different file in case it must be reviewed)

Unit system: MMGS (millimeter, gram, second)
Decimal places: 2
Part origin: Arbitrary
All holes through all unless shown otherwise.
Material: Aluminium 1060 Alloy

Screen shots from the exam.

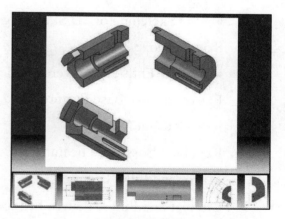

D12801: Intermediate Part (Wheel) - Step 1
Build this part in SolidWorks.
(Save part after each question in a different file in case it must be reviewed)

Unit system: MMGS (millimeter, gram, second)
Decimal places: 2
Part origin: Arbitrary
All holes through all unless shown otherwise.
Material: Aluminium 1060 Alloy

A = 134.00
B = 890.00

Note: All geometry is symmetrical about the plane represented by the line labeled F"" in the M-M Section View.

What is the overall mass of the part (grams)?

Screen shots from the exam.

- Extruded Boss/Base Feature

- Extruded Cut Feature

- Revolved Boss/Base Feature

- Mirror and Fillet Feature

- Circular and Linear Pattern Feature

- Plane Feature

- Modification of Intermediate Part:

  - Sketch, Feature, Pattern, etc.

- Modification of Intermediate part

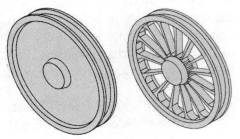

Screen shots from the exam.

- *Advanced Part Creation and Modification:* (Three questions - one multiple choice / two single answers - 15 points each).

  - Sketch Planes:

    - Front, Top, Right, Face, Created Plane, etc.

    - 2D Sketching or 3D Sketching

  - Sketch Tools:

    - Offset Entities, Convert Entitles, etc.

  - Extruded Boss/Base Feature

  - Extruded Cut Feature

  - Revolved Boss/Base Feature

  - Mirror and Fillet Feature

  - Circular and Linear Pattern Feature

  - Shell Feature

  - Plane Feature

  - More Difficult Geometry Modifications

C12801: Advanced Part (Bracket) - Step 1
Build this part in SolidWorks.
(Save part after each question in a different file in case it must be reviewed)

Unit system: MMGS (millimeter, gram, second)
Decimal places: 2
Part origin: Arbitrary
All holes through all unless shown otherwise.
Material: AISI 1020 Steel
Density = 0.0079 g/mm^3

A = 64.00
B = 20.00
C = 26.50

What is the overall mass of the part (grams)?

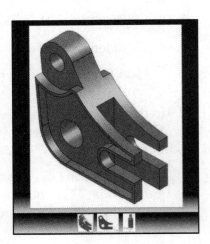

Screen shots from the exam.

- *Assembly Creation and Modification*: (Two different assemblies - four questions - two multiple choice / two single answers - 30 points each).

  - Insert the first (fixed) component

  - Insert all needed components

  - Standard Mates

  - Modification of key parameters in the assembly

Download the needed components in a zip folder during the exam to create the assembly.

Note: To apply for CSWA Provider status for your institution, go to www.solidWorks.com/cswa and fill out the CSWA Provider application. It is as easy as that.

A total score of 165 out of 240 or better is required to obtain your CSWA Certification.

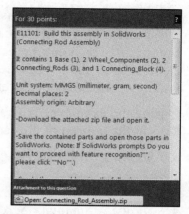

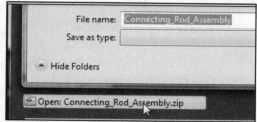

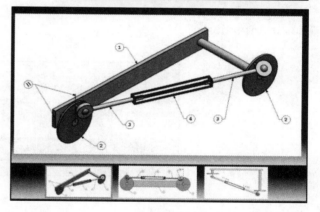

Screen shots from the exam.

You are allowed to answer the questions in any order. Use the Summary Screen during the CSWA exam to view the list of all questions you have or have not answered.

During the exam, use the control keys at the bottom of the screen to:

- *Show the Previous the Question.*
- *Reset the Question.*
- *Show the Summary Screen.*
- *Move to the Next Question.*

Do NOT use feature recognition when you open the downloaded components for the assembly in the CSWA exam. This is a timed exam. Manage your time. You do not need this information.

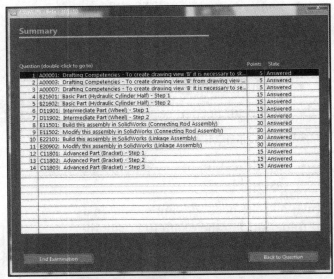

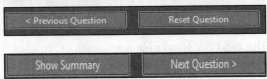

Screen shots from the exam.

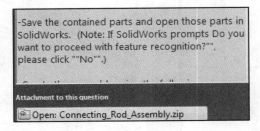

-Save the contained parts and open those parts in SolidWorks. (Note: If SolidWorks prompts Do you want to proceed with feature recognition?"", please click ""No"".)

Attachment to this question

Open: Connecting_Rod_Assembly.zip

During the exam, SolidWorks provides the ability to click on a detail view below (as illustrated) to obtain additional details and dimensions during the exam.

🔆 No Simulation (CosmosXpress) questions are on the CSWA exam.

🔆 No Sheetmetal questions are on the CSWA exam.

🔆 FeatureManager names were changed through various revisions of SolidWorks. Example: Extrude1 vs. Boss-Extrude1. These changes do not affect the models or answers in this book.

🔆 No Surface questions are on the CSWA exam.

About the CSWA exam

Most CAD professionals today recognize the need to become certified to prove their skills, prepare for new job searches, and to learn new skill, while at their existing jobs.

Specifying a CSWA or CSWP certification on your resume is a great way to increase your chances of landing a new job, getting a promotion, or looking more qualified when representing your company on a consulting job.

Exam Day

You will need:

- A computer with SolidWorks installed on it in a secure environment.

- An internet connection.

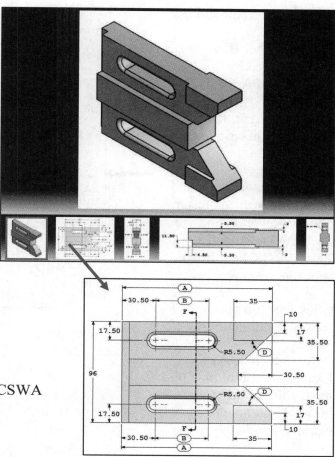

Screen shots from the exam.

- A valid email address.

- A voucher ID code (Provided by the CSWA Provider).

- ID: Student, ID, drivers license, etc.

Log into the Tangix_TesterPRO site (http://www.virtualtester. com/solidworks) and click the Download TesterPRO Client link to start the exam process.

Click the Run button and follow the directions.

After you click the Start Examination button. You have 180 minutes (3 hours) to complete the exam. Good luck!

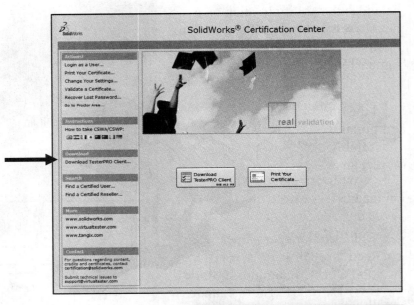

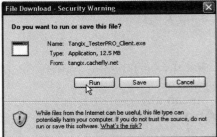

Screen shots from the exam.

The exam generates unique questions for each student.

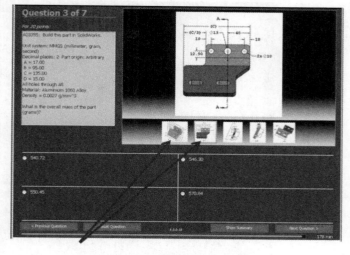

Click on different images to display additional views as illustrated.

Read the questions slowly, view the additional views, model your part or assembly and then select the correct answer.

The exam uses a multi-question single answer format or a fill in the blank format.

Screen shots from the exam.

In the Basic Part Creation and Modification, Intermediate Part Creation and Modification, Advanced Part Creation and Modification, and Assembly Creation and Modification categories, you will be required to read and interpret all types of drawing views.

Click the Next Question button to procedure to the next question in the exam.

You can also click in the image itself to zoom in on that area.

Below are examples for a part and an assembly screen shot in the CSWA exam.

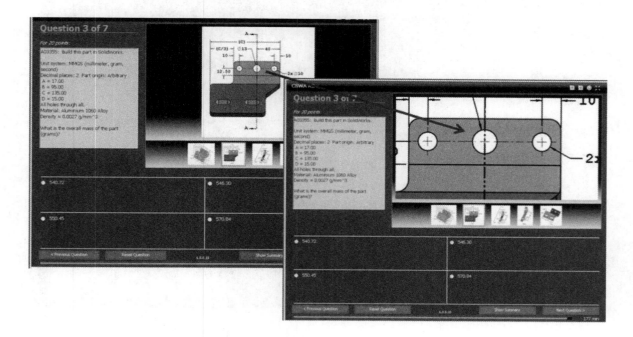

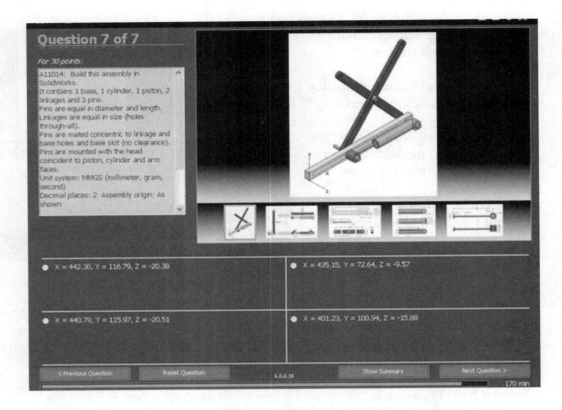

For 30 points:

A11014: Build this assembly in SolidWorks.
It contains 1 base, 1 cylinder, 1 piston, 2 linkages and 3 pins.
Pins are equal in diameter and length.
Linkages are equal in size (holes through-all).
Pins are mated concentric to linkage and base holes and base slot (no clearance).
Pins are mounted with the head coincident to piston, cylinder and arm faces.
Unit system: MMGS (millimeter, gram, second)
Decimal places: 2 Assembly origin: As shown

- X = 442.30, Y = 116.79, Z = -20.38
- X = 435.15, Y = 72.64, Z = -9.57
- X = 440.79, Y = 115.97, Z = -20.51
- X = 401.23, Y = 100.94, Z = -15.88

| < Previous Question | Reset Question | 6.0.0.18 | Show Summary | Next Question > |

170 min

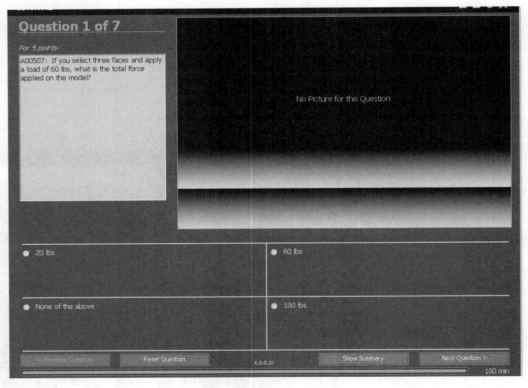

For 5 points:

A00507: If you select three faces and apply a load of 60 lbs, what is the total force applied on the model?

No Picture for this Question

- 20 lbs
- 60 lbs
- None of the above
- 180 lbs

| < Previous Question | Reset Question | 6.0.0.18 | Show Summary | Next Question > |

180 min

Click the Show Summary button to display the current status of your exam.

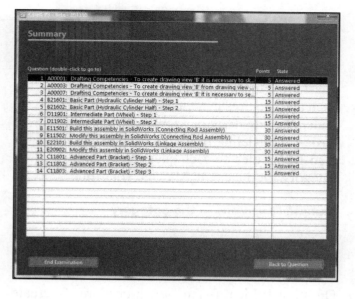

⚡ This is a timed exam. Skip a question and move on it you are stuck. You can also go back to the question you skipped anytime in the exam.

At the completion of the exam, Click the End Examination button. Click Yes to confirm.

Candidates receive a score report along with a score breakout by exam section.

⚡ For detail exam information see the Official Certified SolidWorks Associate Examination Guide book. The primary goal of this book is not only to help you pass the CSWA exam, but also to ensure that you understand and comprehend the concepts and implementation details of the CSWA process.

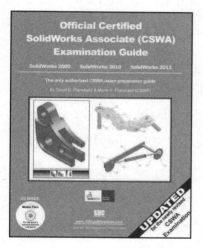

Drafting Competencies

Drafting Competencies is one of the five categories on the CSWA exam. There are three questions - multiple choice format - 5 points each and requires general knowledge and understanding drawing view methods, and basic 3D modeling techniques. Spend no more than 10 minutes on each question in this category for the exam. Manage your time.

> A00006: Drafting Competencies - To create drawing view 'B' it is necessary to select drawing view 'A' and insert which SolidWorks view type?

Screen shots from the exam.

Sample Questions in the category

In the *Drafting Competencies* category, an exam question could read:

Question 1: Identify the view procedure. To create the following view, you need to insert a:

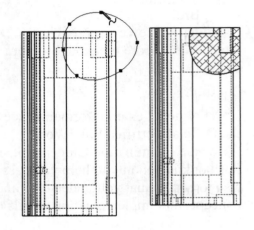

- A: Open Spline

- B: Closed Spline

- C: 3 Point Arc

- D: None of the above

The correct answer is B.

Question 2: Identify the illustrated view type.

- A: Crop view

- B: Section view

- C: Projected view

- D: None of the above

The correct answer is A.

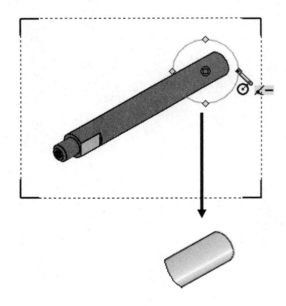

Question 3: Identify the illustrated Drawing view.

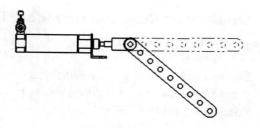

- A: Projected View
- B: Alternative Position View
- C: Extended View
- D: Aligned Section View

The correct answer is B.

Question 4: Identify the illustrated Drawing view.

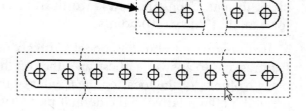

- A: Crop View
- B: Break View
- C: Broken-out Section View
- D: Aligned Section View

The correct answer is B.

Question 5: Identify the illustrated Drawing view.

- A: Section View
- B: Crop View
- C: Broken-out Section View
- D: Aligned Section View

The correct answer is D.

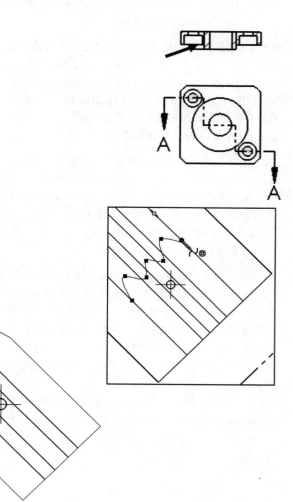

Question 6: Identify the view procedure. To create the following view, you need to insert a:

- A: Rectangle Sketch tool
- B: Closed Profile: Spline
- C: Open Profile: Circle
- D: None of the above

The correct answer is a B.

Basic Part Creation and Modification and Intermediate Part Creation and Modification

Basic Part Creation and Modification and Intermediate Part Creation and Modification is two of the five categories on the CSWA exam.

There are two questions on the CSWA exam in the *Basic Part Creation and Modification* category. One question is in a multiple choice single answer format and the other question (Modification of the model) is in the fill in the blank format. Each question is worth fifteen (15) points for a total of thirty (30) points.

You are required to build a model, with six or more features and to answer a question either on the overall mass, volume, or the location the Center of mass for the created model relative to the default part Origin location. You are then requested to modify the part and answer a fill in the blank format question.

There are two questions on the CSWA exam in the *Intermediate Part Creation and Modification* category. One question is in a multiple choice single answer format and the other question (Modification of the model) is in the fill in the blank format. Each question is worth fifteen (15) points for a total of thirty (30) points.

You are required to build a model, with six or more features and to answer a question either on the overall mass, volume, or the location the Center of mass for the created model relative to the default part Origin location. You are then requested to modify the model and answer a fill in the blank format question.

The main difference between the *Basic Part Creation and Modification* category and the *Intermediate Part Creation and Modification* or the *Advance Part Creation and Modification* category is the complexity of the sketches and the number of dimensions and geometric relations along with an increase in the number of features.

🔆 Spend no more than 40 minutes on the question in these categories. This is a timed exam. Manage your time.

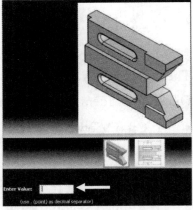

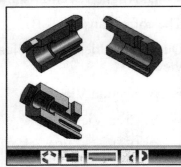

B22001: Basic Part (Hydraulic Cylinder Half) - Step 1
Build this part in SolidWorks.
(Save part after each question in a different file in case it must be reviewed)

Unit system: MMGS (millimeter, gram, second)
Decimal places: 2
Part origin: Arbitrary
All holes through all unless shown otherwise.
Material: Aluminium 1060 Alloy

D12801: Intermediate Part (Wheel) - Step 1
Build this part in SolidWorks.
(Save part after each question in a different file in case it must be reviewed)

Unit system: MMGS (millimeter, gram, second)
Decimal places: 2
Part origin: Arbitrary
All holes through all unless shown otherwise.
Material: Aluminium 1060 Alloy

A = 134.00
B = 890.00

Note: All geometry is symmetrical about the plane represented by the line labeled F"' in the M-M Section View.

What is the overall mass of the part (grams)?

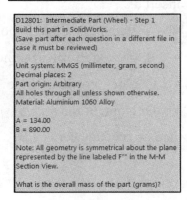

Screen shots from the exam.

Sample Questions in these categories

Question 1: Build the illustrated model from the provided information. Locate the Center of mass relative to the default coordinate system, Origin.

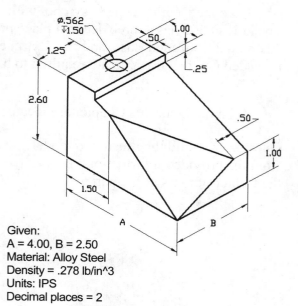

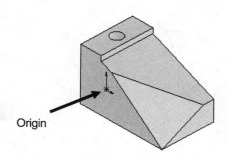

Origin

Given:
A = 4.00, B = 2.50
Material: Alloy Steel
Density = .278 lb/in^3
Units: IPS
Decimal places = 2

- A: X = -1.63 inches, Y = 1.48 inches, Z = -1.09 inches

- B: X = 1.63 inches, Y = 1.01 inches, Z = -0.04 inches

- C: X = 43.49 inches, Y = -0.86 inches, Z = -0.02 inches

- D: X = 1.63 inches, Y = 1.01 inches, Z = -0.04 inches

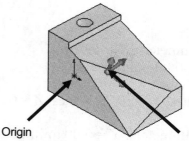

Origin　　　　　　　　　　　　Center of mass relative to the part Origin

The correct answer is B.

In the *Basic Part Creation and Modification and Intermediate Part Creation and Modification* category of the exam; you are required to read and understand an engineering document, set document properties, identify the correct Sketch planes, apply the correct Sketch and Feature tools, and apply material to build a part.

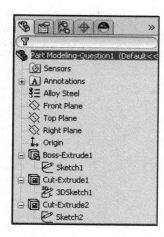

💡 Note the Depth/Deep ⊽ symbol with a 1.50 dimension associated with the hole. The hole Ø.562 has a three decimal place precision. Hint: Insert three features to build this model: Extruded Base, and two Extruded Cuts. Insert a 3D sketch for the first Extruded Cut feature. You are required to have knowledge in 3D sketching for the exam.

💡 All models for this Chapter are located on the DVD in the book.

Question 2: Build the illustrated model from the provided information. Locate the Center of mass of the part?

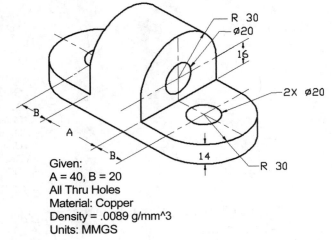

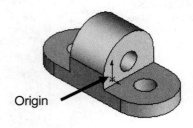

Origin

Given:
A = 40, B = 20
All Thru Holes
Material: Copper
Density = .0089 g/mm^3
Units: MMGS

- A: X = 0.00 millimeters, Y = 19.79 millimeters, Z = 0.00 millimeters

- B: X = 0.00 inches, Y = 19.79 inches, Z = 0.04 inches

- C: X = 19.79 millimeters, Y = 0.00 millimeters, Z = 0.00 millimeters

- D: X = 0.00 millimeters, Y = 19.49 millimeters, Z = 0.00 millimeters

- The correct answer is A.

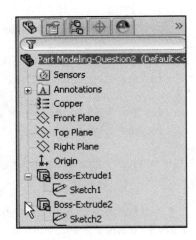

Question 3: Build the illustrated model
from the provided information. Locate the
Center of mass of the part.

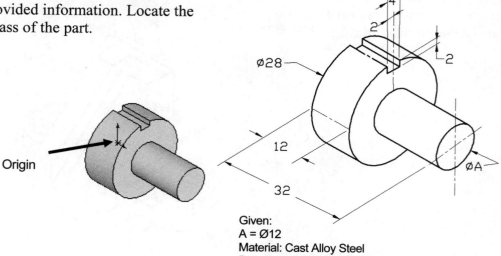

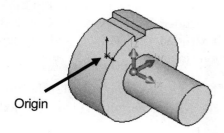

Given:
A = Ø12
Material: Cast Alloy Steel
Density = .0073 g/mm^3
Units: MMGS

- A: X = 10.00 millimeters, Y = -79.79 millimeters,
  Z: = 0.00 millimeters

- B: X = 9.79 millimeters, Y = -0.13 millimeters,
  Z = 0.00 millimeters

- C: X = 9.77 millimeters, Y = -0.10 millimeters,
  Z = -0.02 millimeters

- D: X = 10.00 millimeters, Y = 19.49 millimeters,
  Z = 0.00 millimeters

- The correct answer is B.

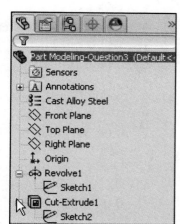

Question 4: Build the illustrated model from the provided information. Locate the Center of mass of the part.

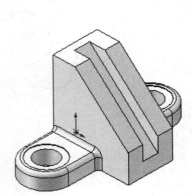

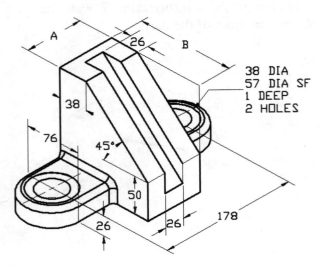

38 DIA
57 DIA SF
1 DEEP
2 HOLES

There are numerous ways to build this model. Think about the various features that create the model. Hint: Insert seven features to build this model: Extruded Base, Extruded Cut, Extruded Boss, Fillet, Extruded Cut, Mirror, and a second Fillet. Apply symmetry.

In the exam, create the left half of the model first, and then apply the Mirror feature. This is a timed exam.

- A: X = 49.00 millimeters,
  Y = 45.79 millimeters,
  Z = 0.00 millimeters

- B: X = 0.00 millimeters,
  Y = 19.79 millimeters,
  Z = 0.04 millimeters

- C: X = 49.21 millimeters,
  Y = 46.88 millimeters,
  Z = 0.00 millimeters

- D: X = 48.00 millimeters,
  Y = 46.49 millimeters,
  Z = 0.00 millimeters

The correct answer is C.

Given:
A = 76, B = 127
Material: 2014 Alloy
Density: .0028 g/mm^3
Units: MMGS
ALL ROUNDS EQUAL 6MM

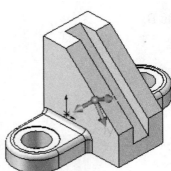

Part Modeling-Question4 (Default<<
- Sensors
- Annotations
- 2014 Alloy
- Front Plane
- Top Plane
- Right Plane
- Origin
- Boss-Extrude1
- Cut-Extrude1
- Boss-Extrude2
- Fillet1
- Cut-Extrude2
- Mirror1
- Fillet2

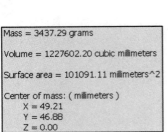

Mass = 3437.29 grams

Volume = 1227602.20 cubic millimeters

Surface area = 101091.11 millimeters^2

Center of mass: (millimeters)
  X = 49.21
  Y = 46.88
  Z = 0.00

Question 5: Build the illustrated model from the provided information. Locate the Center of mass of the part.

Think about the various features that create this model. Hint: Insert five features to build this part: Extruded Base, two Extruded Bosses, Extruded Cut, and Rib. Insert a Reference plane to create the Extruded Boss feature.

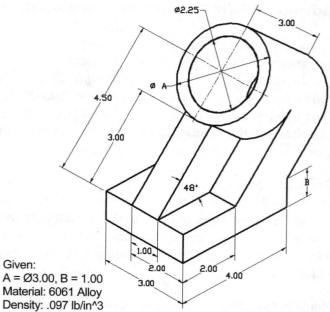

Given:
A = Ø3.00, B = 1.00
Material: 6061 Alloy
Density: .097 lb/in^3
Units: IPS
Decimal places = 2

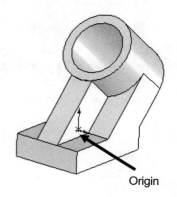

Origin

- A: X = 49.00 inches, Y = 45.79 inches, Z = 0.00 inches

- B: X = 0.00 inches, Y = 19.79 inches, Z = 0.04 inches

- C: X = 49.21 inches, Y = 46.88 inches, Z = 0.00 inches

- D: X = 0.00 inches, Y = 0.73 inches, Z = -0.86 inches

The correct answer is D.

All models for this Chapter are located on the DVD in the book.

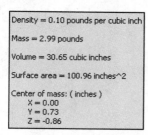

Density = 0.10 pounds per cubic inch

Mass = 2.99 pounds

Volume = 30.65 cubic inches

Surface area = 100.96 inches^2

Center of mass: (inches)
 X = 0.00
 Y = 0.73
 Z = -0.86

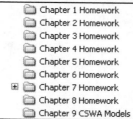

Advanced Part Creation and Modification

Advanced Part Creation and Modification is one of the five categories on the CSWA exam. The main difference between the *Advanced Part Creation and Modification* and the *Basic Part Creation and Modification* category and the *Intermediate Part Creation and Modification* is the complexity of the sketches and the number of dimensions and geometric relations along with an increase number of features.

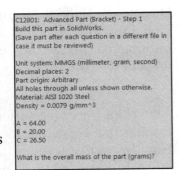

C12801: Advanced Part (Bracket) - Step 1
Build this part in SolidWorks.
(Save part after each question in a different file in case it must be reviewed)

Unit system: MMGS (millimeter, gram, second)
Decimal places: 2
Part origin: Arbitrary
All holes through all unless shown otherwise.
Material: AISI 1020 Steel
Density = 0.0079 g/mm^3

A = 64.00
B = 20.00
C = 26.50

What is the overall mass of the part (grams)?

There are three questions - one multiple choice / two single answers - 15 points each. The question is either on the location of the Center of mass relative to the default part Origin or to a new created coordinate system and all of the mass properties located in the Mass Properties dialog box: total overall mass, volume, etc.

Sample Questions in the category

In the *Advanced Part Creation and Modification* category, an exam question could read:

Screen shots from the exam.

Question 1: Build the illustrated model from the provided information.
Locate the Center of mass of the part.

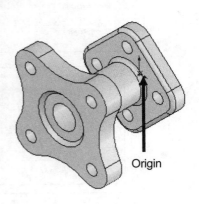

Origin

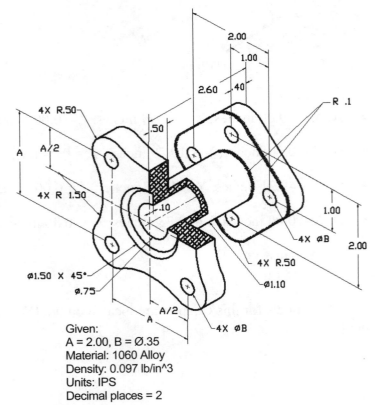

Given:
A = 2.00, B = Ø.35
Material: 1060 Alloy
Density: 0.097 lb/in^3
Units: IPS
Decimal places = 2

Think about the steps that you would take to build the illustrated part. Identify the location of the part Origin.

Start with the back base flange. Review the provided dimensions and annotations in the part illustration.

The key difference between the *Advanced Part Creation and Modification* and the *Basic Part Creation and Modification* category and the *Intermediate Part Creation and Modification* is the complexity of the sketches and the number of features, dimensions, and geometric relations. You may also need to locate the Center of mass relative to a created coordinate system location.

- A: X = 1.00 inches, Y = 0.79 inches, Z = 0.00 inches

- B: X = 0.00 inches, Y = 0.00 inches, Z = 1.04 inches

- C: X = 0.00 inches, Y = 1.18 inches, Z = 0.00 inches

- D: X = 0.00 inches, Y = 0.00 inches, Z = 1.51 inches

The correct answer is D.

All models for this chapter are located on the DVD in the book.

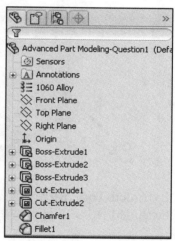

Mass = 0.59 pounds

Volume = 6.01 cubic inches

Surface area = 46.61 inches^2

Center of mass: (inches)
X = 0.00
Y = 0.00
Z = 1.51

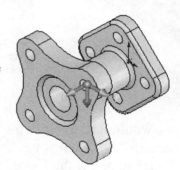

Advanced Part Modeling-Question1 (Def
- Sensors
- Annotations
- 1060 Alloy
- Front Plane
- Top Plane
- Right Plane
- Origin
- Boss-Extrude1
- Boss-Extrude2
- Boss-Extrude3
- Cut-Extrude1
- Cut-Extrude2
- Chamfer1
- Fillet1

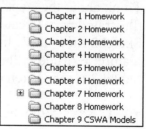

Chapter 1 Homework
Chapter 2 Homework
Chapter 3 Homework
Chapter 4 Homework
Chapter 5 Homework
Chapter 6 Homework
Chapter 7 Homework
Chapter 8 Homework
Chapter 9 CSWA Models

Question 2: Build the illustrated model from the provided information. Locate the Center of mass of the part.

Hint: Create the part with eleven features and a Reference plane: Extruded Base, Plane1, two Extruded Bosses, two Extruded Cuts, Extruded Boss, Extruded Cut,

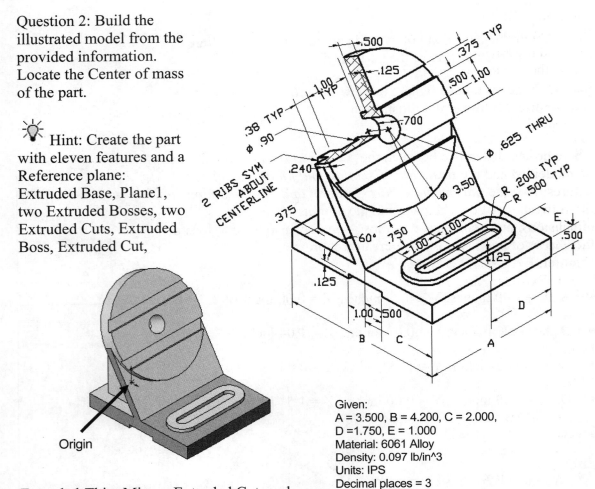

Given:
A = 3.500, B = 4.200, C = 2.000,
D =1.750, E = 1.000
Material: 6061 Alloy
Density: 0.097 lb/in^3
Units: IPS
Decimal places = 3

Origin

Extruded-Thin, Mirror, Extruded Cut, and Extruded Boss.

Think about the steps that you would take to build the illustrated part. Create the rectangular Base feature. Create Sketch2 for Plane1. Insert Plane1 to create the Extruded Boss feature: Extrude2. Plane1 is the Sketch plane for Sketch3. Sketch3 is the sketch profile for Extrude2.

- A: X = 1.59 inches, Y = 1.19 inches, Z = 0.00 inches

- B: X = -1.59 inches, Y = 1.19 inches, Z = 0.04 inches

- C: X = 1.00 inches, Y = 1.18 inches, Z = 0.10 inches

- D: X = 0.00 inches, Y = 0.00 inches, Z = 1.61 inches

The correct answer is A.

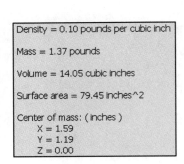

Density = 0.10 pounds per cubic inch

Mass = 1.37 pounds

Volume = 14.05 cubic inches

Surface area = 79.45 inches^2

Center of mass: (inches)
    X = 1.59
    Y = 1.19
    Z = 0.00

Question 3: Build the illustrated model from the provided information. Locate the Center of mass of the part. Note the coordinate system location of the model as illustrated.

Where do you start? Build the model. Insert thirteen features: Extruded-Thin1, Fillet, two Extruded Cuts, Circular Pattern, two Extruded Cuts, Mirror, Chamfer, Extruded Cut, Mirror, Extruded Cut, and Mirror.

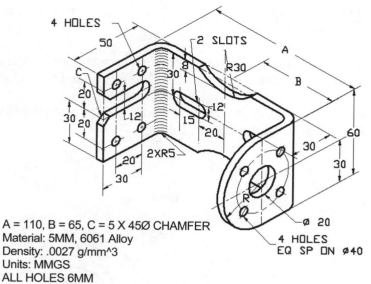

A = 110, B = 65, C = 5 X 45Ø CHAMFER
Material: 5MM, 6061 Alloy
Density: .0027 g/mm^3
Units: MMGS
ALL HOLES 6MM

Think about the steps that you would take to build the illustrated part. Review the provided information. The depth of the left side is 50mm. The depth of the right side is 60mm

Create Coordinate System1 to locate the Center of mass.

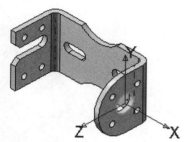

Coordinate system: +X, +Y. +Z

🔅 The SolidWorks software displays positive values for (X, Y, Z) coordinates for a reference coordinate system. The CSWA exam displays either a positive or negative sign in front of the (X, Y, Z) coordinates to indicate direction as illustrated, (-X, +Y, -Z).

- A: X = -53.30 millimeters, Y = -0.27 millimeters, Z = -15.54 millimeters

- B: X = 53.30 millimeters, Y = 0.27 millimeters, Z = 15.54 millimeters

- C: X = 49.21 millimeters, Y = 46.88 millimeters, Z = 0.00 millimeters

- D: X = 45.00 millimeters, Y = -46.49 millimeters, Z = 10.00 millimeters

The correct answer is A.

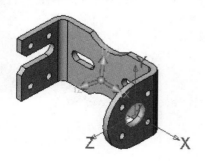

Question 4: Build the illustrated model from the provided information. Locate the Center of mass of the part.

Hint: Insert twelve features and a Reference plane: Extruded-Thin1, two Extruded Bosses, Extruded Cut, Extruded Boss, Extruded Cut, Plane1, Mirror, five Extruded Cuts.

Think about the steps that you would take to build the illustrated part. Create an Extrude-Thin1 feature as the Base feature.

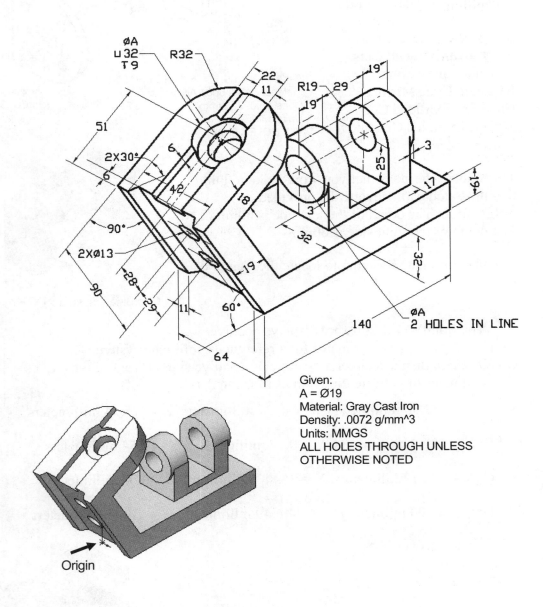

Given:
A = Ø19
Material: Gray Cast Iron
Density: .0072 g/mm^3
Units: MMGS
ALL HOLES THROUGH UNLESS
OTHERWISE NOTED

- A: X = -53.30 millimeters, Y = -0.27 millimeters, Z = -15.54 millimeters

- B: X = 53.30 millimeters, Y = 1.27 millimeters, Z = -15.54 millimeters

- C: X = 0.00 millimeters, Y = 34.97 millimeters, Z = 46.67 millimeters

- D: X = 0.00 millimeters, Y = 34.97 millimeters, Z = -46.67 millimeters

The correct answer is D.

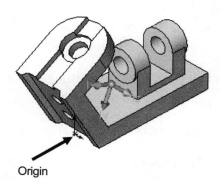

```
Mass = 2536.59 grams

Volume = 352304.50 cubic millimeters

Surface area = 61252.90 millimeters^2

Center of mass: ( millimeters )
    X = 0.00
    Y = 34.97
    Z = -46.67
```

⛯ Due to software rounding, you may view a negative -0.00 coordinate location in the Mass Properties dialog box.

Question 5: Build the illustrated model from the provided information. Locate the Center of mass of the part.

Origin

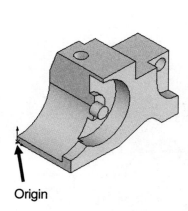

Origin

Given:¶
A·=·63,·B·=·50,·C·=·100¶
Material:·Copper¶
Units:·MMGS¶
Density:·.0089·g/mm^3¶
All·HOLES·THROUGH·ALL¶

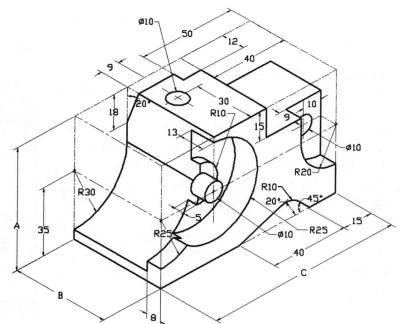

The center point of the top hole is located 30mm from the top right edge.

Think about the steps that you would take to build the illustrated part.

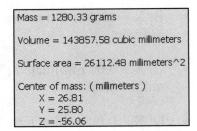

- A: X = 26.81 millimeters, Y = 25.80 millimeters, Z = -56.06 millimeters

- B: X = 43.30 millimeters, Y = 25.27 millimeters, Z = -15.54 millimeters

- C: X = 26.81 millimeters, Y = -25.75 millimeters, Z = 0.00 millimeters

- D: X = 46.00 millimeters, Y = -46.49 millimeters, Z = 10.00 millimeters

The correct answer is A.

This example was taken from the SolidWorks website, **www.solidworks.com/cswa** as an example of an Advanced Part on the CSWA exam. This model has thirteen features and twelve sketches.

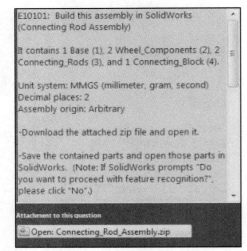

☼ There are numerous ways to create the models in this chapter.

Assembly Creation and Modification

Assembly Creation and Modification is one of the five categories on the CSWA exam. In the last two section of this chapter, a *Basic Part Creation and Modification, Intermediate Part Creation and Modification,* or an *Advanced Part Creation and Modification* was the focus.

The *Assembly Creation and Modification* category addresses an assembly with numerous sub-components.

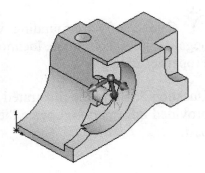

Screen shots from the exam.

Knowledge to insert Standard mates is required in this category.

There are four questions on the CSWA exam in the Assembly Creation and Modification category: (Two different assemblies - four questions - two multiple choice / two single answers - 30 points each).

You are required to download the needed components from a provided zip file and insert them correctly to create the assembly as illustrated. You are then requested to modify the assembly and answer fill in the blank format questions.

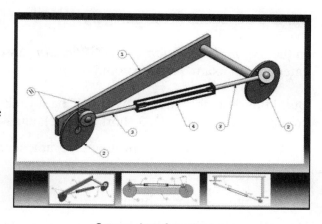

Screen shots from the exam.

Components for the assembly are supplied in the exam.

Do NOT use feature recognition when you open the downloaded components for the assembly in the CSWA exam. This is a timed exam. Manage your time. You do not need this information.

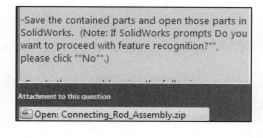

-Save the contained parts and open those parts in SolidWorks. (Note: If SolidWorks prompts Do you want to proceed with feature recognition?"", please click ""No"".)

Attachment to this question

Open: Connecting_Rod_Assembly.zip

Sample Questions in the category

In the *Assembly Creation and Modification* Assembly Modeling category, an exam question could read:

Build this assembly in SolidWorks (Chain Link Assembly). It contains 2 long_pins (1), 3 short_pins (2), and 4 chain_links (3).

- Unit system: MMGS (millimeter, gram, second)
- Decimal places: 2
- Assembly origin: Arbitrary

IMPORTANT: Create the Assembly with respect to the Origin as shown in the Isometric view. (This is important for calculating the proper Center of Mass). Create the assembly using the following conditions:

1. Pins are mated concentric to chain link holes (no clearance).

2. Pin end faces are coincident to chain link side faces.

A = 25 degrees, B = 125 degrees, C = 130 degrees

What is the center of mass of the assembly (millimeters)?

Hint: If you don't find an option within 1% of your answer please re-check your assembly.

A) X = 348.66, Y = -88.48, Z = -91.40

B) X = 308.53, Y = -109.89, Z = -61.40

C) X = 298.66, Y = -17.48, Z = -89.22

D) X = 448.66, Y = -208.48, Z = -34.64

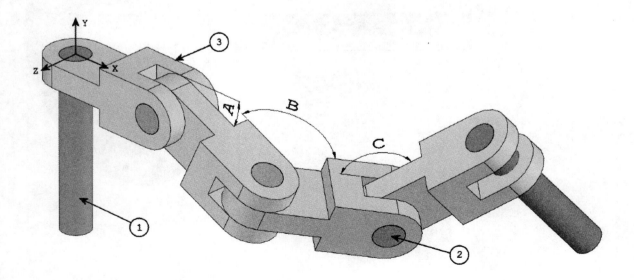

There are no step by step procedures in this section.

Download the *needed components* from the Chapter 9 CSWA Models folder on the DVD in the book to create the assembly.

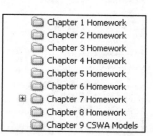

Below are various Assembly FeatureManagers that created the above assembly and obtained the correct answer.

The correct answer is:

A) X = 348.66, Y = -88.48, Z = -91.40

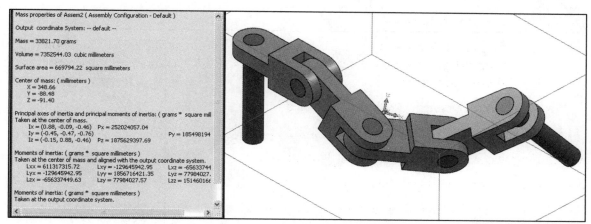

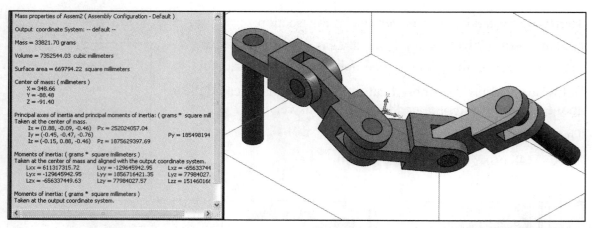

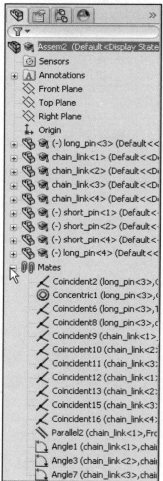

Appendix

Engineering Changer Order (ECO)

D&M	Engineering Change Order		ECO # _____ Page 1 of __

	☐ Hardware		Author
Product Line	☐ Software		Date
	☐ Quality		Authorized Mgr.
	☐ Tech Pubs		Date

Change Tested By

Reason for ECO(Describe the existing problem, symptom and impact on field)

D&M Part No.	Rev From/To	Part Description	Description	Owner

ECO Implementation/Class		Departments	Approvals	Date	
All in Field	☐	Engineering			
All in Test	☐	Manufacturing			
All in Assembly	☐	Technical Support			
All in Stock	☐	Marketing			
All on Order	☐	DOC Control			
All Future	☐				
Material Disposition		ECO Cost			
Rework	☐	DO NOT WRITE BELOW THIS LINE (ECO BOARD ONLY)			
Scrap	☐	Effective Date			
Use as is	☐	Incorporated Date			
None	☐	Board Approval			
See Attached	☐	Board Date			

This text follows the ASME Y14 Engineering Drawing and Related Documentation Practices for drawings. Display of dimensions and tolerances are as follows:

TYPES of DECIMAL DIMENSIONS (ASME Y14.5M)			
Description:	**UNITS: MM**	**Description:**	**UNITS: INCH**
Dimension is less than 1mm. Zero precedes the decimal point.	0.9 0.95	Dimension is less than 1 inch. Zero is not used before the decimal point.	.5 .56
Dimension is a whole number. Display no decimal point. Display no zero after decimal point.	19	Express dimension to the same number of decimal places as its tolerance. Add zeros to the right of the decimal point. If the tolerance is expressed to 3 places, then the dimension contains 3 places to the right of the decimal point.	1.750
Dimension exceeds a whole number by a decimal fraction of a millimeter. Display no zero to the right of the decimal.	11.5 11.51		

TABLE 1 TOLERANCE DISPLAY FOR INCH AND METRIC DIMENSIONS (ASME Y14.5M)		
DISPLAY:	**UNITS: INCH:**	**UNITS: METRIC:**
Dimensions less than 1	.5	0.5
Unilateral Tolerance	$1.417^{+.005}_{-.000}$	$36^{0}_{-0.5}$
Bilateral Tolerance	$1.417^{+.010}_{-.020}$	$36^{+0.25}_{-0.50}$
Limit Tolerance	.571 .463	14.50 11.50

SolidWorks Keyboard Shortcuts

Listed below are some of the pre-defined keyboard shortcuts in SolidWorks:

Action:	Key Combination:
Model Views	
Rotate the model horizontally or vertically:	**Arrow** keys
Rotate the model horizontally or vertically 90 degrees.	**Shift + Arrow** keys
Rotate the model clockwise or counterclockwise	**Alt** + left of right **Arrow** keys
Pan the model	**Ctrl + Arrow** keys
Magnifying glass	**g**
Zoom in	**Shift + z**
Zoom out	**z**
Zoom to fit	**f**
Previous view	**Ctrl + Shift + z**
View Orientation	
View Orientation menu	**Spacebar**
Front view	**Ctrl + 1**
Back view	**Ctrl + 2**
Left view	**Ctrl + 3**
Right view	**Ctrl + 4**
Top view	**Ctrl + 5**
Bottom view	**Ctrl + 6**
Isometric view	**Ctrl + 7**
NormalTo view	**Ctrl + 8**
Selection Filters	
Filter edges	**e**
Filter vertices	**v**
Filter faces	**x**
Toggle Selection Filter toolbar	**F5**
Toggle selection filters on/off	**F6**
File menu items	
New SolidWorks document	**Ctrl + n**
Open document	**Ctrl + o**
Open From Web Folder	**Ctrl + w**
Make Drawing from Part	**Ctrl + d**
Make Assembly from Part	**Ctrl + a**
Save	**Ctrl +s**
Print	**Ctrl + p**
Additional shortcuts	
Access online help inside of PropertyManager or dialog box	**F1**
Rename an item in the FeatureManager design tree	**F2**
Rebuild the model	**Ctrl + b**
Force rebuild – Rebuild the model and all its features	**Ctrl + q**
Redraw the screen	**Ctrl + r**

Cycle between open SolidWorks document	**Ctrl + Tab**
Line to arc/arc to line in the Sketch	**a**
Undo	**Ctrl + z**
Redo	**Ctrl + y**
Cut	**Ctrl + x**
Copy	**Ctrl + c**
Additional shortcuts	
Paste	**Ctrl + v**
Delete	**Delete**
Next window	**Ctrl + F6**
Close window	**Ctrl + F4**
View previous tools	**s**
Selects all text inside an Annotations text box	**Ctrl + a**

 In a sketch, the **Esc** key un-selects geometry items currently selected in the Properties box and Add Relations box. In the model, the **Esc** key closes the PropertyManager and cancels the selections.

Use the **g** key to activate the Magnifying glass tool. Use the Magnifying glass tool to inspect a model and make selections without changing the overall view.

Use the **s** key to view/access previous command tools in the Graphics window.

Windows Shortcuts

Listed below are some of the pre-defined keyboard shortcuts in Microsoft Windows:

Action:	Keyboard Combination:
Open the Start menu	Windows Logo key
Open Windows Explorer	Windows Logo key + E
Minimize all open windows	Windows Logo key + M
Open a Search window	Windows Logo key + F
Open Windows Help	Windows Logo key + F1
Select multiple geometry items in a SolidWorks document	Ctrl key (Hold the Ctrl key down. Select items.) Release the Ctrl key.

Helpful On-Line Information

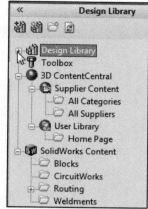

The SolidWorks URL: http://www.solidworks.com contains information on Local Resellers, Solution Partners, Certifications, SolidWorks users groups, and more.

Access 3D ContentCentral using the Task Pane to obtain engineering electronic catalog model and part information.

Use the SolidWorks Resources tab in the Task Pane to obtain access to Customer Portals, Discussion Forums, User Groups, Manufacturers, Solution Partners, Labs, and more.

Helpful on-line SolidWorks information is available from the following URLs:

- http://www.dmeducation.net

    Helpful tips, tricks and what's new in SolidWorks.

- http://www.mechengineer.com/snug/

    News group access and local user group information.

- http://www.nhcad.com

    Configuration information and other tips and tricks.

- http://www.dmeducation.net

    Helpful tips, tricks and what's new in SolidWorks.

- http://www.topica.com/lists/SW

    Independent News Group for SolidWorks discussions, questions and answers.

*On-line tutorials are for educational purposes only. Tutorials are copyrighted by their respective owners.

Notes:

INDEX